Distillery Wastewater Pollution and Bioremediation

Distillery Wastewater Pollution and Bioremediation

Ram Chandra PhD, FAMI, FBRSI, FAEB

Professor and Head
Department of Environmental Microbiology
School for Environmental Sciences
Babasaheb Bhimrao Ambedkar University
(A Central university)
Lucknow 226025, UP, India

Sangeeta Yadav PhD

Postdoctoral Fellow
Department of Environmental Microbiology
School for Environmental Sciences
Babasaheb Bhimrao Ambedkar University
(A Central university)
Lucknow 226025, UP, India

CBS

CBS Publishers & Distributors Pvt Ltd

New Delhi • Bengaluru • Chennai • Kochi • Mumbai • Pune
Hyderabad • Kolkata • Nagpur • Patna • Vijayawada

Distillery Wastewater Pollution and Bioremediation

ISBN: 978-81-239-2503-5

Copyright © Authors and Publisher

First Edition: 2014

Published by Satish Kumar Jain for
CBS Publishers & Distributors Pvt Ltd
4819/XI Prahlad Street, 24 Ansari Road, Daryaganj, New Delhi 110 002, India.
Ph: 23289259, 23266861, 23266867   Website: www.cbspd.com
Fax: 011-23243014   e-mail: delhi@cbspd.com; cbspubs@airtelmail.in.
Corporate Office: 204 FIE, Industrial Area, Patparganj, Delhi 110 092
Ph: 4934 4934   Fax: 4934 4935   e-mail: publishing@cbspd.com; publicity@cbspd.com

Branches

• **Bengaluru:** Seema House 2975, 17th Cross, K.R. Road,
  Banasankari 2nd Stage, Bengaluru 560 070, Karnataka
  Ph: +91-80-26771678/79   Fax: +91-80-26771680   e-mail: bangalore@cbspd.com
• **Chennai:** 20, West Park Road, Shenoy Nagar, Chennai 600 030, Tamil Nadu
  Ph: +91-44-26260666, 26208620   e-mail: chennai@cbspd.com
• **Kochi:** 36/14 Kalluvilakam, Lissie Hospital Road, Kochi 682 018, Kerala
  Ph: +91-484-4059061-65   Fax: +91-484-4059065   e-mail: kochi@cbspd.com
• **Mumbai:** 83-C, Dr E Moses Road, Worli, Mumbai-400018, Maharashtra
  Ph: +91-22-24902340/41   Fax: +91-22-24902342   e-mail: mumbai@cbspd.com
• **Pune:** Bhuruk Prestige, Sr. No. 52/12/2+1+3/2 Narhe, Haveli
  (Near Katraj-Dehu Road Bypass), Pune 411 041, Maharashtra
  Ph: +91-20-64704058, 64704059, 32392277 Fax: +91-20-24300160   e-mail: pune@cbspd.com

Representatives

• **Hyderabad** 0-9885175004   • **Kolkata** 0-9831437309, 0-9051152362
• **Nagpur**   0-9021734563   • **Patna**   0-9334159340
• **Vijayawada** 0-9000660880

Printed at: Paras Offset Pvt. Ltd., New Delhi

Preface

The sugarcane molasses-based distillery waste is well known for its complex and non-biodegradable properties. It pollutes the aquatic resources nearby the distilleries even after long run off and treatment. Further, due to environmental pollution and stringent rule for safe disposal of industrial waste by national and international protection agencies, the safe disposal of postmethanated distillery effluents is still a challenge for industry and threat to the environmentalists also. The researchers and biotechnologists in the recent past have reported some specialized bacterial strains and plants which are capable to degrade the effluent in specific condition. However, our knowledge on properties of distillery wastes and their environmental health hazards is quite fragmentary and insufficient. The behavior of distillery waste pollutants with microbes and other biological system in environment is also not much known. Therefore, the scarcity of knowledge regarding distillery effluents and their colorants (i.e. melanoidins, phenolics, metal sulphides) for safe disposal has inspired me to write this book. This book is an outcome of two decades' extensive research work by authors' team and various scientists in the Indian scenario. The book has been compiled with the physicochemical properties and their health hazards at different trophic level. Further, the bacterial species and their consortium capable to degrade melanoidin in different environmental condition. Moreover, the phytoremediation potential of wetland plants for postmethanated distillery waste and melanoidin degradation has also been evaluated. Furthermore, the biostimulation effects on bacterial degradation of distillery effluent and combination of wetland plants in biphasic treatment have been described with the latest information which is not known so far at any other sources. Most of the findings of this book has been published in high impact international journal as original research work. It is also appreciated by the scientific community globally. The writing of this book is made possible by the extensive research work funded by DBT, DOEn, CSIR and industry.

This book will be very useful to understand the physicochemical properties of effluent by researchers and postgraduate students of environmental sciences, biotechnology and microbiology. Moreover, the content of this book will be a boon for industrial personnel to understand the factual data of distillery waste and their environmental health hazards. Moreover, the prospect of phytoremediation for safe disposal with the novel technology and biotechnological approach will be a sustainable and feasible approach for safe disposal or utilization of wastewater. We also extend our sincere thanks to all PhD students and research scholars who have contributed the knowledge with their hard work during the research and generating the original data.

Ram Chandra
Sangeeta Yadav

Contents

Preface v

1. Introduction 1

2. Physico-chemical Properties of Distillery Wastewater 9

3. Melanoidin as a Major Colourant of PMDE, its Synthesis, Structure and Property 23

4. Bioredegradation of Sucrose-aspartic Acid Maillard as Model Melanoidin Products and its Degradative Metabolites 47

5. Effect of Phenol and Heavy Metals on Biological Decolourization of Melanoidin 71

6. Health Hazards of PMDE on Aquatic and Terrestrial Ecosystem 87

7. Phytoremediation of Heavy Metals from Post Methanated Distillery Effluent 106

8. Phytoremediation Potential of Wetland Plants for Melanoidin Decolourization in Presence of Heavy Metals and Phenol 144

9. Different Bioremediation Technique for the Decolourization and Detoxification of PMDE 169

10. Biocomposting of Distillery Waste for Safe Disposal 208

11. Challenges and Further Suggestions for Decolourization and Detoxification of PMDE in Indian Scenario 224

References 230

Abbreviations 239

Glossary 242

Index 247

1

Introduction

Alcohol distilleries are rated among the 17 most polluting industries in India, the second largest producer of ethanol in Asia. Sugarcane molasses is a basic raw material for the manufacture of ethanol in the country. Other materials include sugar-based corn, wheat, cassava, rice and barley. The main source of wastewater generation is the distillation step. In the distillation process, ethanol is 5–12% hence, the amount of waste is 88–95% by volume of the alcohol distilled. An average molasses based distillery generates 15 L of spentwash per lit of alcohol produced (Beltran et al., 2001). Molasses based distilleries in India are classified as 'Red Category' because of the large volume of high strength wastewater (spentwash) generation. The spentwash is highly acidic in nature and has a variety of recalcitrant colouring compounds as melanoidins, phenolics and metal sulfides that are mainly responsible for the dark colour of distillery effluent. The pH of spentwash increases from 4.5 to 8.5 during the anaerobic treatment process and finally it is called post methanated distillery effluent (PMDE). In the year 1999, there were 285 distilleries in India producing 2.7×10^9 L of alcohol each year (Joshi, 1999) and this number has gone up to 319, producing 3.25×10^9 L of alcohol annually and generating 40.4×10^{10} L of wastewater annually (AIDA, 2005; Uppal, 2004; Fig. 1.1). The spentwash is a cumbersome waste having very high biochemical oxygen demand (BOD; 35,000–40,000 mg/L), chemical oxygen demand (COD; 90,000–1,10,000 mg/L), total solids (TS; 82,480 mg/L), nitrogen (2,200 mg/L), phenolics (4.20 mg/L) and sulphate (3,410 mg/L). Besides, several heavy metals cadmium (Cd), manganese (Mn), iron (Fe), zinc (Zn), nickel (Ni) and lead (Pb) are also present (Chandra et al., 2004). The spentwash primarily undergoes anaerobic treatment process which converts a significant proportion (>50%) of the BOD and COD. However, different biochemical changes in the spentwash occur during anaerobic digestion. There is formation of significant amount of hydrogen sulfide (H_2S) as a result of the reduction of oxidized sulfur compound. Sulfide binds with the heavy metals present in the effluent and forms a colloidal solution of metal sulfide, colourant (dark black). Wastewater generated after anaerobic digestion in sugarcane molasses based distilleries known as PMDE. PMDE retains high BOD (18,000–22,000 mg/L), COD (32,400–35,000 mg/L), colour (1,50,000–1,80,000 mg/L), sulphate (3,100–5,760 mg/L), phenol (4.0–4.2 mg/L), total suspended solids (TSS; 11,920–25,308 mg/L), total dissolved solids (TDS; 10,480–77,776 mg/L) and having heavy metals (Chandra et al., 2004). Apart from high organic content, PMDE also contains nutrients in the form of nitrogen (1,660–4,200 mg/L), phosphorus (225–3,038 mg/L) and potassium (9,600–17,475 mg/L) that can lead to eutrophication of water bodies.

The unpleasant odour of the effluent is due to the presence of skatole, indole and other sulphur compounds, which are not effectively decomposed during fermentation and distillation. Further, its dark colour hinders photosynthesis by blocking sunlight and is therefore deleterious to aquatic life. Water quality of a river contaminated with

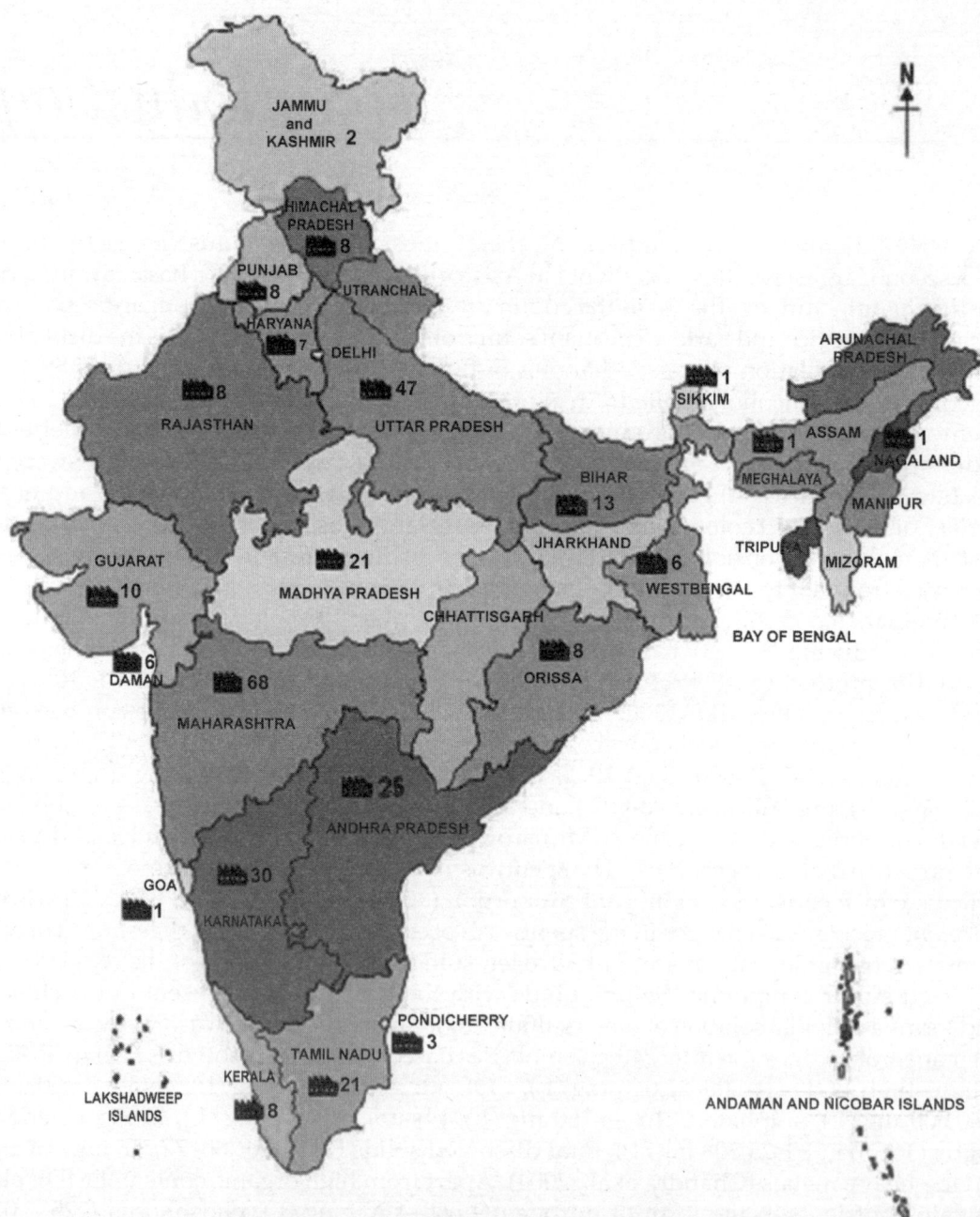

Fig. 1.1: State wise distribution of distilleries in India

distillery effluent is affected up to very large distance due to high BOD, COD and colour. The indiscriminate discharge of PMDE from distilleries is a major source of soil and water pollution which directly or indirectly affects the animals and human beings. Use of effluent for irrigation without proper monitoring can affect groundwater quality by altering its physio-chemical properties such as colour, pH, electrical conductivity etc. due to leaching down of organic and inorganic ions. Therefore adequate treatment is imperative before the effluent is discharged. The increasingly stringent environment regulations are forcing the distilleries worldwide to improve the existing treatment and also explore alternative methods of effluent management.

For instance, Indian distilleries were stipulated to achieve zero discharge of spentwash to inland surface by December 2005 (Uppal 2004). Hence, its colour removal is important prior to its disposal in the environment. The most common decolourisation process being operated at industry in India is either aeration or lagooning. However, these processes are not much effective at the existing pollution load of PMDE due to the presence of various recalcitrant compounds in high quantity. Though some physical and chemical

Table 1.1: Wastewater production by different distilleries in India (AIDA, 2003)

S.No	Names of States	No. of Distilleries	Annual Licenced Capacity	Annual Installed Capacity	Wastewater Production (Litres)
1.	Andhra Pradesh	27	144463	151022	2265.33×10^6
2.	Bihar	13	88512	93910	1408.65×10^6
3.	Daman	4	22360	22360	335.4×10^6
4.	Gujarat	13	143350	141550	2123.25×10^6
5.	Goa	4	5044	5206	78.09×10^6
6.	Haryana	7	64050	64050	960.75×10^6
7.	Himachal Pradesh	2	12485	13600	204.00×10^6
8.	Jammu & Kashmir	5	26650	29450	441.75×10^6
9.	Karnataka	33	243225	282044	4230.66×10^6
10.	Kerala	8	24728	22930	343.95×10^6
11.	Madhya Pradesh	22	469600	463120	6946.80×10^6
12.	Maharashtra	69	766142	683432	10251.48×10^6
13.	Nagaland	1	1350	1350	20.25×10^6
14.	Orissa	9	21585	21405	321.075×10^6
15.	Pondicherry	4	4400	11700	175.50×10^6
16.	Punjab	9	167920	158197	2372.96×10^6
17.	Rajasthan	7	53715	45265	678.98×10^6
18.	Sikkim	2	4600	6580	98.70×10^6
19.	Tamil Nadu	22	292780	315105	4726.58×10^6
20.	Uttar Pradesh	43	645240	689373	10340.60×10^6
21.	West Bengal	5	18240	30660	459.90×10^6
22.	Andhra Pradesh	27	144463	151022	2265.33×10^6

methods, i.e. adsorption, coagulation/precipitation, oxidation and membrane filtration are used for decolourisation of anaerobically digested distillery effluent (Chandra and Pandey, 2001), but the chemical and physical methods generate huge quantity of sludge and secondary pollutants.

Biological methods employing different fungi, bacteria and algae is an environmental-friendly and cost-competitive alternative to chemical decomposition for decolourisation and degradation of distillery effluent. Microorganisms play the vital role in the process of bioremediation and biodegradation because of their great metabolic diversity, which includes the ability to metabolise these pollutants. The degradation of toxic compounds to less harmful forms with the use of biological systems is called bioremediation. There has been limited success in search for bacteria and fungi, which can efficiently degrade melanoidins in order to reduce the colour and COD. It is considered highly desirable to exploit the degradation potential of soil micro-organisms from polluted site exposed to recalcitrant compounds of distillery spentwash for prolonged period.

1.1 SOURCE OF COLOURANTS AND NATURES IN PMDE

Most of the colourants are generated and concentrated from sugarcane juice during the crushing of sugarcane, while some are added during the processing of sugar. Phenolics are more pronounced in cane molasses wastewater whereas melanin is significant colourant in beet molasses. However, heating of glucose and fructose under acidic or basic condition leads to degradation reaction, forming highly reactive intermediates which can undergo further condensation and polymerization reactions to form coloured polymers known as melanoidins as complex of amino-carbonyl. In general colourants can be divided into enzymatic colourants such as melanins and non-enzymatic colourant such as melanoidin, alkaline degradation product of hexoses and caramel. Cane molasses contains around 2% of melanoidins that imparts colour to the effluent. There is high concentration of sulphate also in sugarcane molasses, which is added to molasses during the cleaning of sugar crystals. High level of sulfate can lead to the production of sulfides in anaerobic digestion of the effluent, these precipitate out along with the existing metal as metallic sulfides in PMDE, consequently increasing the total solid (TS) of effluent.

1.2 IN SITU/INTRINSIC BIOREMEDIATION/ NATURAL ATTENUATION

The most effective means of implementing in situ bioremediation depends on the hydrology of the subsurface area, the extent of the contaminated area and the nature (type) of the contamination. In general, this method is effective only when the subsurface soils are highly permeable, the soil horizon to be treated falls within a depth of 8 – 10 m and shallow groundwater is present at 10 m or less below ground surface. The depth of contamination plays an important role in determining whether or not an in situ bioremediation project should be employed. If the contamination is near the groundwater but the groundwater is not yet contaminated then it would be unwise to set up a hydrostatic system. It would be safer to excavate the contaminated soil and apply an onsite method of treatment away from the groundwater. The average time frame for an in situ bioremediation project can be in the order of 12–24 months depending on the levels of contamination and depth of contaminated soil. Due to the poor mixing in this system it becomes necessary to treat for long periods of time to ensure that all the pockets of contamination have been treated. In situ bioremediation is a very site specific technology that involves establishing a hydrostatic gradient through the contaminated area by flooding it with water carrying nutrients and possibly organisms adapted to the contaminants. Water is continuously circulated through the site until it is determined to be clean.

The in situ treatment methods of contaminated soil include the following:

1.2.1 Bioventing

Bioventing is an in situ remediation technology that uses indigenous microorganisms to biodegrade organic constituents adsorbed to soils in the unsaturated zone. Soils in the capillary fringe and the saturated zone are not affected. In Bioventing, the activity of the indigenous bacteria is enhanced by inducing air (or oxygen) flow into the unsaturated zone (using extraction or injection wells) and, if necessary, by adding nutrients. This process combines an increased oxygen supply with vapour extraction. A vacuum is applied at some depth in the contaminated soil which draws air down into the soil from holes drilled around the site and sweeps out any volatile organic compounds. All aerobically biodegradable constituents can be treated by Bioventing. In particular, Bioventing has proven to be very effective in remediating releases of petroleum products including gasoline, jet fuels, kerosene, and diesel fuel. Bioventing is most often used at sites with mid-weight petroleum products (i.e. diesel fuel and jet fuel), because lighter products (i.e. gasoline) tend to volatilize readily

1.2.2 Biosparging

Biosparging is an in situ remediation technology that uses indigenous microorganisms to biodegrade organic constituents in the saturated zone. The objective of a biosparging system is to promote contaminants biodegradation by the injection of air or gas (ex: oxygen, gaseous nutrients) and to minimize the volatilization of volatile and semi-volatile organic compounds. In some instances air injections are replaced by pure oxygen to increase the degradation rates. However, in view of the high costs of this treatment in addition to the limitations in the amount of dissolved oxygen available for microorganisms, hydrogen peroxide (H_2O_2) was introduced as an alternative, and it was used on a number of sites to supply more oxygen. Each liter of commercially available H_2O_2 (30%) would produce more than 100 L of O_2, and was more efficient in enhancing microbial activity during the bioremediation of contaminated soils and groundwater. The H_2O_2 put into the soil would supply ~ 0.5 mg/L of oxygen from each mg/L of H_2O_2 added, but a disadvantage comes from its dangerous toxicity to microorganisms even at low concentrations. Biosparging requires high pressure, whereas bioventing uses low pressure.

1.2.3 Phytoremediation

Phytoremediation is a form of bioremediation and applies to all chemical or physical processes that involve plants for degrading or immobilizing contaminants in soil and ground water. While, the technology is not new, current trends suggest its popularity is growing. Phytosequestration, rhizodegradation, phytohydraulics, phytoextraction, phytovolatilization and phytodegradation are six different types of phytoremediation mechanism of plant. Phytosequestration is also called phytostabilization. Many different processes fall under this category which can involve absorption by roots, adsorption to the surface of roots or the production of biochemicals by the plant that are released into the soil or ground water in the immediate vicinity of the roots, and can sequester, precipitate, or otherwise immobilize nearby contaminants. Further, rhizodegradation takes place in the soil or ground water immediately surrounding the plant roots. Exudates from plants stimulate rhizosphere bacteria to enhance biodegradation of soil contaminants. Furthermore, in phytohydraulics we use deep-rooted plants (usually trees) to contain, sequester or degrade ground water contaminants that come into contact with their roots. In addition, phytoextraction is also known as phytoaccumulation. Plants take up or hyperaccumulate contaminants through their roots and store them in the tissues of the stem or leaves. The contaminants are not necessarily degraded but are removed from the environment when the plants are harvested. This is particularly useful for removing metals from soil/effluent and, in some cases; the metals can be recovered for reuse, by incinerating the plants, in a process called phytomining. However, plants take up volatile compounds through their roots, and transpire the same compounds, or their metabolites, through the leaves, thereby releasing them into the atmosphere are called phyto-volatilization. Finally, in phytodegradation contaminants are taken up into the plant tissues where they are metabolized, or biotransformed. Where the transformation takes place depends on the type of plant, and can occur in roots, stem or leaves

1.3 ON SITE (EX SITE) BIOREMEDIATION

In this process the contaminated soil is excavated and placed into a lined treatment cell. Thus, it is possible to sample the site in a more thorough and, therefore, representative manner. On site treatment involves land treatment or land farming, where regular tilling of the soil increases aeration and the supplement area is lined and dammed to retain any contaminants that leak out. The use of the liner is an added benefit, since the liner prevents migration of the contaminants and there is no possibility of contaminating the groundwater. However, excavation of the contaminated soil adds to the cost of a bioremediation project as does the liner and the landfarming equipment. In addition to these costs, it is necessary to find enough space to treat the excavated soil on site. This process allows for better control of the system by enabling the engineering firm to dictate the depth of soil well as the exposed surface area. As a consequence of the depth and exposed surface area of the soil being determined, one is able to better control the temperature, nutrient concentration, moisture content and oxygen availability. There are two techniques for utilizing bacteria to degrade petroleum in the aquatic and terrestrial environments. One method, biostimulation, uses the indigenous bacteria which are stimulated to grow by introducing nutrients into the soil or water environment and thereby enhancing the biodegradation process. The other method, bioaugmentation, involves culturing the bacteria independently and then adding them to the site.

1.3.1 Biostimulation

This process involves the stimulation of indigenous microorganisms to degrade the contaminant. The microbial degradation of many pollutants in aquatic and soil environments is limited primarily by the availability of nutrients, such as nitrogen, phosphorus, and oxygen availability. The addition of nitrogen and phosphorus containing substrates has been shown to stimulate the indigenous microbial populations.

1.3.2 Bioaugmentation

This process involves the introduction of pre-selected organisms to the site for the purpose of increasing the rate or extent, or both, of biodegradation of contaminants. It is usually done in conjunction with the development and monitoring of an ideal growth environment, in which the selected bacteria can live and work. The selected microorganisms must be carefully matched to the waste contamination present as well as the metabolites formed. Effective seed organisms are characterized by their ability to degrade most petroleum components, genetic stability and viability during storage, rapid growth following storage, a high degree of enzymatic activity and growth in the environment, ability to compete with indigenous microorganisms, nonpathogenicity and inability to produce toxic metabolites. Mixed cultures have been most commonly used as inocula for seeding because of the relative ease with which microorganisms with different and complementary biodegradative capabilities can be isolated.

Biodegradation and decolourisation of PMDE is very challenging and important due to its complexity and huge volume being discharge in the environment. The PMDE become dark black viscous during anaerobic digestion, which further retains dark brown colour even after extended aeration. Moreover, it contributes high BOD, COD, TDS and colour. Microorganisms (bacteria/fungi/actinomycetes) due to their versatile inherent property to metabolize a variety of complex compounds have been utilized since long for biodegradation of complex, toxic and recalcitrant compounds present in various industrial wastes for environmental safety. Microbial degradation and decolourisation of industrial wastes is an environment-friendly and cost-competitive method of industrial wastes minimization and clean-up. The utility of microbes in industrial wastes treatment process largely depends on the enzymatic setup, nutrient requirement of microbes as well as nature and chemical structure of recalcitrant compounds and environmental conditions.

Therefore, this book has been emphasized on distillery wastewater pollution and bioremediation and chapter two described the evaluation of pollution load of distillery effluent from physico-chemical properties of PMDE. The main content of book has focused on the generation of distillery effluent and its physico-chemical properties. Besides, chapter has also highlight the recent knowledge of persistent organic pollutant present in PMDE.

Chapter three has described the synthesis, chemical structure and detail of melanoidin properties. Further, chapter four has focused on biodegradation of synthetic melanoidin and its degradative metabolites while chapter five shows effect of different pollutant (phenol and heavy metals) present in PMDE on melanoidin degradation. These findings are very important to understand the mechanism of bacterial degradation and decolourisation of melanoidins in presence of other co-pollutants present in PMDE. This will be much close to understanding the problem of microbial decolourisation of PMDE. Furthermore, this book involves health hazards of PMDE on aquatic and terrestrial ecosystem in chapter six. However, chapter seven summarized the role of plants on remediation of heavy metals, a major constituent of PMDE. In addition, chapter eight also explores the effect of phenol and heavy metals on biodegradation of melanoidins. Chapter nine described the different techniques like conventional, physico-chemical, biological and novel techniques for treatment of PMDE. The traditional aim of wastewater treatment is to enable wastewater to be disposed safely, without being a danger to public health and without polluting watercourses or causing other nuisance. Increasingly another aim of wastewater is to recover energy, nutrient, water and other valuable resources from wastewater. The treatment methods include anaerobic digesters which not only substantially reduce the organic load of the effluent but also produce methane gas. The gas is used as a fuel to compensate the energy needs of the industry. Since the conventional aerobic processes for primary treatment of distillery waste are not cost effective and require large land area, the thrust has been mainly on anaerobic processes as these have dual advantages of pollution control and production of fuel. Similarly chapter ten has described a very interesting and novel technique for treatment of PMDE as biocomposting which will be a very prosperous in future for environmental safety. The last chapter of the book has described the challenges and future need in the area of distillery effluent treatment for environmental safety. Therefore, this book has a major significance in the prevention of environmental pollution.

2

Physico-chemical Properties of Distillery Wastewater

Alcohol distilleries are rated among the 17 most polluting industries in India, fourth largest producer of ethanol in the world and the second largest in Asia. That's why distilleries have been included in the red category list by the Central Pollution Control Board, India. There are above 319 distilleries in India, mostly concentrated in Maharashtra, Uttar Pradesh, Andhra Pradesh, Karnataka, Tamil Nadu, Gujarat and Madhya Pradesh. The First distillery in the country was set up at Crwnpore (Kanpur) in 1805 by Carew & Co. Ltd. for manufacture of Rum for the army. The technique of fermentation, distillation and blending of alcoholic beverages was developed in India on the lines of practices adopted overseas particularly in Europe. Distilleries manufacture rectified spirit and extra neutral alcohol for human consumption and for industrial utilization. The distillery industry today consists broadly of two parts, potable liquor and the industrial alcohol. The potable distillery producing Indian Made Foreign Liquor and Country Liquor has a steady but limited demand with a growth rate of about 7 – 10 per cent per annum. The industrial alcohol industry on the other hand, is showing a declining trend because of high price of Molasses which is invariantly used as substrate for production of alcohol. Molasses is the one of the by-products of cane sugar production and is used in various fermentation processes, biofertilizer production, and feed for domestic animal molasses is commonly used as a raw material in alcohol distillery industries because of its low cost, availability, and suitability for fermentation processes. The alcohol produced is now being utilized in the ratio of approximately 52 per cent for potable and the balance 48 percent for industrial use. Apart from its use for beverage, medicinal, pharmaceutical and flavouring, alcohol constitutes the feedstock for large number of organic chemicals, which are used in manufacturing a wide variety of intermediates, drugs, rubber, pesticides, solvents, etc. with the advent of ethanol blending with petrol/motor fuel, the requirement of ethanol/industrial alcohol has increased manifold in the country to the extent that in case 5% blending, if made mandatory all over the country, the sugar factory molasses available in the country shall not prove to be adequate for meeting the total requirement of ethanol including its use for potable liquors and other industrial uses. Due to government promoting ethanol to mix in petrol there will be drastic demand for ethanol, which could overcome the existing unutilized capacity and thus creating an excess demand. Looking to its wide use, it can be inferred that the demand for alcohol is likely to increase in the country and so is the number of distilleries producing alcohol. All India Distillers Association (AIDA) and Ethanol India are predicting the birth of many new distilleries along with major expansion in capacity of existing distilleries (AIDA, 2008; Ethanol India, 2008). But, huge amount of wastewater is generated after alcohol production i.e. approx 88 – 95% by volume of the alcohol distilled. Different terminologies are used for this type of effluent in different countries viz., stillage, vinasse, slop, dunder, still residue, etc. An average molasses based distillery generates 15 L of spentwash per liter of alcohol produced. The effluent is a potential water pollutant in two ways. Firstly, effluent block out sunlight from entering the bottom layers of rivers and streams, thus reducing oxygenation of the water by blocking photosynthesis and becomes detrimental to aquatic life. Secondly, it has a high pollution load which leads to the eutrophication of water courses. Due to the presence of putriciable organics like skatole, indole and other sulphur compounds, the effluent that is disposed in canals or rivers produces obnoxious smell. Undiluted effluent has toxic effect on fishes and other aquatic organisms. The estimated LC_{50} for distillery spentwash was found to be 0.5% using a bio-toxicity study on fresh water fish *Cyprinus carpio* var. communis. The respiratory process in *C. carpio* under distillery effluent stress was affected resulting in a shift towards anaerobiosis at organ level during sublethal intoxication. Soil pollution and acidification are the other pollution issues caused by spentwash and it is reported to inhibit seed germination, reduce soil alkalinity, cause soil manganese deficiency and damage agricultural crops. Hence, this is hazardous for environment and human beings. The following major steps are involved for manufacture of alcohol, viz. feed preparation, fermentation, distillation and packaging.

Alcohol manufacturing process:

A. Feed preparation and fermentation

Ethanol (CH_3CH_2OH; or EtOH for short) is produced by fermentation of sugars. Fermentation alone does not produce beverages with alcohol content greater than 12 to 15% because the fermenting yeast is destroyed at high alcohol concentrations. To produce beverages of higher alcohol content the aqueous solution must be distilled. The fermentation of carbohydrates into alcohol is one of the oldest known chemical processes. Fermentation can be represented as:

$$Sugar \longrightarrow Alcohol + Carbon\ Dioxide$$

The reaction is catalyzed by yeast enzymes called zymases. A balanced chemical reaction for this process, assuming the sugar is table sugar or sucrose, is:

$$C_{12}H_{22}O_{11} + H_2O \longrightarrow 4\ CH_3CH_2OH + 4CO_2$$

The fermentation process is started by mixing a source of sugar, water and yeast and allowing the yeast to act in an oxygen free environment. The process of culturing yeast under alcohol- producing conditions is referred to as brewing. The same process produces CO_2 in situ, and may be used to carbonate the drink. Taste, texture, or both are improved by carbonation of drink. This anaerobic environment forces the yeast to shut down the "burning" of sugar and allows them to; instead, ferment alcohol (Fig. 2.1).

Overall, fermentation is a process that uses yeast to break down sugar molecules into carbon dioxide gas and ethyl alcohol. This chemical change is achieved because the yeast rapidly reproduces itself in any solution that contains sugar. In chemistry, alcohol is a

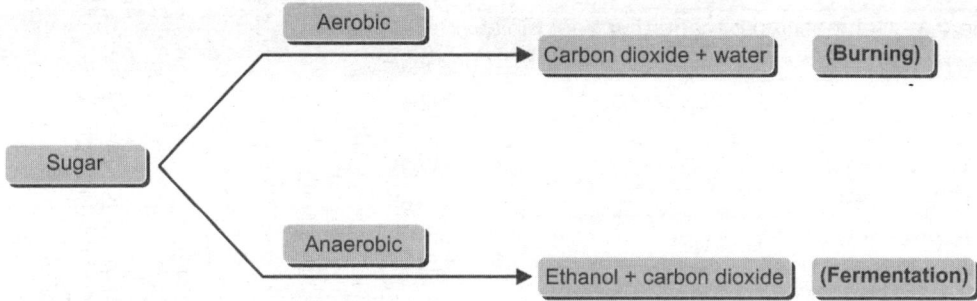

Fig. 2.1: Sugar fermentation process

general term for any organic compound in which a hydroxyl group (–OH) is bound to a carbon atom, which in turn may be bound to other carbon atoms and further hydrogens. Ethanol is classified as a primary alcohol, meaning that the carbon to which its hydroxyl group is attached has at least two hydrogen atoms attached to it as well. The chemistry of ethanol is largely that of its hydroxyl group. (CH_3CH_2OH), the active ingredient in alcoholic drinks, for consumption purposes is always produced by fermentation. Drinks with a concentration of more than 50% ethanol by volume (100 US proof) are flammable liquids and easily ignited. Spirits with higher ethanol content can be ignited with ease by heating slightly, e.g. adding the spirit to a warmed short glass. Other alcohol such as propylene glycol and the sugar alcohols may appear in food or beverages regularly, but these alcohols do not make them alcoholic. Methanol (one carbon), the propanol (three carbon giving two isomers), and the butanol (four carbons, four isomers) are all commonly found alcohols, and none of these three should ever be consumed in any form. Alcohol are toxicated into the corresponding aldehydes and then into the corresponding carboxylic acids. These metabolic products cause a poisoning and acidosis. In the case of other alcohols other than ethanol, the aldehydes and carboxylic acids are poisonous and the acidosis can be lethal. In contrast, fatalities from ethanol are mainly found in extreme doses and are related to induction of unconsciousness or chronic addiction (alcoholism). Ethanol can be produced from a wide range of feedstock. These include sugar-based (cane and beet molasses, cane juice), starch-based (corn, wheat, cassava, rice, barley) and cellulosic (crop residues, sugarcane bagasse, wood, municipal solid wastes) materials. In general, sugar-based feedstock containing readily available fermentable sugars are preferred since starch and cellulosic substrates involve an additional pre-treatment step to convert starch into fermentable sugars. Thus, cane juice is a commonly used substrate in Brazil, while Indian distilleries almost exclusively use sugarcane molasses named vinasses. Overall, nearly 61% of world ethanol production is from sugar crops. The composition of molasses varies with the variety of cane, the agro climatic conditions of the region, sugar manufacturing process and handling and storage (Godbole, 2002). Table 2.1 summarizes the chemical composition of beet and cane molasses.

Molasses, a by-product of sugar industry is used as raw material by most of the distilleries for production of alcohol by fermentation and distillation processes. Molasses is a thick viscous, acidic in nature (pH 5), contains about 40–50% sugar, rich in salts, dark brown in colour and it also contains sugar which could not be crystallized. For manufacturing alcohol, the Molasses is diluted with water into a solution contents

Table 2.1: Composition of cane and beet molasses

Properties	Cane molasses	Beet molasses
Brix (%)	79.5	79.5
Special gravity	1.38–1.52	1.41
Total solids (%)	75–88	77.0
Total sugars (%)	44–60	48.0
Crude protein (%)	2.5–4.5	6.0
Ash (%)	7–15	8.7
Calcium (%)	0.8	0.2
Phosphorus (%)	0.08	0.03
Potassium (%)	2.4	4.7
Sodium	0.2	1.0
Chlorine (%)	1.4	0.9
Sulfur (%)	0.5	0.5

10–15% of sugars or about 20–25 Brix (measurement of sugar concentration in a solution) due to its high osmotic pressure, and its pH adjusted (pH 5), if required, before fermentation. This solution is then inoculated with yeast strain and is allowed to ferment at room temperature. The feed is inoculated with about 10% by volume of yeast (*Saccharomyces cerevisiae*) inoculum. The fermented wash is distilled in a series of distillation columns to obtain alcohol of adequate/requisite strength and quality/specification. This alcohol is used for various purposes including potable and industrial. In India, about 90% of the molasses produced in cane sugar manufacture is consumed in ethanol production. Yeast culture is prepared in the laboratory and propagated in a series of fermenters, each about 10 times larger than the previous one. This is an anaerobic process carried out under controlled conditions of temperature and pH wherein reducing sugars are broken down to ethyl alcohol and carbon dioxide. The reaction is exothermic. To maintain the temperature between 25 and 32°C plate heat exchangers are used; alternatively some units spray cooling water on the fermenter walls. Fermentation can be carried out in either batch or continuous mode (CPCB, 2003). Fermentation time for batch operation is typically 24–36 h with an efficiency of about 95%. Continuous operation, involving higher sugar concentration and an osmotolerant variety of yeast, is faster (16–24 h fermentation time) but the efficiency is marginally lower (T.R. Sreekrishnan, pers. comm.). The resulting broth contains 8–9% alcohol. The sludge (mainly yeast cells) is separated by settling and discharged from the bottom, while the cell free fermentation broth is sent for distillation. Apart from yeast, the bacterium *Zymomonas mobilis* has also been investigated for ethanol production. The organism follows a simple catabolic pathway and has advantages over *S. cerevisiae* because of its higher sugar uptake rate, lower biomass yields and higher ethanol production (Lin and Tanaka, 2006).

Distillation of fermented molasses feed and generation of spentwash

To produce beverages of higher alcohol content the aqueous solution after fermentation must be distilled. Distillation is a separation process for a mixture of liquids. It relies on differences in the boiling points of the component liquids to be separated. Lower boiling components will preferentially vaporize first. Hence, initially low boiling components are collected followed by higher boiling components. In distilleries it is a two-stage process and is typically carried out in a series of bubble cap fractionating columns. The first stage consists of the analyzer column and is followed by rectification columns. The cell free fermentation broth (wash) is preheated to about 90°C by heat exchange with the effluent ("spentwash") and then sent to the degasifying section of the analyzer column. Here, the liquor is heated by live steam and fractionated to give about 40–45% alcohol. The bottom discharge from the analyzer column is the spentwash. The alcohol vapors are led to the rectification column where by reflux action, ~96% alcohol is tapped, cooled and collected. The condensed water from this stage, known as "spentlees" is usually pumped back to the analyzer column. The process of ethanol production, spentwash generation and treatment are shown in Fig. 2.2.

The molasses spentwash has very high biological oxygen demand (BOD, 35,000 to 40,000 mg/L), Chemical oxygen demand (COD, 90000–110000 mg/L), total solids (TS, 82480 mg/L), Nitrogen (2200 mg/L), phenolics (4.20 mg/L), sulphate (3410 mg/L). The wastewater generated from distillation of fermented mash is in the temperature range 70–80°C, deep brown in colour, acidic in nature (low pH), and has high concentration of organic materials and solids. It is a very complex, caramelized and cumbersome agro industrial waste. However, the pollution load of the distillery effluent depends on the quantity of molasses, unit operations for processing of molasses and process recovery of alcohols. Distillery spentwash has very high BOD, COD and high BOD/COD ratio. The amount of organic substances such as nitrogen, potassium, phosphates, calcium, sulphates is also very high. High COD, total nitrogen and total phosphate content of the influent may result in eutrofication of the natural water body. Disposal of the distillery spentwash on land is equally hazardous to the vegetation it is reported to reduce soil alkanity and manages availability, thus inhabiting seed germination. Application of distillery spentwash to soil without proper monitoring, seriously affects the ground water quality by altering its physico-chemical properties such as colour, pH, electric conductivity due

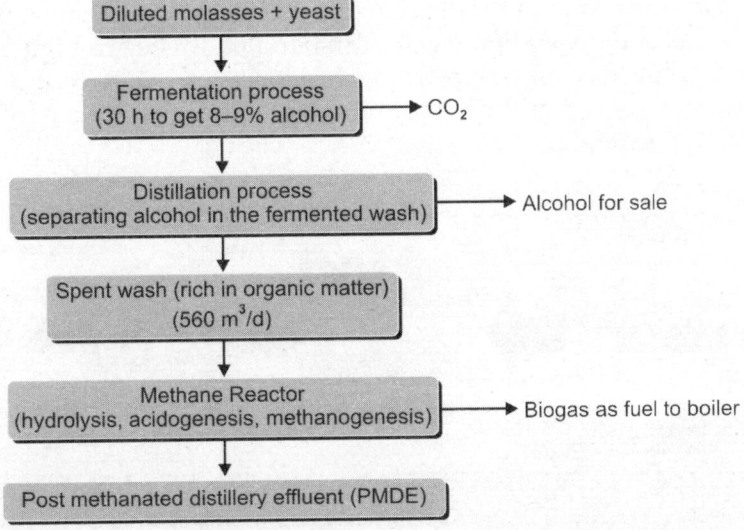

Fig. 2.2: Process description of ethanol and effluent generation

to leaching down of organic and inorganic ions. Besides these pollutants there are some heavy metals, i.e. cadmium (Cd), manganese (Mn), iron (Fe), zinc (Zn), nickel (Ni) and lead (Pb) are detected (Sangeeta and Chandra, 2013). In addition, cane molasses spentwash contains low molecular weight compounds such as lactic acid, glycerol, ethanol and acetic acid. Cane molasses also contains around 2% of a dark brown pigment called melanoidins that impart colour to the spentwash (Kalavathi et al., 2001). Melanoidins are low and high molecular weight polymers formed as one of the final products of Maillard reaction, which is a nonenzymatic browning reaction resulting from the reaction of reducing sugars and amino compounds. This reaction proceeds effectively at temperatures above 50 °C and pH 4–7. The structure of melanoidins is still not well known. Apart from melanoidins, spentwash contained other colourants such as phenolics, caramel and melanin. Phenolics have been more pronounced in cane molasses wastewater, whereas melanin was significant in beet molasses.

Spentwash treatment and generation of Post Methanated Distillery Effluent (PMDE)

Spentwash treatment is proposed by three different routes currently viz; (a) Concentration followed by incineration (b) Direct wet oxidation of stillage by air at high temperature with generation of steam followed by aerobic polishing and (c) Anaerobic digestion with biogas recovery followed by aerobic polishing. All of these processes are capital intensive. The incineration process involves an investment of the order of 400% of the distillery cost, whereas the other two processes along with the secondary treatment require an investment of 200–300% of the distillery cost. In recent years, anaerobic wastewater treatment has become a technology of growing importance, especially for highly polluted wastewater from the sugar and distillery industries. In India very common treatment method for spentwash is anaerobic digestion with biogas recovery. In spite of the fact of that there is the negative environmental impact associated with industrialization, the effect can be minimized and energy can be tapped by means of anaerobic digestion of the wastewater. Biological treatment of the distillery spentwash is either aerobic or anaerobic but in most cases the combination of both is used. A typical COD/BOD ratio of 1.8 to 1.9 of effluent indicates the suitability of influent of biological treatment. This anaerobic reactor known as second generation reactor can handled waste at a high organic loading rate of 24 kg COD/m^3 day and high up flow velocity of 2 mm/h at a low hydraulic retentions time. Anaerobic digestion is the most suitable option for the treatment of high strength organic effluents (Fig. 2.3).

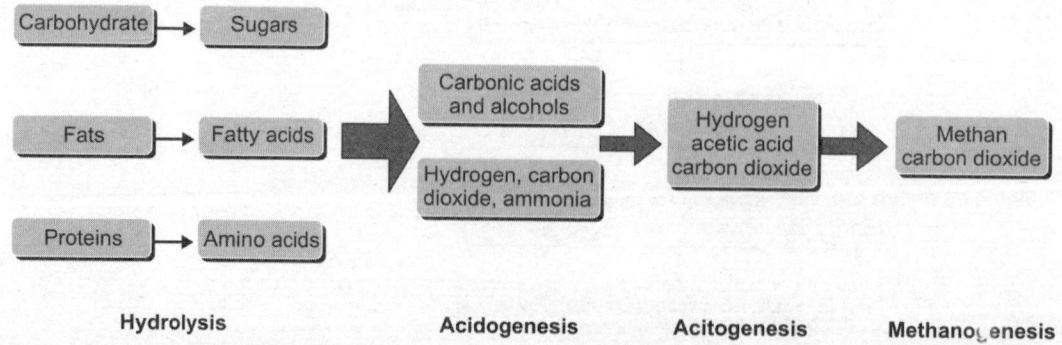

Fig. 2.3: The key process stages of anaerobic digestion of spentwash

The first step is the hydrolysis and liquefaction of the large organic molecules by extracellular enzymes. Carbohydrases catalyse the hydrolysis of glycidic bonds. These enzymes attack polysaccharides from the non-reducing end of the chain cleaving alternate glycidic bonds. In spentwash which comes out fresh from distillery, the temperature remains 60–80°C with acidic pH. Therefore, before processing into anaerobic digestion, the spentwash is cooled and neutralized with lime, and even sometimes the treated distillery effluent which becomes alkaline after anaerobic digestion is recycled for neutralization. Thereafter, actual hydrolysis starts where carbohydrases catalyse the monosaccharides and disaccharides. The disaccharides are then cleaved to mono-saccharides by a glycosidase. The specific enzyme involved depends on the nature of the glycosidase bond and the monosaccharide involved, i.e. whether the configuration is alpha or beta, dextra or hexo and the size of the heterocyclic ring. Cellulase and chitinase degrade the structural polysaccharides, cellulose and chitin; lipases and esterases hydrolyse fats and lipids when the substrate is a mixed of triglyceride of long chain fatty acids similar to those found in natural fats and oils. Starch and glycogen are hydrolysed to a disaccharide by the action of enzymes called amylases. The hydrolysis proceeds in a step-wise manner with the rapid formation of di and monoglycerides fallowed by slow hydrolysis of the monoglycerides. Proteolytic enzymes, proteases, catalyse the hydrolytic cleavage of the peptide bonds of proteins. These enzymes are somewhat specific exopeptidases, restricted to terminal peptide bonds; and endopeptidases those attack peptide bond located centrally in the peptide chains. The enzyme trypsin is specific for bond involving the amino acids argenine or lysine.

After the hydrolysis and liquefaction the initial degradation products are in a form that can be utilized by the microorganisms provided that they are able to pass into the cell through the cytoplasmic membrane. The membranes selectively regulate the flux of nutrients, ions and waste products in and out of the cell. It is composed of lipids and proteins. The lipids provide the structural properties and the proteins provide distinctive functional properties to the membrane. It is be-specific substances or types of molecules. This explains the ability of the membrane to pump against the head of an osmotic pressure gradient, a phenomenon called active transport. The lactose transport system of *E coli*, for example, is capable of producing a lactose concentration 500 times greater inside the cell than outside. The operation of these protein gates requires a certain amount of membrane fluidity which is largely determine by the relative amounts of saturated and unsaturated fatty acids in the lipid portion. Bacteria grown at a low temperature have membranes with a greater proportion of unsaturated fatty acids than those grown at a higher temperature. A typical metabolic pathway for the degradation of monosaccharides is the Embden-Meyerno and Parnas Pathway of glycolysis. Note that the end product of glycolysis in yeast is ethanol and in bacteria is acetic acid. Carbon dioxide is evolved in either case. Energy from exothermic transformation is stored in high energy phosphate bonds of ATP to ADP. Nicotinamide Adenine Dinucleotide (NAD) is a coenzyme, often involved in hydrogen transfer reactions, which must be regenerated with a coupled reaction. In recent years, carbon-14 tracers have been used in substrate to aid in defining of the metabolic pathways. It is generally the anaerobic degradation of long chain fatty acids proceeds primarily via beta oxidation in which two carbon atoms at a time are split from the chain in the sequence. It shows the removal of one acetate unit which combines with reduced coenzyme A (CoASH). Acetic acid can then either be liberated from CoA in a subsequent reaction, or the acetate can be transferred to other functional compounds. Flavin Adenine Dinucleotide (FAD) is a prosthetic group for a wide variety of enzymes (E-FAD) associated with hydrogen transfer reactions. In nature as well as effluent anaerobic biological processes of waste treatment, it is most likely that proteolytic species of *Clostridium* are largely responsible for the anaerobic decompositions of amino acids. They can accomplish this in one of the two ways. Some amino acids may be fermented individually by pathways specific to that compound as revealed from the studies using corbon-14 tracers during the anaerobic digestion of glutamic acid. The second method is the stick and reaction whereby pairs of amino acids are fermented. One amino acid acts as the electron donor (is oxidised) while the other is reduced. During these hydrolysis and fermentation steps very little COD is removed, however, some carbon dioxide formed in the process, results from the one of two carbon units broken down from large molecules. The growth yield of fermentative cells is low due to the small net production of ATP by substrate phosphorylation. Finally the low molecular weight acids produced in the acid production stage are further degraded to methane and carbon dioxide by a highly specialized group of bacteria commonly referred to as the methane producing bacteria. These organisms have the unique ability to couple organic oxidation to reduction of carbon dioxide. In this process carbon dioxide is the terminal hydrogen accepter.

Four genera of strictly anaerobic bacteria are known to produce methane.

1. Methanobacterium : a nonspore forming rod
2. Methanobacillus : a spore forming rod
3. Methanococcus : a nonspore forming coccus
4. Methanosarcina : a nonspore forming coccus in packets

It was believed earlier that a variety of low molecular weight alcohols and acids could be used as substrate by particular methanogens. More recent studies suggest that the cultures previously used were not pure and that the methanogens are able to utilize only a few are able to form methane from species of Methanosarcina are able to form methane from methanol or acetic acid.

In the absence of hydrogen the reaction is:

$$CH_3COOH \longrightarrow CH_4 + CO_2$$

If hydrogen is available the carbon dioxide is reduced to methane.

$$CO_2 + 4H_2 \longrightarrow CH_4 + 2H_2O$$

In the overall process of the anaerobic conversion of organic material into methane there is interaction between the fermentative (acid forming) bacteria and the methanogens in the form of interspecies hydrogen transfer. The utilisation of hydrogen by the methanogens enhances the acid production stage by, for example, making the conversion of acetyl-CoA to acetic acid thermodynamically under anaerobic conditions.

In general, biochemical sequences for the anaerobic degradation of organic compounds are through pyruvic acid and acetic acid. They serve as building blocks for the synthesis of more complicated organic molecules by getting converted into the intercellular storage product poly-beta hydrobutyric acid (PBH), or degraded to ethanol by yeast or to methane and carbon dioxide by the methane bacteria. In hydrolysis, fermentation and methane production only a small amount of oxidation has occurred. Net energy generated in the form of ATP has been minimal and new cell yields are low. The large organic molecules have been converted to methane (50–70%), and carbon dioxide (25–45%) and small amount of hydrogen, nitrogen and hydrogen sulphide but from thermodynamic prospective much of the chemical energy of the original organic molecule are transferred to methane. The methane generated in continued conditions can be collected and used as fuel or simply destroyed by torching. In anaerobic treatment process significant proportion (>50%) of the COD is converted in to biogas. The overall 50% BOD, COD is reduced during anaerobic digestion.

Methanogenesis and sulfate reduction are terminal steps in the anaerobic degradation of organic matter. The anaerobic digestion of organic material is accomplished by the concerted action of various trophic groups of bacteria (methanogens or sulfate-reducing bacteria, SRB). These reactions are known to compete with each other for electron donors, i.e. H_2 and acetate, in various environments. SRB are believed to outcompete methanogens in the presence of nonlimiting sulfate concentrations because they compete better for common substrates. This is explained on the basis of their kinetic properties (Ks and µmax) and given the favourable thermodynamic conditions. In contrast, methanogens tend to dominate in low-sulfate environments. However, in anaerobic digestion various biochemical changes occur. There is formation of significant amount of H_2S as a result of the reduction of oxidized sulfure compound in anaerobic digestion along with other biochemical changes. An inhibitory effect of sulfide, the end-product of sulfate reduction, on anaerobic biogas production is known. Thus, sulfate reduction is generally thought to be obstructive to the methanogenic digestion. Besides, this sulfide has binding properties with the existing heavy metals present in effluent. This, converts as colloidal solution of metal sulfide, as colourant (dark black). This makes effluent more complex and recalcitrant.

The spentwash remains highly acidic in nature and has a variety of recalcitrant colouring compounds as melanoidins, phenolics and metal sulfides that are mainly responsible for the dark colour of distillery effluent. The pH of spentwash increases from 4.5 to 8.5 during the anaerobic treatment process and finally it is called as post methanated distillery effluent (PMDE). Generation of different stages of effluent from distillery are shown in Fig. 2.4. The PMDE retains high BOD (18000–22000 mg/L), COD (32400–35000 mg/L), colour (150000–180000 mg/L), sulphate (3100–5760 mg/L), phenol (4.0–4.2 mg/L), TSS (11920–25308 mg/L), TDS (10480–77776 mg/L) and heavy metals (Cu, Mn, Fe, Ni and Zn). Apart from high organic content distillery wastewater also contains nutrient in the form of nitrogen (1660–4200 mg/L) phosphorus (225 3038 mg/L) and potassium (9600–17475 mg/L) that can also lead to eutrophication of water bodies. The comparative physico-chemical properties of spentwash and PMDE are shown in table 2.2.

The colour of spentwash is found dark brown. The colour of PMDE becomes dark black. Odour of the distillery effluent remains offensive. Odourous compounds from distillery wastewater mainly consist of volatile fatty acids such as butyric and valeric acids that have a high odour index. Due to the presence of putriciable organics like skatole, indole and other sulphur compounds, the PMDE that is disposed in canals or rivers produces obnoxious smell (Mahimaraja and Bolan, 2004). Distillery has distinct organic compositions. Various anaerobic bacteria ferment these compounds and generate products such as volatile fatty acids for example glycerol is fermented into butyric acid by *Clostridium butyricum*. The total solids of spentwash were 42400.2 mg/L, but when it was biomethanated, the value decreased. Total solids are the residues that include both dissolved solids and suspended solids. Distillery effluents contain huge amount of solids. PMDE is converted more dark and viscous than original spentwash in anaerobic digestion

Fig. 2.4: Generation of different types of wastewater after alcohol production, Spentwash (a) Biometha-nated reactor (b) distillery effluent after treatment in aerobic reactor (c) showing final discharge of PMDE (d).

with alkaline pH. The pH increase in methane reactor from 4 to 8.5 due to oxidation of organic acid to CO_2 and the reaction between CO_2 to basic compounds and generates carbonates and bicarbonates this increase the pH of the effluent. The electrical conductivity of spentwash is high but when it is biomethanated, value decreased. The total hardness of the PMDE remains also higher. The term total hardness indicates the concentration of calcium and magnesium ions.

The calcium, magnesium and chloride of the spentwash content 2070.0, 2260.5 and 8530.2 mg/L, respectively. The calcium, magnesium and chloride content of distillery effluent remains 872.0, 1742.0 and 5352.6 mg/L, respectively. Dissolve Oxygen (DO) become nil in spentwash and the value increases to 2.6 mg/L when PMDE is biomethanated. Hence, the PMDE is a potential water pollutant in two ways. First, the highly coloured nature of PMDE can block out sunlight from rivers and streams, thus reducing oxygenation of the water by photosynthesis and hence becomes detrimental to aquatic life. Secondly, it has a high pollution load which would result in eutrophication

Table 2.2: Characteristics of wastewater generated from distillery

S.No.	Parameters	Distillery Spentwash	Post Methanated Distillery Effluent	*Permissible Limits for Discharge
1.	Colour (Co-pt)	Dark brown (150,000)	Dark Brown (70,000)	Not clear
2.	Odour	Jaggery smell	Mild sulphur smell	Not clear
3.	pH	3.9 0.25	8.3 0.310	5–9
4.	Temperature(°c)	90 2.0	35 1.2	40
5.	T.S.	1,03,084 5.50	34317 455	2200
6.	T.D.S	77,776 3768	20022 438	2100
7.	T.S.S.	25,308 1201	14276 16	100
8.	COD	90,000 231	58018 185	120
9.	BOD	42,000 123	29120 265	40
10.	Chloride	2,200 105	1,300 60.5	1500
11.	Phenol	4.20 1.8	1.65 0.76	0.5
12.	Sulphate	5,760 260	13656 21.23	1500
13.	Phosphate	5.36 0.168	1.16 0.15	Not clear
14.	Total nitrogen	2,800 130	568 23	25
15.	Total organic carbon	25,368 1.060	10,904 0.34	Not clear
16.	Metals			
i.	Cadmium	0.02 0.00	2.281 0.067	0.01
ii.	Chromium	0.192 0.008	0.440 0.013	0.05
iii.	Iron	6.312 0.210	84.01 1.980	2.0
iv.	Nickel	0.1706 0.006	1.241 0.037	0.10
v.	Copper	0.961 0.001	0.955 0.022	0.50
vi.	Lead	0.945 0.002	4.446 0.064	0.05
vii.	Zinc	2.012 0.001	4.631 0.108	2.00
viii.	Manganese	0.214 0.001	2.112 0.045	0.20

Mean± SD n = 3; Values are reported in mg/L except pH, temperature, odour and colour; *Effluent discharge standards (EPA 2002)

of contaminated water courses. Undiluted effluent has toxic effect on fishes and other aquatic organisms. The estimated [LC_{50}] for distillery spentwash is found to be 0.5% using a bio-toxicity study on fresh water fish *Cyprinus carpio var. communis* (Mahimaraja and Bolan, 2004). Impacts of distillery effluent on carbohydrate metabolism of freshwater fish, *C. carpio* have been studied recently by Ramakritinan et al. (2005).

The respiratory process in *C. carpio* under distillery effluent stress was affected resulting in a shift towards anaerobiosis at organ level during sublethal intoxication. PMDE also leads to significant levels of soil pollution and acidification in the cases of inappropriate land discharge. It is reported to inhibit seed germination, reduce soil alkalinity, cause soil manganese deficiency and damage agricultural crops. However, effect of distillery effluent on seed germination is governed by its concentration and is crop-specific. In a study by Ramana et al. (2002a) the germination percent in five crops decreased with increase in concentration of the effluent. The germination was inhibited in all the five crops studied with concentration exceeding 50%. At the same time, organic wastes contained in distillery effluent are valuable source of plant nutrients especially N, P, K and organic substrates if properly utilized (Pathak et al., 1999). For instance, distillery effluent in combination with bioamendments such as farm yard manure, rice husk and Brassica residues was used to improve the properties of sodic soil (Kaushik et al., 2005). The use of fungi for bioconversion of distillery waste into microbial biomass or some useful metabolites has been recently reviewed by Friedrich (2004). The end products of bioconversion are fungal biomass, ethanol, enzymes, etc. and substantially purified and decolourised effluents. Recently enhanced production of oyster mushrooms (*Pleurotus sp.*) using distillery effluent as a substrate amendment have been reported (Pant et al., 2006). The details about impact of distillery effluent are described in chapter 6.

Other chemical constituents and colourant of PMDE

Most of the colour contributing substances are generated and concentrated during the processing of sugarcane juice while some are added during the processing of sugar in sugar industries. Heating of glucose and fructose under acidic or basic condition in sugar industries leads to the degradation reaction forming the highly reactive intermediates which undergo condensation and polymerization reactions forming more complex and coloured polymers known as melanoidins. The colourants formed from sucrose can be divided into enzymatic colourants such as melanins and non-enzymatic colourant such as melanoidins, alkaline degradation product of hexoses and caramel. Cane molasses contains around 2% of melanoidin that imparts colour to the effluent. There is high concentration of sulphate also in sugarcane molasses, which is added to molasses during the cleaning of sugar crystals. High level of sulfate can lead to the production of sulfides in anaerobic digestion of the effluent, these precipitate out along with the existing metal as metallic sulfides in PMDE, consequently increasing the total solid (TS) of effluent. The colour is hardly degraded by the conventional treatment methods and can even be increased during anaerobic treatments, due to repolymerization of compounds. Phenolics (tannic and humic acids) from the feedstock, melanoidins from Maillard reaction of sugars (carbohydrates) with proteins (amino groups), caramels from overheated sugars, and furfurals from acid hydrolysis mainly contribute to the colour of the effluent (Table 2.3).

Table 2.3: Major colourants and their characteristics present in PMDE

Colourant	Characteristics
Melanoidin	Browning reaction products of sugars and amino acids with high molecular weight.
Caramel	Process colourant, thermal degradation of sugar, high molecular weight low net charge.
Phenolics	Plant pigments, low molecular weight, may be attached to polysaccharides, pH sensitive, darker at high pH, pale yellow to orange colour, react with iron to produce very dark colour, may dimerize or oxidize to form darker colour.
Alkaline degradation product of fructose	Process colourants, reddish to dark brown in colour, low molecular weight up to polymeric depending on degree of degradation.
Sulfide	Process colourant, toxic to microorganisms and creating foul smell, strong binding tendency with metals.
Heavy Metals	Process colourants, toxic to microorganisms, animals and humans.

Table 2.4: Organic compounds detected in PMDE after Pyrolysis

S. No.	Compounds	Molecular Weight
1.	Acetic acid ($C_2H_4O_2$)	60
2.	3-pyrroline (C_4H_7N)	69
3.	Hydroxypropanone	74
4.	2, 3-dihydro-5-methylfuran (C_5H_8O)	84
5.	Butenoic acid	86
7.	Methyl benzene (C_6H_8)	92
8.	Phenol	94
9.	2,5 Dimethylfuran	96
10.	2,4-Dimethylfuran	96
11.	Furfural	96
12.	Furfuryl alcohol	98
13.	2-hydroxymethylfuran	98
14.	2, 3-Dihydro-5-methylfuran-2-one	98
15.	2-Methylhexane	100
16.	Vinylbenzene (styrene)	104
17.	2, 3-dimethyl-pyrazine ($C_6H_8N_2$)	108
18.	Methylphenol	108
19.	2-Ethyl-5-methylfuran (isomer)	110
20.	5-methyl-2-furfuraldehyde	110
21.	2-Ethyl-5-methylfuran	110
22.	Indole (C_8H_7N)	117
23.	Methylbenzaldehyde	120
24.	2-Methoxyphenol (guaiacol)	124
25.	5-Hydroxymethyl-2-furfuraldehyde (impure)	126
26.	Methylindole	131
27.	2, 2'-bifuran ($C_8H_6O_2$)	134
28.	p-chloroanisol (C_7H_7ClO)	142
29.	4-Vinyl-2-methoxyphenol (4-vinylguaiacol)	150
30.	2,6-Dimethoxyphenol (syringol)	154
31.	2-nitroacetophenone ($C_8H_7NO_3$)	165
32.	3,4-Dimethoxybenzyl alcohol (veratryl alcohol)	168
33.	2, 2'-bifuran-5-carboxylic acid ($C_9H_6O_4$)	178
34.	Dihydroxyconiferyl alcohol ($C_{10}H_{14}O_3$)	182
35.	2-Hydroxy-3-methoxyphenyl propanol (dihydroconiferyl alcohol)	182
36.	trans-2-tridecenal	196
37.	1-Hexadecanol	242
38.	Palmitic acid	256

Besides these colour contributing components, during heat treatment, the Maillard reaction (non-enzymatic reaction) is accompanied by the formation of a class of compounds known as Maillard reaction products (MRPs) in different fractions of a distillery effluent. The distillery effluent comprises many other compounds according to recent reports which are also listed in table 2.4. The pyrolysis/gas chromatography/ mass spectrometry (Py/GC/MS) shows different products in suspended solids (SS) which may be as hydroxypropanone, methylbenzene, phenol and methylphenol. Most have been reported as pyrolysis products of different amino acids: methylbenzene from phenylalanine; indole and methylindole from tryptophan; and, finally, phenol and methylphenol from the amino acid tyrosine. These compounds could be derived from the amino acids present in the original molasses or from the rest of the yeast cells that remain in the distillery effluent after the alcoholic fermentation. Despite hydro-xypropanone being a low-molecular weight compound it is not a good diagnostic marker; it has been reported as a dehydration product of glycerol and as a typical carbohydrate pyrolysis product. Glycerol represents 5–6% of the dry weight of vinasses. The existence of carbohydrate-derived compounds such as furfuryl alcohol and 2,3-dihydro-5-methylfuran-2-one or lignin-derived moieties such as 4-vinylguaiacol and syringol could be ascribed residual sugarcane fragments present in the suspended solids fraction of the distillery effluent. In addition, the presence of furans could also be explained from both, their presence in molasses, as well as from pyrolysis degradation of the non-fermented sugars present in the original molasses or even from thermal degradation of melanoidins. Moreover, 4-vinylguaiacol has been described as a pyrolysis product of ferulic acid, one of the two cinnamic acids present in the vegetal cell wall of gramineous plants as is the case of sugarcane (*Saccharum officinarum* L.) leads to sugarcane molasses.

Palmitic acid could come from yeast cells or from the original sugarcane molasses. A series of diagnostic methoxyphenols such as guaiacol, 4-vinylguaiacol and syringol were also found in the polysaccharide free fraction of effluent (PF). In a minor proportion, some long-chain alkyl compounds were represented in the PF fraction by trans-2-tridecenal, 1-hexadecanol and palmitic acid. These compounds have been described as derived from lipids. Styrene, a pyrolysis product of phenylalanine is also present in PF fraction. It has also been reported in the Py/GC/MS of humic acids. Furan-related moieties, i.e. 2,5-dimethylfuran, 2-hydroxymethylfuran, 2,4-dimethylfuran, 2, 3-dihydro-5-methylfuran-2-one and 5-methyl-2-furfuraldehyde are major compounds present in Polysaccharide-rich fraction (PR) of effluent. The comparative LC-MS-MS and other spectrophotometric analysis of distillery effluent has shown the presence of dihydroxyconiferyl alcohol, 2, 2'-bifuran-5-carboxylic acid, 2-nitroacetophenone, p-chloroanisol, 2, 3-dimethylpyrazine, 2-methylhexane, methylbenzene, 2, 3-dihydro-5-methylfuran, 3-pyrroline, and acetic acid (Bharagava and Chandra, 2010a). Many unknown persistent organic pollutants (POPs) still are unknown.

3

Melanoidin as a Major Colourant of PMDE, its Synthesis, Structure and Property

3.1. INTRODUCTION

Melanoidins are natural condensation products of sugar and amino acids produced by non-enzymatic Maillard amino-carbonyl reaction taking place between the amino and carbonyl groups in organic substances. The **Maillard reaction** is named after the French scientist Louis Camille Maillard (February 4, 1878–May 12, 1936), who studied the reactions of amino acids and carbohydrates in 1912, as part of his PhD thesis, which was published in 1913. The **Maillard reaction** is not a single reaction, but a complex series of reactions between amino acids and reducing sugars, usually at increased temperatures. Like caramelization, it is a form of non-enzymatic browning. Non-enzymatic browning reactions depend on many parameters, such as, temperature, water activity (a_w), pH, moisture content and chemical composition. In general, maximum browning occurs at aw between 0.60 and 0.85 and the browning rate increases with increasing pH, up to a pH of around 10. In the process, hundreds of different flavour compounds are created. These compounds in turn break down to form yet more new flavour compounds, and so on. Melanoidins extensively exist in food products, drinks and wastewaters released from distilleries and fermentation industries. Melanoidins have commercial, nutritional and toxicological significance as these have significant effect on the quality of food, since colour and favours are important food attributes and key factor in consumer's acceptance. Food and drinks as bakery products, cofee and beer having brown coloured melanoidins exhibited antioxidant, antiallergenic, antimicrobial and cytotoxic properties as in vitro studies have revealed that Maillard reaction products (MRPs) may offer substantial health promoting effects as they can act as reducing agents, metal chelators and radical scavengers. Besides, these health-promoting properties, in vitro studies have also revealed some harmful effects of melanoidins as mutagenic, carcinogenic and cytotoxic effects. The wastewaters released from distilleries and fermentation industries are the major source of soil and aquatic pollution due to presence of water-soluble melanoidins. Melanoidins are very important from the nutritional, physiological and environmental aspects and due to their structural complexity, dark colour and offensive odor, these pose serious threat to soil and aquatic ecosystem that release of melanoidins increased load of recalcitrant organic material to natural water bodies. This then causes the problems, like reduction of sunlight penetration, decreased photosynthetic activity and dissolved oxygen concentration whereas on land it causes reduction in soil alkalinity and inhibition of seed germination. Further, due to the possibility of complexation reactions of introduced melanoidins with metal ions, they could influence the biogeochemical cycle of many constituents in natural waters. Melanoidin which are highly resistant to microbial attack and conventional biological processes such as activated sludge treatment process are insuficient to treat these melanoidins containing wastewater released from distilleries and fermentation industries. Hence, these wastewaters require pretreatment before its safe disposal into the environment. Degradation and decolourisation of these wastewaters by chemical methods, focculation treatment and physico-chemical treatment such as ozonation and activated carbon adsorption have been accomplished, but these methods are not economically feasible on large scale due to cost limitation whereas biological decolourisation by using fungi such as *Coriolus*, *Aspergillus*, *Phanerochaete* and certain bacterial sp. as *Bacillus*, *Alkaligenes* and *Lactobacillus* have been successfully achieved and thus can be applied as a bioremediation techniques. However, the biological decolourisation of wastewaters containing melanoidins depends on pH, temperature, concentration of nutrients, oxygen and inoculum size while the enzymatic system responsible for the degradation of melanoidins consists mainly sugar oxidases and peroxidases as sarbos oxidase, glucose oxidase, manganese dependent and independent peroxidases (MnP and MnIP). Since, MnP and MnIP showed melanoidin decolourizing activity in presence of H_2O_2 and the decolourizing activity of both sugar oxidases and peroxidases were found optimum at a particular pH, temperature and substrate specific. The details of melanoidin containing distillery effluent degradation are described in chapter nine. The structure of melanoidins is still not completely understood but it is assumed that it does not have a definite structure as its elemental composition and chemical structures largely depend on the nature and molar concentration of parent reacting compounds and reaction conditions as pH, temperature, heating time and solvent system used. Carbon isotope ratios support the stoichiometric ratios for the combination of sugars with amino acids, which are based on the elemental composition data of melanoidins. In this chapter we described the synthesis, chemical structure and properties of melanoidins.

3.2. MELANOIDIN FORMATION

General Chemistry of Melanoidin formation/Maillard reaction

The Maillard reaction occurs in three main steps:

A. Initial stage: Products colourless, without absorption in the ultraviolet (280 nm). The first stage involves the sugar-amine condensation and the Amadori rearrangement. The reaction steps have been well-defined and no browning occurs at this stage.

Reaction A : Sugar-Amine condensation

Reaction B : Amadori rearrangement

The Maillard reaction is initiated by a condensation reaction between the carbonyl group of the aldose and the free amino group of an amino acid to give an N-substituted aldosylamine (Fig. 3.1). This is the result of a nucleophilic attack group by the NH group of the amino acid on the electrophilic carbonyl groups of sugar. It is basically an amine assisted dehydration reaction of sugar. The condensation product rapidly loses water as a product and is converted into a Schiff base. This reaction is reversible and acid-base catalyzed. The Schiff base is cyclizes into the aldosylamine. The Amadori rearrangement follows for aldoses; Heyns rearrangement for ketoses to form a ketosamine. If the aldose is glucose and amino acid glycine, then the Amadori product is 1-amino-1-deoxy-2-fructose (monofructoseglycine, MFG). The Amadori rearrangement is considered to be the key step in the formation of major intermediates for the browning reaction. Ketoses, such as fructose, react with amines to form aminoaldoses, this is called the Heyns reaction. The intermediates to this reaction are imines. Aminoaldoses are not very stable and readily react forming the Amadori compound.

B. Intermediate stage: Products colourless or yellow, with strong absorption in the ultraviolet. The second stage involves sugar dehydration and fragmentation, and amino acid degradation via the Strecker reaction especially at high temperatures as used in candy manufacture. At the end of stage two, there is a beginning of flavor formation depending on which flavor is studied.

Reaction C: Sugar dehydration

Reaction D: Sugar Fragmentation

Reaction E: Amino acid degradation (Stecker degradation)

The Amadori product degrades by one of three main pathways depending on the conditions. Sugar dehydration occurs in two ways: under acid conditions, furfurals are produced,

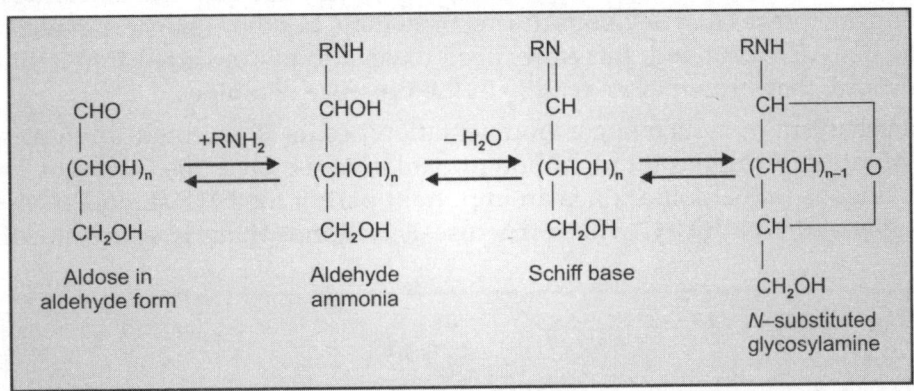

Fig. 3.1: Sugar-amine condensation to form N-substituted glycosylamine

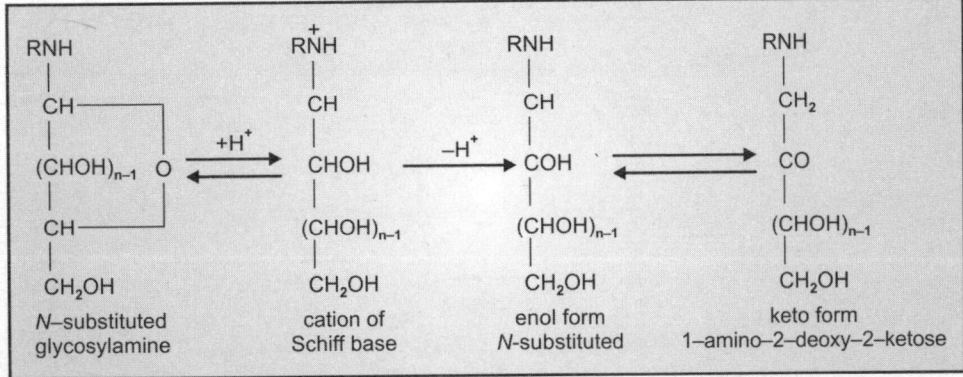

Fig. 3.2: Amadori rearrangement leading to the Amadori compound, the N-substituted 1-amino-2-deoxy-2-ketose

whereas in neutral or alkaline conditions and/or in the presence of amines in nearly anhydrous systems, 6-carbon and other reductones are favoured. The two pathways are illustrated in Fig. 3.3, where it can be seen that the first proceeds via 1,2-enolisation and the 3-deoxyosone, whereas the second does so via 2,3-enolisation and the 1-deoxy-2,3-dicarbonyl.

1. **Furfural formation:** Various compounds can accelerate furfural formation; for example, glycine accelerates both the conversion of xylose to furfural and glucose to hydroxymethylfurfural (HMF). The reason seems to be that the Amadori product dehydrates more readily than the original aldose or N-Substituted glycosylamine, giving the Schiff base of furfural, which is then hydrolysed, re-liberating part of the amine, but also condensing to melanoidins. It is generally thought that HMF is of low browning potential and does not lie on the main pathway to melanoidins. Furfural formation is aften used as a relatively simple means of following the deterioration of food during storage.

2. **Reductone formation:** Reductons can be thought of as products formed from sugar by the loss of only two molecules of water, as compared with loss of three that leads to furfurals. Reductones are compounds which contain the group- C(OH):C(OH)-, as in ascorbic acid, and a hexose can readily be converted on paper into the vinylogue of a reductone. Compounds such as reductones explain the reducing power that develops during browning, but they take part in browning in the dehydro form and, therefore, need oxygen to be converted into it. Similar to furfural, they brown more readily in the presence of amines.

The mechanism by which sugar fragmentation occurs is accepted to be principally retroaldolisation (dealdolisation), although oxidative fission is also thought to play a role. Vivylogous retroaldolisation is an important part of the EMP (Embden-Meyerhof-Parnas) glycolytic pathway, where fructose-1, 6-diphosphate is split into dihydro-

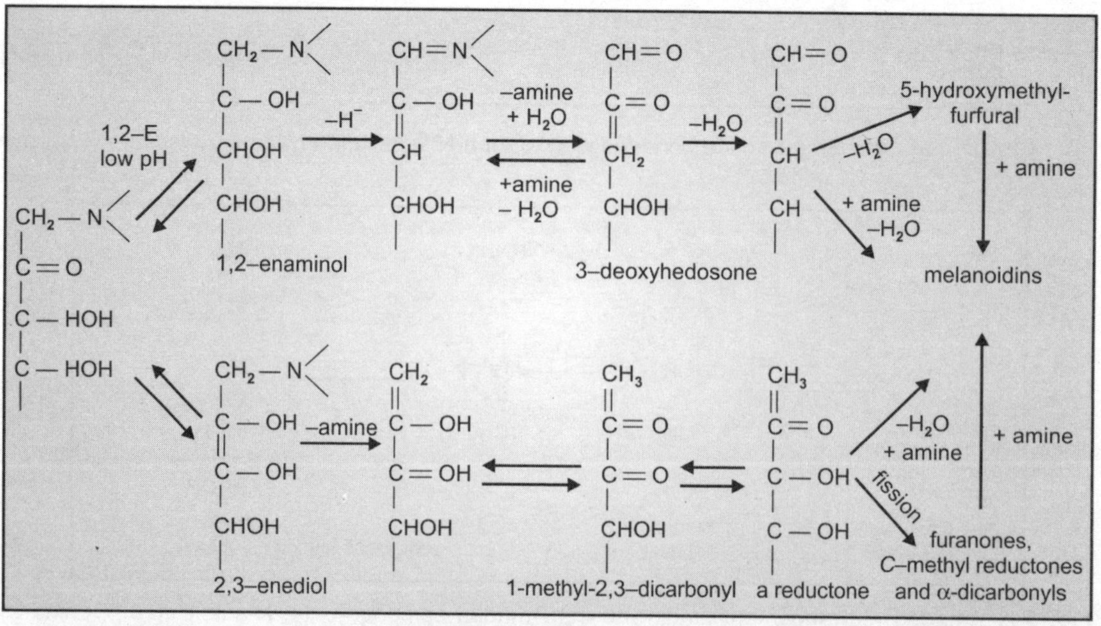

Fig. 3.3: Maillard reactions: the two major pathways from Amadori compounds to melanoidins

xyacetone-phosphate and glyceraldehydes-3-phosphate. The cleavage of hexose derivatives can be C_5/C_1, C_4/C_2, or C_3/C_3. The sorts of reactions which occur are illustrated in Fig. 3.4.

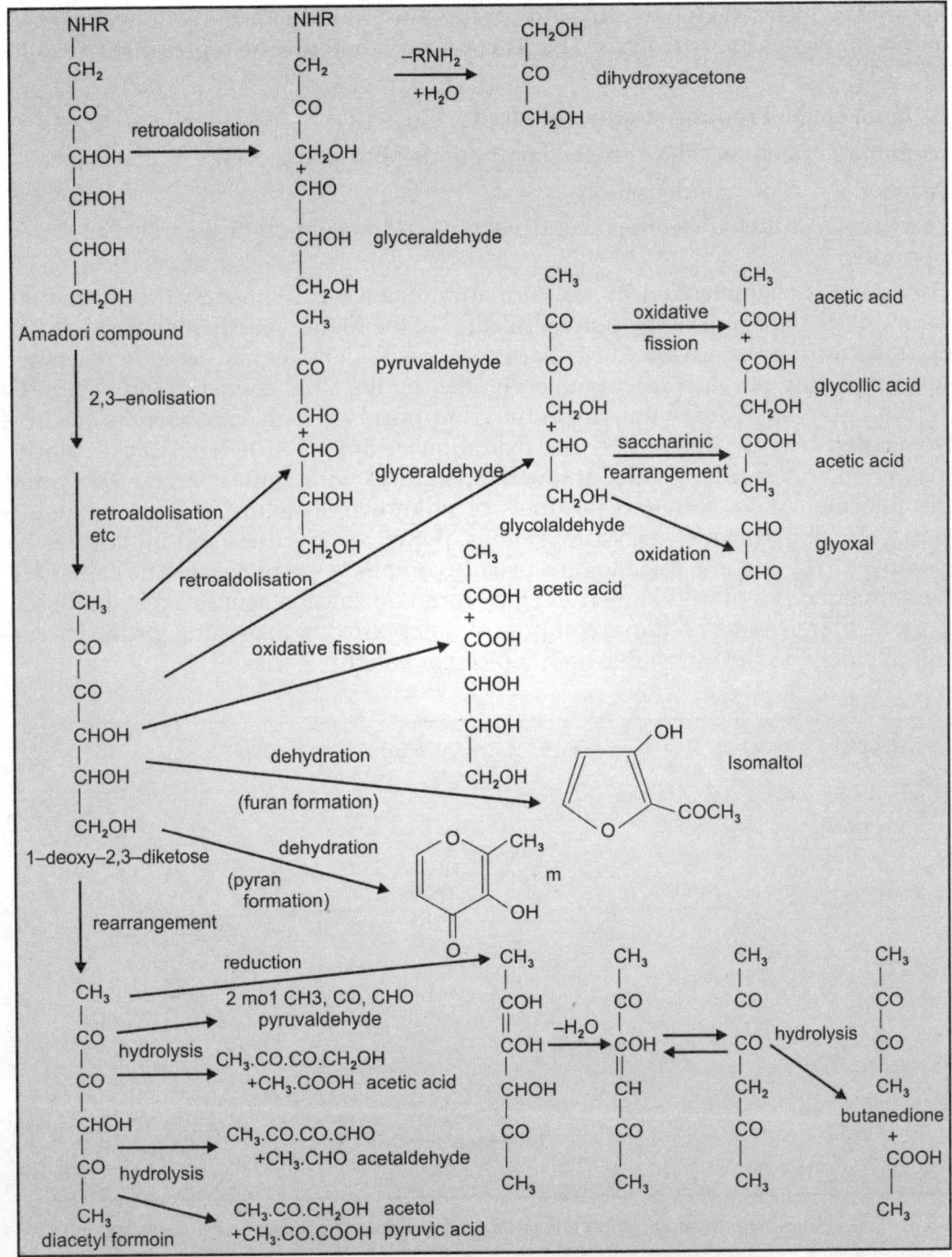

Fig. 3.4: Examples of sugar fragmentation

In Strecker degradation, this is a reaction of α-amino acids, in which they are oxidized to the corresponding aldehydes, giving off carbon dioxide and ammonia is transferred to other components of the system, very little being liberated as such. The reaction is initiated by compounds, such as α-dicarbonyl compounds and their vinylogues, or compounds which can give rise to them readily, such as reductones by dehydrogenation or imino analogues by hydrolysis. The reaction may therefore be represented as follows in Fig. 3.5.

C. Final stage: Products highly coloured

Formation of heterocyclic nitrogen compounds. Browning occurs at this stage.

Reaction F : Aldol condensation

Reaction G : Aldehydes-amine condensation and formation of heterocyclic nitrogen compounds.

This stage is characterized by the formation of brown pigments. The formation of melanoidins is the result of the polymerization of the highly reactive intermediates that are formed during the advanced Maillard reaction. Aldehydes can arise by reactions C, D and E and they can then react with each other by the aldol condensation. Amines are effective catalyst. Additional carbonyl compounds which can participate in the condensation may be derived by the oxidation of lipids. Aldehydes, particularly α, β-unsaturated ones, react readily at low temperatures with amines to give "polymeric" high molecular mass, coloured products of unknown structure, called melanoidins. Heterocyclic ring systems, such as pyridines, pyrazines, pyrroles and imidazoles have been shown to be present. Melanoidins usually contain 3–4% nitrogen. The chemistry of these compounds is not well-known and their formation mechanism also remains obscure. The molecular weight of these compounds increases as browning proceeds, until eventually they become insoluble high molecular weight species.

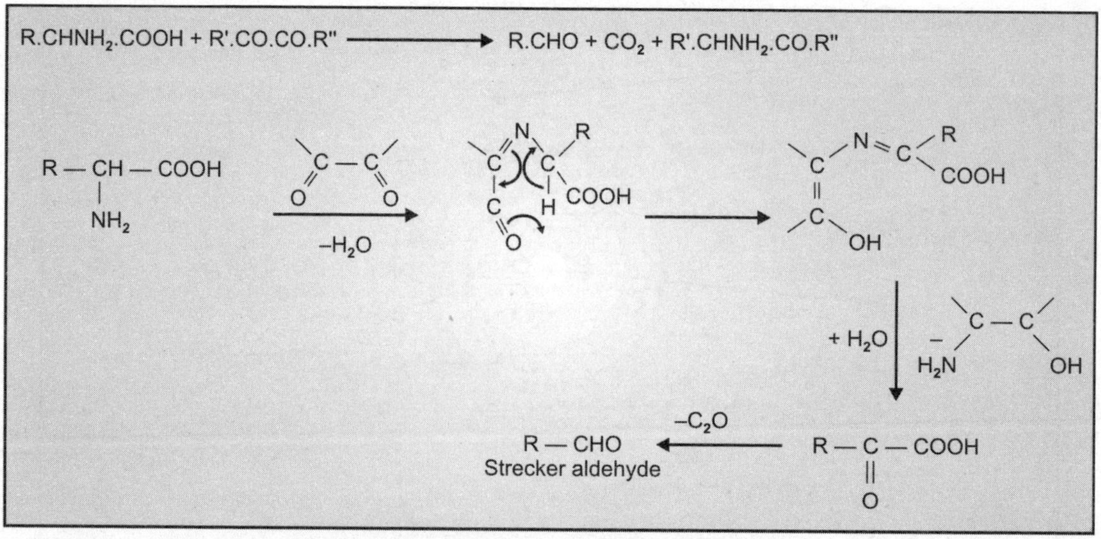

Fig. 3.5: The ketosamine products then either dehydrates into reductones and dehydro reductones, which are caramel, or products short chain hydrolytic fission products such as diacetyl, acetol or pyruvaldehyde which then undergo the Strecker degradation.

The Hodge Diagram

The classical scheme of the chemical reaction is that of Hodge as shown in Figure 3.6. This is still used today to describe the reaction. The Maillard reaction was first described as between reducing sugars with amino acids, but now is extended to include many other carbohydrate and amine groups. Sugar sources include dextrose, fructose, high fructose corn syrup, sucrose, corn starches and maltodextrins. Protein (-NH) sources for candy may include milk solids, cream, egg solids, nuts and nut fragments, cocoa solids, butter (small source of nitrogen), fruits and fruit juices provide free amino acids, gelatin, whey proteins and emulsifiers such as lecithin. The mechanism of the Maillard reaction as shown in Fig. 3.6 is very complicated. However, it is generally divided into three stages:

(A) The initial reaction between a reducing sugar and amino group forms an unstable Schiff base; (B) The Schiff base slowly rearranges to form the Amadori product; (C) Degradation of the Amadori product; (D) Formation of reactive carbonyl and dicarbonyl compounds; (E) Formation of Strecker aldehydes of amino acids and aminoketones; (F) Aldol condensation of furfurals, reductones, and aldehydes produced in Steps C, D, and E without intervention of amino compounds; (G) Reaction of furfurals, reductones, and aldehydes produced in Steps C, D, and E with amino compounds to form melanoidins; (H) Free radical-mediated formation of carbonyl fission products from the reducing sugar (Namiki pathway).

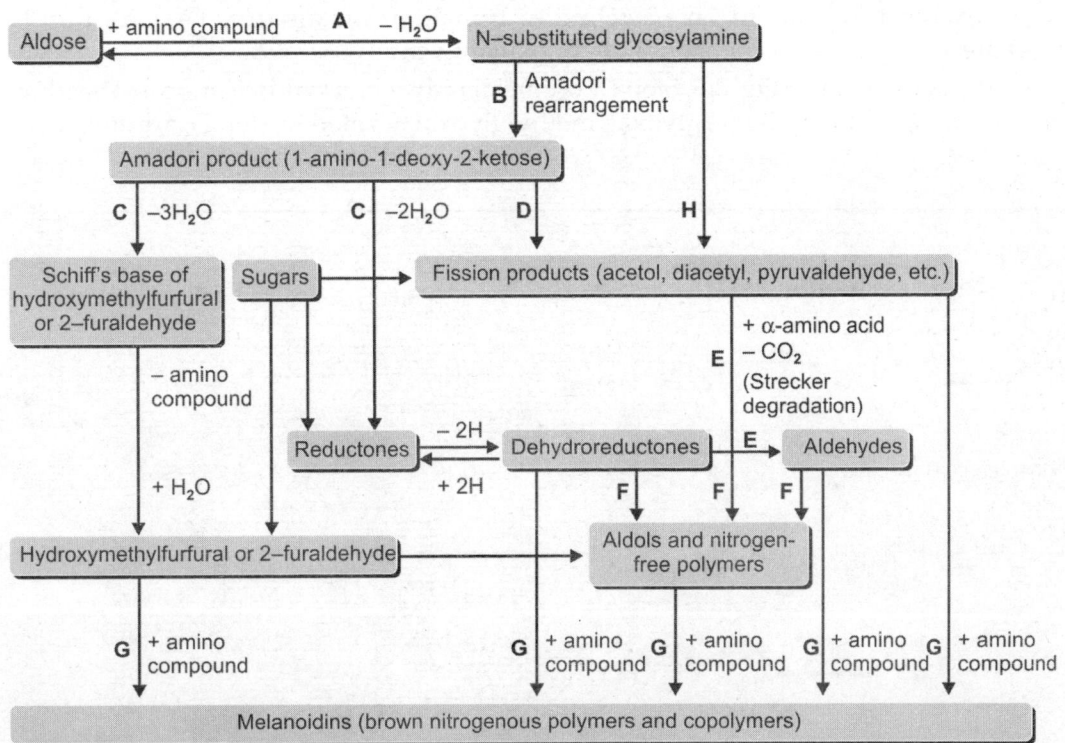

Fig. 3.6: Hodge Diagram

The initial stage of the Maillard reaction (Fig. 3.6, Step A) involves the condensation of a carbonyl group, for example from a reducing sugar such as glucose, with a free amino group, typically the epsilon amino group of lysine residues within proteins. This glycation reaction results in the formation of an unstable Schiff base (aldimine) that spontaneously rearranges (Fig. 3.6, Step B) to form the more stable 1-amino-1-deoxy-2-ketose (ketoamine), which is also known as the Amadori product (Fig. 3.6) after the Italian scientist Mario Amadori. When the initial sugar is glucose, the Amadori product is commonly known as fructoselysine (FL). The equilibrium constant for the formation of the Amadori product lies strongly in the forwards direction at physiological pH, 2 but reversibility increases with increasing pH and also in the presence of phosphate.

Reaction between glucose and amino group of protein to form the Amadori product Nucleophilic attack by a free amino group of protein on the aldehyde of glucose initially forms a carbinolamine, which subsequently dehydrates to a Schiff base (Fig 3.7). The Schiff base then undergoes a slow rearrangement to form the Amadori product. While only a single cyclic isoform of the Amadori product is shown, it is important to note that it exists as a mixture of several isoforms. Amadori products are degraded via various pathways (Fig. 3.6, Steps C and D), leading to the formation of furfurals, reductones and fragmentation products (carbonyl and hydroxycarbonyl compounds). All of these intermediates can also form directly from the sugar in uncatalyzed reactions, i.e. without the intervention of an amino compound (Fig. 3.6). In Step C, furfural formation is favoured under acidic conditions, while alkaline media favor the production of reductones. These conjugated enediol intermediates possess moderate reducing power; they may catalyze redox reactions dependent on recycling of transition metals (e.g. Fe, Cu), but, like ascorbate, they may also contribute to antioxidant activity.

Sugar fragmentation (Fig. 3.6, Step D) occurs mainly by retroaldolisation. α-Dicarbonyl compounds, e.g. butanedione, glyoxal, methylglyoxal, formed in Step D, are able to react

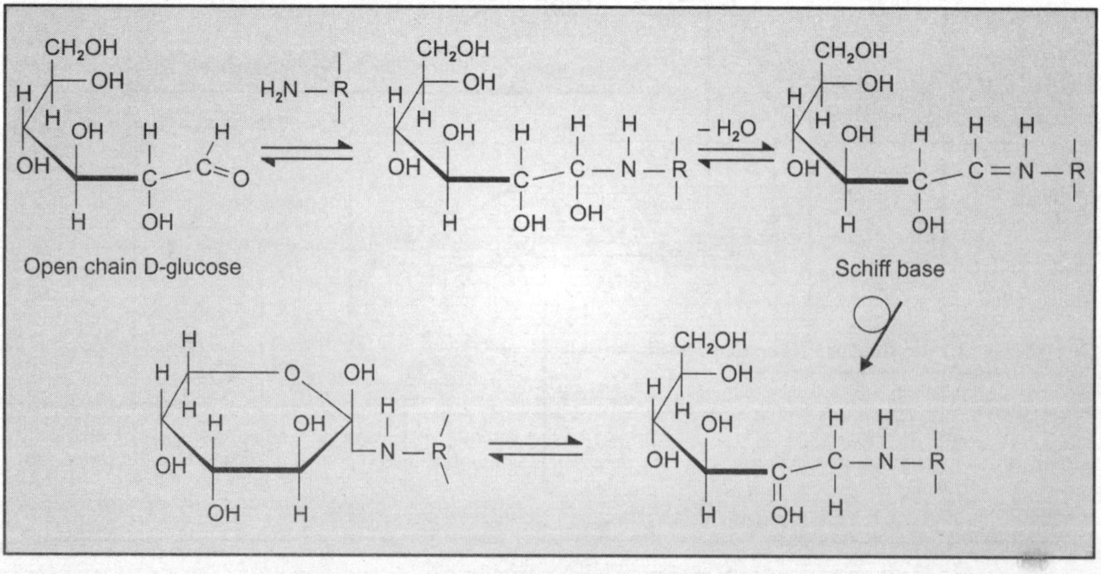

Fig. 3.7: Reaction between glucose and amino group of protein to form the Amadori product

with amino acids via the Strecker degradation (Fig. 3.6, Step E) (named after the German chemist Adolph Strecker) to give Strecker aldehydes of the amino acids and aminoketones; the latter subsequently condense to form pyrazines. Strecker aldehydes and pyrazines contribute to aroma in heated foods.

Steps F and G in figure 3.6 summarize the final stage of the Maillard reaction, and it is here that the majority of the compounds contributing colour are formed. These may be relatively small molecules or much larger polymeric materials. Step F involves aldol condensation of the furfurals, reductones and aldehydes produced in steps C, D and E without the intervention of amino compounds. Step G represents reactions between the same intermediates with amino compounds and that lead to the ultimate reaction products, known in food science as melanoidins. These poorly defined compounds chelate redox-active, transition metal ions and thereby possess antioxidant activity. Hodge defined melanoidins as brown, nitrogenous polymers and copolymers. Polymers are generally considered to contain repeating units but, since the structures of melanoidins are unknown and the presence of true repeating units uncertain, melanoidins are sometimes described more generally as macromolecular materials.

The constitution of the melanoidins differs somewhat depending on how they have been produced. These are temperature and duration of heating, pH and presence of weak acids and bases, water content, type of reactant, amino acid to sugar ratio and oxygen.

3.2 STRUCTURAL PROPERTIES AND THE LIKELY FORMATION MECHANISM OF MELA-NOIDINS

Although numerous attempts have been undertaken to isolate and purify melanoidins from foods such as bread crust and coffee, as well as from sugar-amino acid model systems, little is known about the overall structural properties of melanoidins. A few preliminary studies on melanoidin structure have been conducted, though, and high molecular weight (HMW) melanoidin structures from water free model systems (glucose-gluten heated to 150°C for 45 min and glucose-glycine/glutamic acid heated to 125°C for 2 h) and foodstuffs have been reported. The melanoidins from model systems, bread crust, coffee, and tomato sauce, elucidated from the volatile profiles after thermal degradation, are shown to be comprised primarily of furans accompanied by carbonyl compounds, pyrroles, pyrazines and pyridines. Low molecular weight (LMW) chromogenic compounds and coloured substructures of melanoidin-type compounds (Fig. 3.8A-D), produced by refluxing neutral aqueous solutions of xylose/glucose and alanine/proline/ lysine for less than 5 h, were shown to consist of a similar chemical mixture, that contained furan, pyrrole and pyrrolinone structures and derivatives (Fig. 3.8A-D). Blue and red pigments with polymerising activities have been isolated from alkali aqueous xylose-glycine, glucose-glycine and xylose-alanine reactions conducted under nitrogen (pH 8.1, 60% ethanol, 26.5°C, 48 h). The blue pigments (pyrrolopyrrole dimer) were determined to have a novel chemical structure consisting of two pyrrolopyrrole rings combined with a methane bridge (Fig. 3.8 E), and the red pigments (pyrrolopyrrole azepine) were shown to have unique structures of pyrrolopyrrole, pyrrole and azepin rings (Fig. 3.8F). Yellow compounds (acetyl pyridinone and acetyl azepinone), consisting of either a ring of pyridinone (Fig. 3.8 G) or azepinone (Fig. 3.8 H), are also isolated from the acidic xylose-glycine and

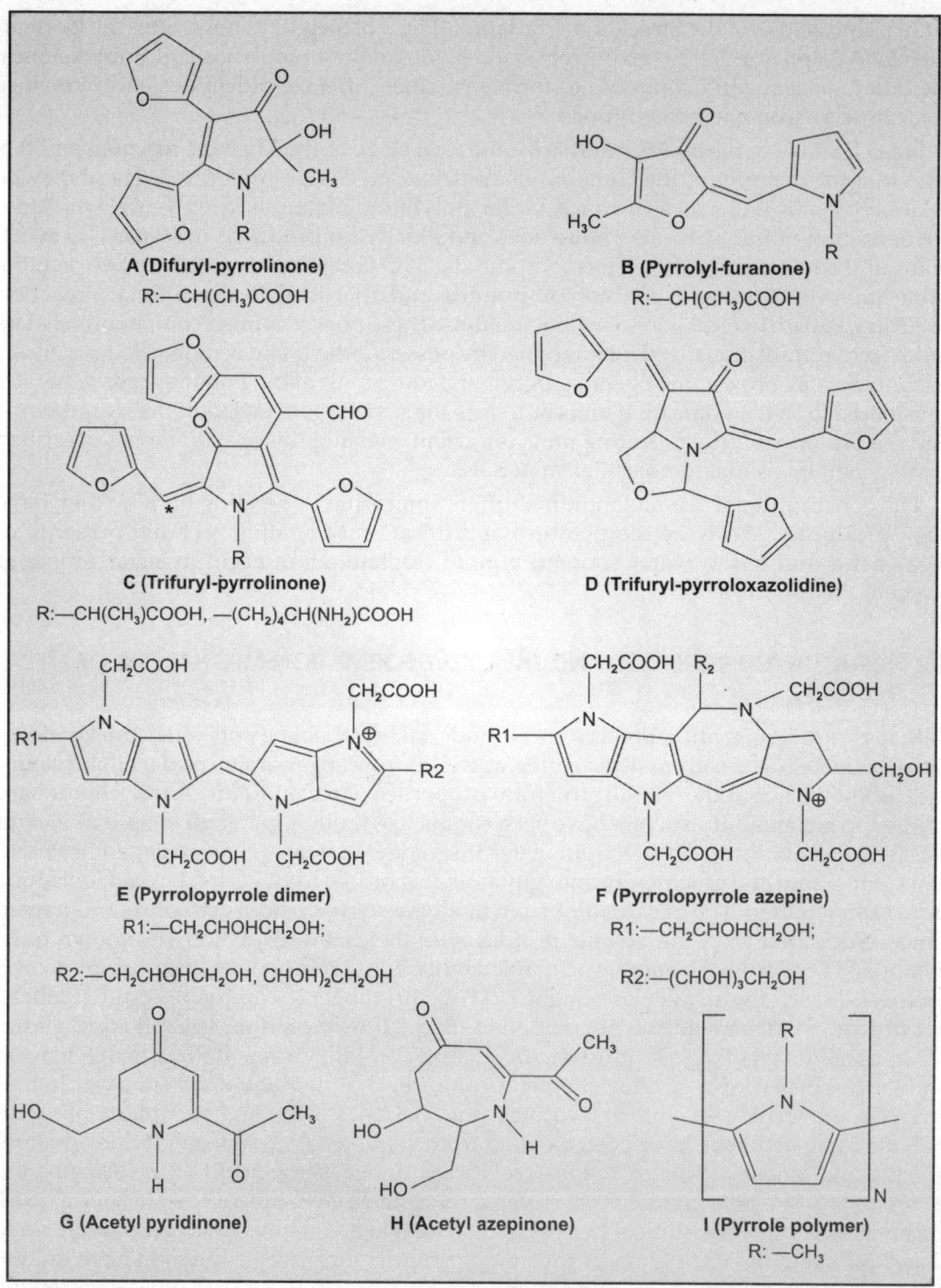

Fig. 3.8: Some LMW compounds with chromophores of substructures for melanoidin *asymmetric carbon for isomer

glucose-glycine systems (refluxing for 2 h at pH 5), respectively. Moreover, Tressl et al. (1998) reported an N-substituted pyrrole polymer chromophore with extraordinary polycondensation activity (Fig. 3.8I), formed from neutral pentose and methylamine in an aqueous MR model system (120°C, 1 h). The polymerising activities of these LMW colourants suggests they are important reaction intermediates in the formation of melanoidins. In model MR systems, pH appears to be an important factor that influences the structure of the chromophoric melanoidins, while the reaction temperature and time dictate the molecular weight of melanoidins. Three theories on HMW melanoidin formation have been described. One theory is that HMW coloured structures are formed by polymerisation of LMW Maillard reaction intermediates, such as furans, pyrroles, pyrrolopyrroles, and/or their derivatives, in the later stages of the reaction. The polymerising activities of the LMW colourants mentioned above suggest these compounds could function as reaction intermediates during polycondensation, making the molecular weight of the final melanoidins dependant on the heating time. Hofmann (1998a, b) has postulated that HMW melanoidins are derived from cross-linking chromophoric LMW MRPs and reactive amino acid side chains such as lysine, arginine, or cysteine. Several melanoidin-amino acid chromores have been described.

An intense red-brown chromophoric compound with the substructure of furyl-imidazolidine dimer linked with two arginine moieties (Fig. 3.9A) has been detected in a heated solution of Na acetyl-L-arginine with glyoxal and furan-2-carboxaldehyde (90 °C for 80 min, at pH 7.0), an MR model system that contains arginines. Brown-orange melanoproteins were detected in heated aqueous reactions of casein and furan-2-carboxaldehyde (70 °C for 1 h, pH 7.0). These compounds were formed by linking trifuryl-pyrrolinone chromophores (Fig. 3.8C) to the e-amino group of a lysine residue (Fig. 3.9B). Another melanoidin structure, protein-bound pyrrolinone reductonyl-lysine, also known as pronyl-L-lysine (Fig. 3.9C), has been identified in bread crust, malt and MRPs from an aqueous model system (100 °C for 25 min, pH 5.5) containing the lysine side chain. This compound is responsible for the dark colour of these foods. The chromophores of melanoproteins usually contain pyrroles, imidazoles, and their nitrogen-containing derivatives. The nitrogen atoms of these structures can be cross-linked to proteins via the amino and guanidine groups of amino acids, as well as to the furan substructures produced upon heating the carbohydrates. The third and final theory proposes that the melanoidin skeleton is built primarily from sugar degradation products branched with amino compounds, such as amino acids. It has been demonstrated in aqueous systems that melanoidins are formed by aldol condensations of highly reactive a-dicarbonyl compounds, which are the main intermediates during the early stages of the MR, and partially branched by amino compounds. In water-free reaction conditions, the transglycosylation reactions that form the carbohydrate-skeleton of melanoidins have been shown to take place during the early stages of MRs. When using di or oligosaccharides as carbonyl components, the glycosidic bonds remain unchanged both in aqueous and water-free systems, though a small fraction of the glycosidic bonds could be broken.

Therefore, the aldol condensations of a-dicarbonyl compounds in aqueous conditions, in addition to the transglycosylation reactions of saccharides in water-free conditions produced the postulated carbohydrate-based skeleton shown in Fig. 3.10A. Amino acids could react with the unsaturated carbonyl structures of this structure to form melanoidins

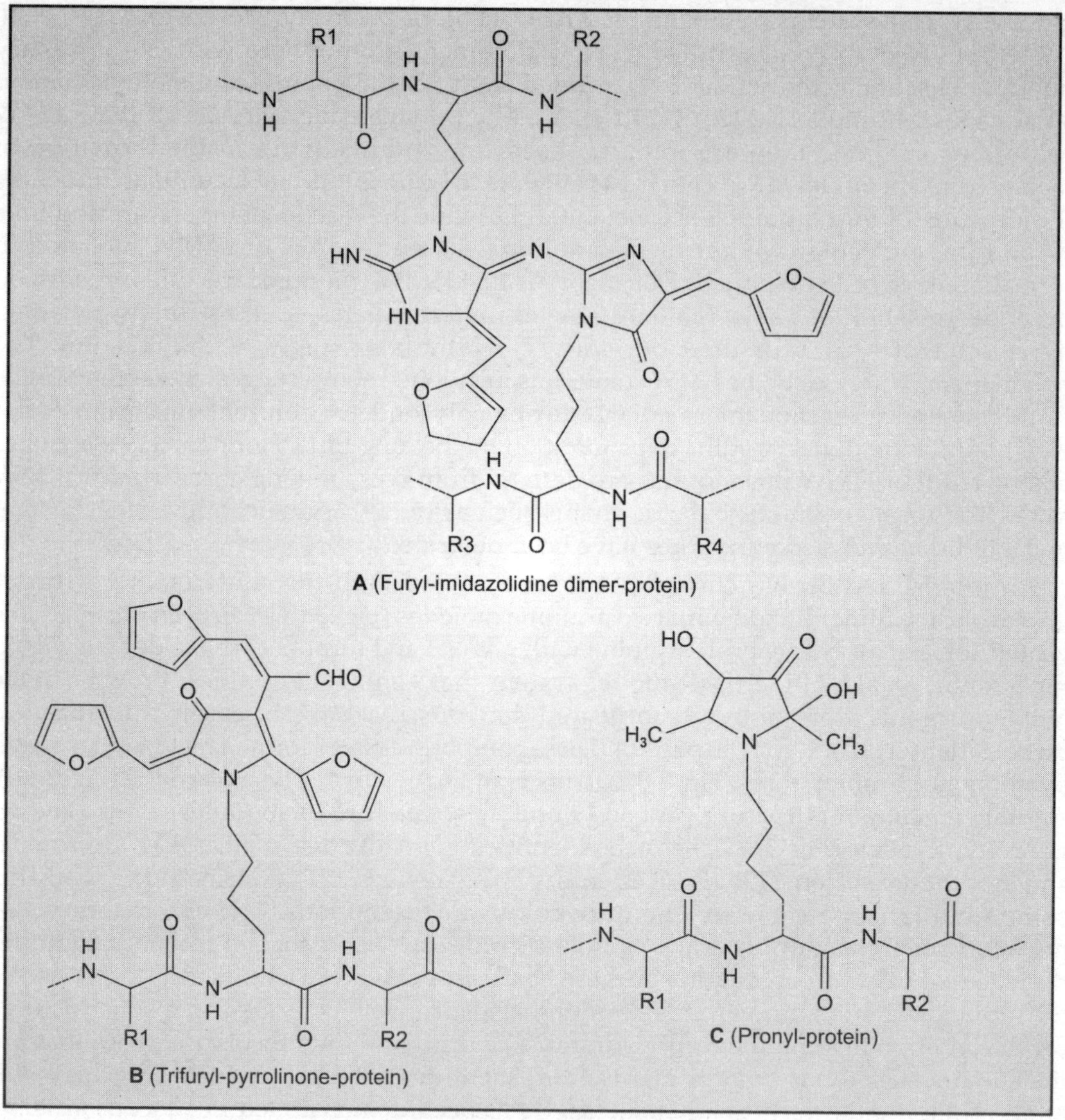

Fig. 3.9: The proposed melanoidin cross-linking structures of chromophoric LMW with HMW protein

branched with amino compounds (Fig. 3.10B). Cosovic et al. (2010) investigated the adsorption behaviour of melanoidins prepared from glucose and amino acids via an electrochemical method. The observed similarity of the adsorption isotherm for melanoidins prepared from glucose compared with those prepared from glucose and amino acids supports the idea that melanoidins have a sugar derived "backbone". In the early stages of the MR, reducing sugars react with amino compounds to form a Schiff base adduct which is then stabilized by an Amadori rearrangement. During the intermediate stages, a-dicarbonyls, aldehydes, furaldehydes and furanone, are generated, which rapidly react with each other in an aldol-type condensation. Also, degradation of aldose sugars leads to the formation of furaldehydes and furanones too, such as furan-2-

carboxaldehyde and furan-2-aldehyde. A range of reactions takes place in the advanced stages of the MR, including cyclisations, dehydrations, retro-adolisations, rearrangements, isomerisations and additional condensations. Ultimately, these intermediate products react with amino acids to form LMWs, leading to the production of HMWs by polymerisation. HMW MRPs can also be produced via the modification of proteins by the intermediate products. Melanoidins with browning and AGEs with fluorescence are two main components of advanced MRPs, but the choromophores of them are much different. It has been reported that fluorescence can be observed in the MR before browning occurs, even though AGEs and melanoidins are generated at similar stages in the MR. Thus, the formation of LMW and HMW melanoidins depends on the degree of polymerisation among the LMW intermediates and/or the degree of cross-linking between LMW chromophores and non-coloured HMW biopolymers. In the production of real foods such as bread, coffee, roasted malt, cocoa and biscuits, the MR often takes place at lower moisture contents, at temperatures higher than 150°C, and usually for reaction times of less than 2 h. Under these conditions, the oligo and polysaccharides preferentially react as complete molecules at the reducing end. Additional side chains could thus be formed by transglycosylation reactions, producing HMW melanoidins with carbohydrate skeletons. Alternatively, sugar or carbohydrate degradation products can react with protein to form HMW melanoidins with a protein skeleton. LMW melanoidins can also be formed by the reaction of sugar or carbohydrate degradation products with amino acids. In foods with higher moisture content such as beer, sweet wine and grape syrup, the MR often takes place during the fermentation and storing, usually at a temperature lower than 50 °C for a period of more than 30 days. Under these conditions, LMW melanoidins produced by the reaction of sugars and amino acids at lower temperatures and long reaction times allow for extensive polymerisation and formation HMW melanoidins. Thus in food melanoidins, HMW compounds are usually more

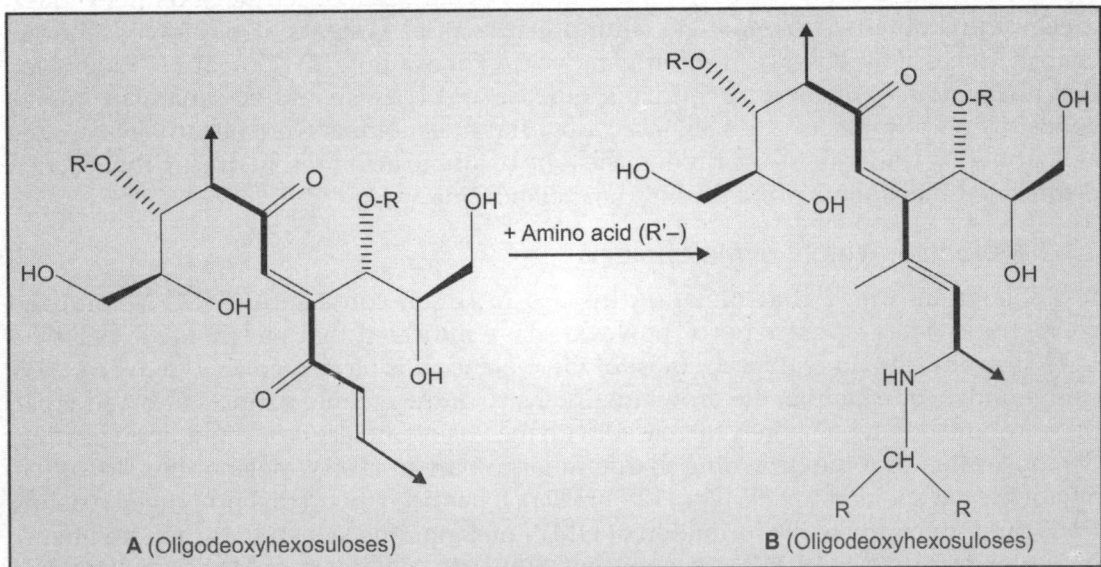

Fig. 3.10: Proposed carbohydrate-based melanoidin structure

enriched compared to LMW compounds. Melanoidin composition in real foods is more complex than in model systems due to the more diverse pool of reactants. Phenolics are constituents of coffee melanoidins, largely due to the incorporation of chlorogenic acids. Arabinogalactan proteins (AGP) were present as AGP-melanoidin complexes in the HMW fraction of coffee melanoidins, too. Similarly, total melanoidins and HMW melanoidins of grape syrup and sweet wine have been shown to be dependent on the presence or absence of polyphenols, as the darkest pigments were formed in the presence of polyphenols.

3.3 CHEMICAL AND STRUCTURAL PROPERTIES OF MELANOIDINS

Melanoidins are formed by cyclizations, dehydrations, retroaldolizations, rearrangements, isomerizations, and condensations of low molecular weight MRPs. The composition of melanoidin chemical structures is relatively unknown, however, due to the complexity of the products that are generated in the reaction.

3.3.1 The Negative Charge of Melanoidins

Although melanoidins are chemically diverse, many studies reported that melanoidins are negatively charged in both real foods and in model systems. Migo et al., (1997) synthesized the model melanoidin of glucose and glycine and found that at pH 2.5, melanoidin has a net zero charge and no migration was observed on 0.8% agarose gel. However, as the pH is increased, the distance travelled towards the positive electrode increases i.e. the net electrical charge of melanoidin becomes more negative. The isoelectric point of melanoidin is approximately 2.5. Melanoidins found in coffee were determined to be negatively charged but heterogeneous with respect to their polyanionic behavior: high molecular weight (HMW) melanoidins in coffee were found to be more negatively charged than low molecular weight (LMW) melanoidins too. Chlorogenic acids have been hypothesised to be the source of the negative charges in coffee melanoidins, though melanoidins obtained from sugar-amino acid model systems also showed anionic characteristics in the absence of chlorogenic acids. For example, Kwak et al. (2005) showed that melanoidins prepared by refluxing glucose and lysine could be separated into 14 bands over a pI range of 3.5–4.85, indicating that melanoidins were negatively charged at neutral pH. Under these conditions, the type of amino acid present during the reaction determined the anionic properties of the melanoidins.

3.3.2 Molecular Weight of Melanoidins

It has been assumed that melanoidins are nitrogen-containing, HMW, coloured compounds. More recent reports, however, have indicated that melanoidins include a LMW fraction, too. In real foods, most of the melanoidins have been shown to be HMW compounds. For example, the browning intensity of the ethanol extracts of bread crust, manuka, and dandelion honey increased with increasing molecular weight upon heating, demonstrating that the browning is due in large part to HMW melanoidins. In coffee, 59% of the melanoidins are HMW (>12–14 kDa). It has been found that prolonged roasting leads predominantly to the formation of HMW melanoidins, which show a more intense brown colour than the LMW melanoidins. Similarly, polymers of HMW melanoidins isolated from sweet wines (>12 kDa), roasted malt (>60 kDa) and roasted cocoa beans

(>5 kDa), have been shown to give rise to the brown colour of these foodstuffs. Many studies have shown that melanoidins produced in model Maillard Reaction systems heated for more than 24 h are predominantly of molecular weights greater than 10 kDa. These studies suggested that melanoidin pigments were predominantly HMW. Some other reaction mixtures and conditions, however, such as glucose and fructose-amino acids heated to 100°C for 2, lactose and glycine under refluxing conditions for 5 h, and glucose and an alanine/glycine mixture heated to 95°C for 4 h, produced primarily LMW coloured compounds (<3.5 kDa). However, the reaction between casein and glucose generates a sizable HMW fraction of pigmented melanoidins. Melanoidins produced in the Maillard reactions between proteins and sugars in real foods are predominantly HMW compounds. Considering that the molecular weight of melanoidins produced from Maillard Reacions is highly dependent on the heating intensity, and HMW melanoidins are produced at longer reaction times (>24 h), it is possible that in the initial stages of the Maillard Reaction, LMW chromophoric melanoidins are formed, which subsequently polymerise or cross-link with other Maillard Reaction Products to produce HMW melanoidins during the later stages of the Maillard Reaction.

3.3.3 Genotoxicity and Cytotoxicity

It has been reported that LMW brown Maillard Reaction Products isolated from dry heated glucose-glycine systems (125°C, 2 h), and HMW brown Maillard Reaction Products from aqueous sugar-lysine model systems (121°C, 1 h, pH 9.0) have genotoxic and cytotoxic effects on cells in high doses. Bread crust and coffee melanoidins were also shown to impair the viability of lung epithelial H358 cells in high doses. Jing and Kitts (2000) observed that HMW Maillard Reaction Products from fructose lysine reactions with lower absorption values (at 420 nm) were more cytotoxic than those from glucose-lysine reactions, which suggested that the melanoidin component of Maillard Reaction Products might not be responsible for the observed toxicity. It has been reported that brown glucose-glycine Maillard Reaction Products and cocoa melanoidin fractions are not mutagenic to *Salmonella typhimurium* strains TA98, TA100 and TA102 (Ames test) at low or high concentrations. It has also been shown that Maillard Reaction Products containing melanoidins produced from proteolyzed products of the dry and wet gluten-glucose model systems, the casein-sugar system, and bread and biscuit displayed no cytotoxicity towards hepatocytes, HepG2 or Caco-2 cells. In conclusion, though brown Maillard Reaction Products containing melanoidins may show modest genotoxic and cytotoxic effects in high doses and after long incubation times, the low concentrations found in foods do not pose a health risk.

3.3.4 Antioxidative Activity

The beneficial antioxidative effects of melanoidins are currently gaining much attention. Studies have primarily focused on the antioxidant properties of melanoidins or brown Maillard Reaction Products from different model systems and melanoidin fractions in foods such as coffee, bread crust, beer, biscuit, roasted barley, roasted cocoa, vinegar, honey and meat flavouring. These studies demonstrated that brown chromophoric MRPs, independent of the reaction conditions, displayed in vitro antioxidant protective effects on lipids, hepatocytes, HepG2 cells and human lymphocytes against oxidation challenge. An in vivo study (Somoza et al., 2005) showed that bread crust, caraffa malt, or pronyl bovine serum albumin (which contained pyrrolinone reductonyl-lysine, a food melanoidin) decreased oxidative stress levels and thiobarbituric acid reactive substances levels, and increased tocopherol in the plasma of rats.

Although the specific components responsible for the protective antioxidant behaviour observed for melanoidins are not yet known, melanoidins are thought to be important in preventing oxidative damage and diseases related to free radicals. The antioxidant properties of melanoidins have been partly ascribed to the metal chelating capacity of these compounds. The anionic nature of melanoidins enables them to chelate transition metals. The nitrogen atoms in melanoidins were proposed to be responsible for the chelation of copper ion. The ketone and/or hydroxyl groups of the pyranone or pyridone components may act as chelation donor groups in melanoidins. Though no relationship was observed between browning and the iron binding ability of melanoidins in model systems, Wen et al. (2005) showed that the zinc-chelating activity of brewed coffee decreased as the intensity of the roasting increased, suggesting that the chromophoric groups of melanoidins may not be the main co-ordination sites for metal complexation, but the metal chelating activity of melanoidins may be due to other, as yet unknown, structures of them. The other main mechanism for the antioxidant activity of melanoidins is their radical-scavenging activity. It has been reported that melanoidin fractions in food and model systems can scavenge a variety of reactive oxygen species. Some studies have shown a positive correlation between the radical-scavenging activity and colour or heated degree of melanoidins, while others have not. A strong linear relationship between the peroxyl, 2, 2'-azobis-2-methyl-propanimidamide, dihydrochloride (AAPH) and 2, 2'-azo-bis(3-ethylbenzothiazoline-6-sulphonic acid) or ABTS radical scavenging activity and the colour intensity of the melanoidin fractions (derived from either model systems or commercial samples) has been established. The ABTS radical-scavenging activity of the non-phenolic fraction of brewed coffee and bread extract was observed to increase with prolonged roasting or baking and the hydroxyl radical scavenging activity of vinegar melanoidins increased during the decoction, storing and aging of the vinegar. These results indicate that the brown compounds formed during the heating of foods or model systems, which are primarily melanoidins, are the compounds most likely responsible for the radical-scavenging activity. Other studies have reported very different results. For example, no positive linear correlation was found between the hydroxyl radical-scavenging activities and browning. In two studies, the antiradical activity measured from ABTS and N, N-dimethyl phenylenediamine dihydrochloride (DMPD) assays of coffee melanoidin fractions was in fact observed to decrease as the intensity of the roasting increased. The conflicting data on the radical-scavenging activity of melanoidins may be a result of the complicated components of melanoidin fractions, as well as the assays used to measure these activities. Although the antiradical activity of MRPs from both model systems and browned foods has been measured several times, most investigations reported only the activity of the bulk browned water extracts or the HMW fraction of the extracts, rather than correlating these effects with chemically distinct melanoidins. The HMW fractions isolated from model systems and food exhibited significant antioxidative activity when compared with the LMW fractions. The variety of different chelation sites potentially present in melanoidins, as well as non-chromophore structures such as phenolic residues, may also introduce some variety among the measured scavenging activity of different radicals by melanoidins.

3.3.5 Antimicrobial Activity

Rufian-Henares's group studied the antimicrobial activity of melanoidins against pathogenic microorganisms and found melanoidin fractions from both aqueous model systems (100°C, 24 h, pH 7) and foods (coffee, beer, sweet wine) exhibited higher antimicrobial activity towards gram-positive microorganisms (e.g. *Staphylococcus aureus*) compared to gram-negative microorganisms (e.g. *Escherichia coli*). The HMW fractions of coffee and biscuit melanoidins with darker colour displayed higher antimicrobial activity against *E. coli* than LMW fractions, and melanoidins from more strongly heated samples such as roasted coffee, displayed a higher activity. It was also determined that the bounded melanoidin compounds (BMCs) fraction contributed to the antimicrobial activity of melanoidins. Strong antimicrobial activity are observed in both pure coffee melanoidins, as well as the BMCs linked to these melanoidins, though pure melanoidins isolated from a glucose-amino acid model system were observed to exert a lower antimicrobial activity than that of the BMCs. Therefore, pure melanoidins do have antimicrobial activity and the chromophore may be responsible for it. Antimicrobial activity has been determined to be bacteriostatic at low concentrations of melanoidins, but bactericidal at higher concentrations. The dependence of the microbicidal mechanism on the melanoidin concentration could be related to their metal-chelating properties. Three different mechanisms for the antimicrobial activity of melanoidins have been proposed. At low concentrations, melanoidins exert a bacteriostatic activity mainly mediated by the chelation of iron from the culture medium. In bacterial strains able to produce siderophores for iron acquisition, the chelation of the siderophore Fe^{3+} complex by melanoidins has been observed, which could decrease the virulence of pathogenic bacteria. Finally, melanoidins at high concentrations can disrupt both the outer and inner membranes by chelating Mg^{2+} ions from the outer membrane, leading to a destabilization of the inner membrane. The higher antimicrobial activity exhibited towards gram-positive microorganisms was ascribed to the absence of a protective outer membrane, which makes them more susceptible to antimicrobial sustains.

3.3.6 Effect of Melanoidins on Xenobiotic Enzymes Activity

It is as yet not clear how non-enzymatic browning products influence xenobiotic enzymes in the human body. Most chemopreventive, non-endogenously formed agents act through enzyme systems by modulating Phase-I and Phase-II enzymes.

The Phase I detoxification system, composed mainly of the cytochrome P450 supergene family of enzymes, is generally the first enzymatic defense against foreign compounds. Most pharmaceuticals are metabolized through Phase I biotransformation. In a typical Phase I reaction, a cytochrome P450 enzyme (CypP450) uses oxygen and, as a cofactor, NADH, to add a reactive group, such as a hydroxyl radical. As a consequence of this step in detoxification, reactive molecules, which may be more toxic than the parent molecule, are produced. If these reactive molecules are not further metabolized by Phase II conjugation, they may cause damage to proteins, RNA, and DNA within the cell. Several studies have shown evidence of associations between induced Phase I and/or decreased Phase II activities and an increased risk of disease, such as cancer, systemic lupus erythematosus, and Parkinson's disease. Phase II conjugation reactions generally follow Phase I activation, resulting in a xenobiotic that has been transformed into a water-soluble compound that can be excreted through urine or bile. Several types of conjugation reactions are present in the body, including glucuronidation, sulfation, and glutathione and amino acid conjugation. These reactions require cofactors which must be replenished through dietary sources. Phase-I metabolic transformations include reduction, oxidation and hydrolytic reactions while Phase-II transformations generally act through conjugation reactions of the parent xenobiotics or of Phase-I metabolites. The conjugation reactions facilitate transport and enhance elimination of the inactive compounds via the renal and biliary routes. Therefore, the main determinant of whether exposure to xenobiotics will result in toxicity is the balance between the activities of Phase-I and Phase-II enzymes. Although, Phase-II enzymes are hypothesized to facilitate the metabolic transit of food-derived Maillard reaction products formed in heat treated proteins it is still an open question whether non-enzymatic browning products formed in foods require specific detoxifying mechanisms or contribute to the chemo preventive potential of the organism via induction of Phase-II enzymes.

Bread crust and pronyl-bovine serum albumin containing large amounts of pronyl-L-lysine were shown to enhance phase II enzymes, e.g. UDP-glucuronyl-tansferase (UDP-GT) and glutathione S-transferase (GST), in rats. Melanoidin fractions from bread crust induced a significant increase in GST activity, but lead to a decrease in the activity of phase I NADPH cytochrome c reductase (CCR) in Caco-2 cells, indicative of their protective effects. Melanoidin fractions from biscuits, however, were observed to decrease GST activity and roasted malt melanoidins were observed to elevate CCR activity. Similarly, the effect of model melanoidin fractions on xenobiotic enzymes is not consistent too. Thus, the effects of melanoidin fractions on phase I and II enzymes are dependent on their source. The relationship between the melanoidin phase I and II enzyme modulating activity and melanoidin colour intensity is not known, so the xenobiotic enzyme-modulating activity described above cannot be linked to specific choromphoric or structural properties of melanoidins. It is known that MRPs contain high concentrations of Ne-carboxymethyllysine (CML), a compound known to induce CCR and inhibit GST, which may partially explain the effects of melanoidin fractions on CCR and GST. However, this xenobiotic enzyme-modulating activity cannot be entirely explained by the presence of CML, as it seems various reaction products formed during the heat-treatment of food and model systems, such as MRPs and native polyphenols, also contribute to this effect, which could help to explain the discrepancies in the above investigations.

3.3.7 Prebiotic Activity

Bifidobacteria and Lactobacilli in the gastrointestinal tract have been shown to confer beneficial health effects. Food components that promote the growth of beneficial versus pathogenic bacteria are known as prebiotics. Data from in vitro and in vivo studies suggest that most melanoidins escape digestion in the upper gastrointestinal tract and are mainly recovered in the faeces. Similar to dietary fibre, Borrelli and Fogliano (2005) demonstrated that melanoidin fractions from bread crust appeared to selectively enhance the growth of Bifidobacteria in the gut, and potentially have a prebiotic activity similar to that of dietary fibres. However, an in vitro study investigating the effects of the melanoidin fractions from glucose-lysine MRPs (refluxing at pH 5 for 2 h) on colonic bacteria demonstrated that melanoidins could cause a non-specific increase in all anaerobic bacteria of the human large-bowel, including Bacteroides, Clostridia, Bifidobacteria and Lactobacilli. Thus, it may be difficult to ascribe prebiotic effects to melanoidins.

3.3.8 Antihypertensive Activity

The potential antihypertensive activity of pure melanoidins isolated from coffee, beer, sweet wine, and purified model melanoidins has been evaluated by investigating the in vitro ACE-inhibitory activity. The in vitro ACE-inhibitory activity of food melanoidins was shown to be similar to that reported for well known antihypertensive peptides. The ACE-inhibitory activity of coffee melanoidins was significantly higher with coffee that had been heated for prolonged periods of time, which indicated that the melanoidins are likely responsible for this activity, though the ACE-inhibitory activity has also been linked to the presence of BMCs as well. Rufian-Henares and Morales (2007) also demonstrated that the in vitro ACE-inhibitory activity of pure melanoidins extracted from aqueous glucose-amino acids systems (100°C, 24 h, pH 7) was related to melanoidin structure, and the chromophores of melanoidins were not involved.

3.3.9 Other Biological Effects

Hiramoto et al. (2004) investigated the effects of food melanoidin fractions on the prevention of *Helicobacter pylori* (*H. pylori*) infection. Urease proteins located on the extracellular surface of *H. pylori* are adhesions that target gastric surfaces. These proteins play an important role in infection and colonisation of the host. It was found that the HMW MRPs of alkali lactose/glucose-protein systems (85°C, 4 h) and brown fractions isolated from the crust of an English muffin inhibited the binding of urease to gastric mucin by competing with gastric mucin for urease-binding sites. Moreover, in an animal model and in human subjects infected with *H. pylori*, these melanoidin fractions were shown to suppress *H. pylori* colonisation. This indicated that food melanoidins may have antibiotic activities and might prevent *H. pylori* infection. Synthesised and purified chromophoric MRPs, identified in thermal food processing products, were reported to inhibit the growth of human carcinoma cells in vitro by interfering with the proliferation of gastric carcinoma cells, causing cell cycle arrest and apoptosis. These results have suggested that some melanoidins may inhibit the growth of human tumour cells. HMW MRP fractions from coffee have been demonstrated to have antiadhesive properties and inhibited the ability of *Streptococcus mutans* (*S. mutans*) to adsorb to saliva-coated hydroxyapatite beads. This is considered an important virulence property of *S. mutans* and a cause of tooth decay. The antiadhesive activity may be largely due to small molecules that occur naturally in coffee (e.g. chlorogenic acid and trigonelline), however, so it was proposed that coffee could minimise *S. mutans* colonisation and might be effective in preventing *S. mutans* induced tooth decay. Takahama and Hirota (2008) reported that in addition to chlorogenic acid and its isomers, coffee melanoidins also reacted with salivary nitrite and SCN at pH 2, mimicking the mixture of coffee, saliva, and gastric juices in the gastric lumen. This produces nitric oxide (NO), which may inhibit microbial growth, and regulate mucosal flow, mucosal formation and gastric mobility. HMW brown MRPs obtained from dry glucose-arginine model systems (125°C, 3 h) were tested for their ability to influence the contractility of gastric smooth muscles. Brown MRPs at low concentrations provoked concentration-dependent contraction, whereas brown MRPs at high concentrations induced muscle relaxation.

3.3.10 In-vivo Effects of Melanoidins and the Effects of Melanoidins on Foods

At present, not much is known about the bioavailability of melanoidins. The amount of protein-bound LMW melanoidins absorbed and excreted in the urine of rats was found at levels ranging from 1% to 5%, which suggested that melanoidins escape digestion in the upper gastrointestinal tract and are mainly recovered in the faeces. Finot and Magnenat (1981) investigated the metabolic transit of LMW and HMW fractions of casein/glycine-14C-glucose mixture MRPs in rats. They observed that 61% of the LMW fractions (<10 kDa) of the casein-14C-glucose mixture were excreted in the faeces. For the HMW fractions (>10 kDa), however, 87% of the ingested radiolabeled casein-14C-glucose and 93% of the ingested radiolabeled glycine-14C-glucose were excreted in the faeces, respectively. These results indicated that HMW melanoidins are absorbed to a much lesser extent than LMW compounds. On the other hand, much research has been done investigating the digestibility and bioavailability of AGEs, which are advanced MRPs formed in reaction pathways that are analogous to the reaction pathways that form melanoidins. It is known that dietary AGEs, particularly LMW AGEs, can be absorbed by humans and rats, with elevated levels observed in the plasma and organs. It can be concluded that about 30% of LMW melanoidins and/or their intestinal degradation products are absorbed and are likely to be taken up in the blood stream. The low dose of absorbed melanoidins is not likely to be genotoxic or cytotoxic to humans. Owning to the much lower dose of digestible melanoidins compared with in vitro studies; however, they may exert limited in vivo beneficial effects, such as antioxidative activities, modulation of phase I and phase II enzymes and antihypertensive activity. The more important in vivo beneficial effects can probably be ascribed to the indigestible melanoidins. As they pass through the intestine, they will behave as antioxidants, display antimicrobial activity against pathogenic microorganisms and prebiotic activity, provided they survive gastric conditions for a long enough period of time. Melanoidins are produced during the processing and storage of foods. In foods, the antioxidant properties of melanoidins can inhibit the oxidation of unsaturated lipids and functional food ingredients, such as vitamins, polyphenols and flavonoids. Moreover, the antimicrobial activity can inhibit the growth of microorganisms and prevent the spoilage and deterioration of foods. Thus, melanoidins may preserve the quality and safety of foods.

3.3.11 Quantification of Model Melanoidins

Although it is difficult to quantify melanoidin composition from a Maillard reaction given the complexity of these reactions, quantification is critical to optimising the browning in processed foods. The extinction coefficient can be used to spectrophotometrically estimate browning (absorbance values), as related to the melanoidin concentration. According to the Lambert-Beer equation $A = \varepsilon cl$, there is a direct linear relation between absorbance (A) and concentration (c). In order to quantify melanoidins using this equation, the extinction coefficient must be known. The extinction coefficient of the melanoidins does not appear to vary with pH, temperature or reaction time. The extinction coefficient of melanoidins at 420 nm (ε_{420}) produced from the glucose-casein and fructose-casein reactions (120°C, 90 min, pH 6.8) were similar: 0.48 ± 0.05 and 0.53 ± 0.04 l mmol/cm, respectively. The ε_{420} of nondialyzable (>3.5 kDa) glucose-glycine melanoidins (both at 120 °C, pH 6.8 and at 100 °C, pH 5.5) was found to be 1.01 ± 0.02 l mmol/cm and those of HMW (>12.4 kDa) melanoidins from glucose/fructose- glycine/asparagine/lysine basic reactions (at 145 °C for 2 h) fell between 0.50 and 0.76 l mmol/cm. Moreover, Mundt and Wedzicha (2003) determined a ε_{420} of 0.48 ± 0.02 l mmol/cm using the kinetic model for melanoidins produced from the fructose-glycine reaction, which is half the value determined for melanoidins from the glucose–glycine reaction (0.96 ± 0.04 l mmol/cm). Bekedam et al. (2006) demonstrated that use of the specific extinction coefficient (K) was preferable to the molar extinction coefficient (ε), since the molecular weight of melanoidins is unknown and probably variable. When using K, the concentration parameter used in the law of Lambert-Beer is expressed in l g/cm. Deduced from the slope of the curve representing the absorbance vs. melanoidin concentration, the K values of glucose-amino acids melanoidins ranged from 0.225 to 4.315 l g/cm for glucose-tryptophan and glucose-lysine melanoidins, respectively.

3.4   EFFECTS ON MELANOIDIN FORMATION

3.4.1   Effect of Reactant type on Melanoidin Formation

Maillard reactants are not just the reducing sugars and amino acids. The common reactive reducing sugars are xylose, ribose, glucose, fructose, lactose and maltose and sucrose under acidic conditions or high temperature conditions where it forms invert sugar. Other reactive reducing moieties include carbonyl compounds formed during lipid peroxidation (glyoxal, methylglyoxal, formaldehyde). The amines may be comprised of free amino acids, the N-terminal amine of proteins or peptides. For sugars, the rate of the reaction depends on the rate at which the sugar ring opens to the reducible, open-chained form and this increases with increasing pH. Pentose sugars react more rapidly than hexoses. For hexoses, the order of reactivity is D-galactose > D-mannose > D-glucose. Reducing disaccharides are considerably less reactive than their corresponding monomers. Kroh and Westphal (1989) found that basic amino acids are more reactive than neutral or acid amino acids. Ashor and Zent (1984) suggested a classification of amino acids into three groups depending on the extent of browning when reacted with glucose at different pH, and 121°C for 10 min. The most reactive were lysine, glycine, tryptophan and tyrosine. These results should be of applicability to candy because this is a typical processing range. Other results show that comparisons can only be used if the same pH and buffering conditions are used. For example, the effect of pH is especially significant due to the different pKa values of the amino acids. Lysine appears to be the most reactive amino acid (Ashor and Zent, 1984; O'Brien and Morrissey, 1989) due to the fact that is has two available amino groups. However, the reactivity of lysine would appear to be dependent on the conditions of the reaction being studied. The reactivity of peptides has hardly been studied and, unfortunately, many of the studies comparing peptide reactivity with that of amino acids measured the rate of browning to quantify the Maillard reaction. In general, the amount of browning has been shown to increase with chain length of the peptide. However, the amount of browning is not, per se, proportional to the conversion of amino acid or peptides, as the degree of browning depends on the type of melanoidin formed during the Maillard reaction (De Kok and Rosing, 1994). Motai (1974) concluded that melanoidins from peptides exhibited a darker degree of colour than those from amino acids. When the degree of sugar decomposition was measured it appeared that small peptides (up to 3 amino acid groups) are more reactive than the corresponding amino acids. The extent of browning seems to vary according to the sugar:amine ratio. The effect of increasing the amino concentration show a greater increase in browning than that of increasing the sugar content on a molar basis and the increase for both is greater than the relative concentration increase (2 times greater for sugar and 2 to 3 fold times greater for amine).

3.4.2 Effect of Temperature and Time on Melanoidin Formation

The effect of temperature and duration of heating on browning are actually studied by Maillard, who reported that the rate of browning increases with temperature. This has been confirmed many times. The Strecker degradation is favored by high temperatures which are helpful to create colour.

3.4.3 Role of water content on melanoidin formation

For the first stage of the Maillard reaction to occur, water is essential. Thus, the rate of this reaction is dependent on the amount of free water available as related to water activity. The rate of brown pigment formation is also dependent on the water content of the.

3.4.4 Effect of pH on melanoidin formation and extraction

The pH has a significant effect on the Maillard reaction. In general, the rate and extent of browning increases with increasing pH. The reaction generally has a minimum is at pH 3. At a pH < 3 and > 9, other nonenzymic interactions (i.e. sugar-sugar and protein-protein) compete with the Maillard reaction. Thus, the Maillard reaction has an optimum above pH 7. A change in pH also leads to a change in the mechanism of the reaction and, hence, to the formation of different volatile and coloured products. The pH dependence of the initial step of the reaction can be related to the amount of unprotonated amine present, which is controlled by the following equilibrium:

$$- NH_3^+ \longleftarrow pH > 7 \longrightarrow - NH_2 + H^+$$

At the pKa of the amine group, by definition, half the amine is present as the protonated NH_3^+ state preventing electron transfer. Thus, the rate of the Maillard reaction is lower at a pH lower than the pKa of the reactive amino group. Theoretically for a pure system a unit change in pH causes a change of 10 fold in rate. From this it can be seen that lysine is about 10 times more reactive than glycine. Schnickles et al (1976) showed that the rate of browning for proteins during storage was directly correlated with the lysine content. If ingredients are changed, such as in changes made to lower calories, it is expected that the pH and buffering capacity will also change, thus there can be significant impact on final colour.

Melanoidin extraction by isopropanol at different pH showed variable results (Sangeeta and Chandra, 2013). According to this study maximum extraction was noted at pH 11.0 (2.87% w/v), while the least was observed at pH 4.0 (0.41% w/v). The extraction of melanoidin resulted into increase decolourisation of distillery effluent. The PMDE sample without melanoidin extraction does not show any significant reduction in physico-chemical parameters even after bacterial treatment. However, the PMDE sample after melanoidin extraction showed significant reduction of pollution parameters. This showed the color contribution of melanoidin in PMDE. It was maximum at pH 11.0 (61.10%) and minimum at pH 4.0 (30.40%) thereby causing extraction of more melanoidin-pigment at pH > 7.0. The removal of melanoidins from effluent showed direct correlation between melanoidin extraction and decolourisation of effluent from pH 4 to 11. However, the melanoidin extraction above pH 11.00 declined and color was increased. This indicated that melanoidin was as major coloring constituent of PMDE. Since the lowering of pH causes oxidation so it might be possible that oxidative decomposition of melanoidin could have occurred thereby leading to decrease in the color intensity of distillery effluent at low pH 4.0–6.0. Whereas, reduction at alkaline pH may have resulted in increased color intensity as during reduction – OH group is released which is one of the most effective auxochrome that results into increase in color intensity at basic pH.

3.4.5 Role of buffer on melanoidin formation

The Maillard reaction forms H^+ ions, so decreasing the pH of the system. As mentioned above, as the pH falls the rate of the Maillard reaction decreases. Consequently, buffers are necessary when studying the Maillard reaction in a model system. It is not, however, clear as to whether or not confectionery products contain natural buffers (for example, protein and ascorbate have been shown to act as a buffer). Buffers have been shown to increase the rate of browning. It is possible that the buffer mops up the H^+ ions keeping the pH and the reaction rate constant or may interact with the reactants in some way enhancing the Maillard reaction.

Table 3.2: Analytical techniques used for the analysis of Maillard reaction products (MRPs)

Maillard reaction products (MRPs)	Analytical techniques
Amadori compounds (Direct analysis)	Column chromatography, HPLC differential refractometry detection; HPLC involving derivatisation; HPAEC coupled electrochimemical and/or DAD FAB-MS; ESI coupled HPLC and EC; EC coupled MS; MALDI-TOF; NMR; LC-MS; NBT; ELISA Immunoblotting (lactosylated proteins)
Indirect analysis (2-FM-AA)	Ion-pair RP-HPLC; CEC UV-detection; HPLC-MS
Unreactive lysine	Colorimetric and fluorimetric methods
Advanced Maillard products	FAST
General AGEs	HPLC-DAD
CML	RP-HPLC; RP-HPLC o-phthalaldehyde; precolumn derivatisation; GC-MS; ELISA; Immunoblotting
Pyrraline	Amino acid analysis with PAD; RP-HPLC
Cross linking products Lysine dimmers Argnine-lysine	LC-MS with ESI; Ion-exchange chromatography FAB-MS
Other amino acid derivates Argypirimidine OMA; PIO	HPLC-coupled GC-MS ELISA RP-UPLC/LC-ESI-TOF-MS/NMR HPLC with UV and fluorescence detection
Pyrazinones	HPLC with UV and fluorescence detection
Lysine aminoreductone	HPLC-DAD
Final stage MRP's General melanoidins	HPLC, NMR, MS, UV, IR spectrometry MALDI-TOF mass spectrometry
Pronyl-L-lysine	GC-MS chemical ionization

3.4.6 Effect of oxygen on melanoidin formation

Recent studies on the Maillard reaction in-vivo have shown that glucose autoxidation plays a significant role in the glycation of proteins. However, it is not clear as to whether oxygen is important for the Maillard reaction. Comparatively few studies have been carried out on the effect of metal ions on the Maillard reaction and browning; and as with many other studies the results are conflicting. A recent study on the effect of the addition of Cu^{2+} ions showed that their influence was concentration dependent and dependent on the pH of the system. At pH 3, the addition of Cu^{2+} increased the absorbance of a glucose-glycine reaction mixture. It also seemed copper became bound to the melanoidins at the highest concentration of 100 ppm. At pH 6.18–6.86, the addition of Cu^{2+} only had a small effect on the absorbance of a glucose-glycine model system heated at 50°C. However, these systems were unbuffered so there was a tendency for the pH to decrease most rapidly in the systems with the greatest copper content.

3.5. RECENT TECHNIQUES AVAILABLE FOR THE ANALYSIS OF MRPS

The qualitative and quantitative determination of the MRPs in foods or physiological samples is difficult since they are converted during the acid hydrolysis of proteins, thus making it impossible to detect them with routine amino acid analysis. This is caused by the large amount of products formed during the Maillard reaction and the difficulties encountered in their purification, identification and quantification of pure compounds. Yet several direct and indirect methods have been developed as mentioned in Table 3.2 which can be of helpful to identify the reaction steps involved in the formation of biologically active Maillard reaction products (MRPs).

4

Bioredegradation of Sucrose-aspartic Acid Maillard as Model Melanoidin Products and its Degradative Metabolites

4.1 INTRODUCTION

Enzymes are biological catalysts that facilitate the conversion of substrates into products by providing favorable conditions that lower the activation energy of the reaction. An enzyme may be a protein or a glycoprotein and consists of at least one polypeptide moiety. The regions of the enzyme that are directly involved in the catalytic process are called the active sites. An enzyme may have one or more groups that are essential for catalytic activity associated with the active sites through either covalent or noncovalent bonds; the protein or glycoprotein moiety in such an enzyme is called the apoenzyme, while the nonprotein moiety is called the prosthetic group. The combination of the apoenzyme with the prosthetic group yields the holoenzyme. Enzyme names apply to a single catalytic entity, rather than to a series of individually catalyzed reactions. Names are related to the function of the enzyme, in particular, to the type of reaction catalyzed. All known enzymes fall into one of these six categories. The six main divisions are (1) the oxidoreductases, (2) the transferases, (3) the hydrolases, (4) the lyases, (5) the isomerases, and (6) the ligases (synthetases). Oxidoreductases catalyze the transfer electrons and protons from a donor to an acceptor. Transferases catalyze the transfer of a functional group from a donor to an acceptor. Hydrolases facilitate the cleavage of C–C, C–O, C-N, and other bonds by water. Lyases catalyze the cleavage of these same bonds by elimination, leaving double bonds (or, in the reverse mode, catalyze the addition of groups across double bonds). Isomerases facilitate geometric or structural rearrangements or isomerizations. Finally, ligases catalyze the joining of two molecules.

Among several environmental pollutants melanoidins (Maillard products) is major pollutant present in distillery effluent. The first investigations on the Maillard Reaction were performed around 90 years ago by Maillard (1912). Subsequently, Amadori reported on the formation of a rearranged stable product from glucose and amino acids which were named after him and Heyns reported a similar compound from fructose. One of the most detailed descriptions of the pathways that lead to the main Maillard Reaction Products (MRPs) are described in chapter 3. Only some relevant points will be discussed here. Reducing sugars are necessary for this reaction; generally they are a monosaccharide, glucose or fructose, or a disaccharide, maltose or lactose, and in some case a pentose. Non-reducing disaccharides, such as sucrose, or bound sugars as in glycoproteins, glycolipids, and flavonoids, react only after hydrolysis, a process often facilitated by fermentation. The counterparts are free amino acids or proteins. Small amounts of ammonia can be produced from amino acids during the Maillard reaction and large amounts are added for the preparation of a particular kind of caramel colouring. The mechanism of this reaction has been studied very seldom, because it is too complex and the separation of non-volatile products is very difficult. The initial step is the condensation of the aldehydic group of the sugar with an amino group to give a relatively unstable glycosylamine which undergoes a reversible rearrangement to give an aminoketose (Amadori compounds) (described in chapter 3). These intermediates have been fully characterised in model systems and detected in many foods. The equivalent rearrangement of the fructose + amino acid adduct produces an amino aldose which is called Heins product.

The wastewaters released from sugar and sugarcane molasses based fermentation industries are the major source of melanoidins in the environment. The empirical formula of melanoidin has been worked out to be $C_{17-18} H_{26-27} O_{10}N$ and the molecular weight is between 5000 and 40000. It consists of acidic polymeric and highly dispersed colloids, which are negatively charged due to the dissociation of carboxylic acids and phenolic group. It is present at a concentration of 2.0% in the effluent discharged from distilleries, most polluting industries generating enormous amount of wastewater with an average of 12–15 L/L of alcohol production. Distillery effluent has high pH (8.0–8.5), COD, BOD, TDS with objectionable odour. Highly coloured components present in distillery effluent reduces the sunlight penetration in river, lakes and lagoons leading to decrease in photosynthetic activity and dissolved oxygen concentration of an aquatic environment. Hence, accordingly the adequate methods for distillery effluent treatment and melanoidins degradation prior to discharge into the environment are necessary. Several physical and chemical methods either adsorption or precipitation have been reported, but the colourants remains unchanged. Moreover, these methods are expensive and generates huge amount of secondary pollutants as sludge. Thus, the feasibility of these treatment methods is defeated. Due to inadequate techniques for distillery effluent treatment, a better alternative technique i.e. biological treatment is warranted, which can degrade melanoidins by enzymes metabolic process and generate less sludge. Hence, recent research has focused on the development of enzymatic processes for the treatment of wastewaters, solid wastes, hazardous wastes and soils in recognition of these potential advantages. Enzymatic treatment has technological advantages and requires economical considerations to apply it on a large scale for industrial wastewater treatment. It has some potential advantages over the conventional treatment. These includes: applicability to biorefractory compounds, operation either at high or low contaminant concentrations, operation over a wide range of pH, temperature and salinity, absence of shock loading effects, absence of delays associated with the acclimatization of biomass, reduction in the sludge volume and the ease and simplicity of controlling the process. A large number of enzymes (e.g. peroxidases, oxidoreductases, cellulolytic enzymes cyanidase, proteases, amylases, etc.) from a variety of different sources have been reported to play an important role in an array of waste treatment applications. Problems that are encountered during the biological treatment of wine distillery wastewater is because of high toxicity and the inhibition of biodegradation due to the presence of polyphenolic compounds, these waters also demonstrates the antibacterial activity reported in the earlier literature. Polyphenol concentrations in some distillery wastewaters vary considerably and can range from 29 to 474 ppm which is responsible for strong inhibitory effects on microbial activity and must be removed during wastewater treatment, owing to the environmental and public health risks they pose. Humans exposed to phenol at 1300 ppm of concentration exhibit significant increases in diarrhoea, dark urine, mouth sores and burning of the mouth. Several workers have reported different kinds of microorganisms such as fungi (*Penicillium decumbens, Aspergillus sp., Aspergillus niger, Flavadon flavus, Phanerochaete sp., Phanerochaete chrysosporium, Trametes versicolor, Coriolus sp., Pleurotus florida, Aspergillus flavus, Alternaria gaisen and Fusarium monoliforme*) and yeast (*Citeromyces sp.*) for melanoidins removal. The degradation of melanoidins has been reported due to the prevalence of manganese peroxidase (MnP) as decolourising enzyme in fungus, which metabolizes melanoidins. But, the large scale application of these techniques has own limitation due to slow growth rate, unfavorable submerged aquatic environment and low pH range (3.0–5.0) for growth of fungus. The removal of excess nitrogen has been reported for enhancement of decolourisation process of biomethanated molasses spentwash in a two stage treatment. Besides, some bacteria *Bacillus sp., Bacillus thuringiensis, Bacillus brevis, Pseudomonas aeruginosa* PAO1, *Stenotrophomonas maltophila., Proteus mirabilis, Lactobacillus hilgardii* W-NS and acetogenic bacteria BP103 has also been reported for melanoidins degradation. Recently few bacteria have been reported for the metabolization of melanoidins by MnP activity, but the decolourisation was achieved only up to certain limit. In addition, bioremediation or decolourisation by bacteria may be improved by biostimulation. Biostimulation involves the modification of the environment to stimulate existing bacteria capable of bioremediation. This can be done by addition of various forms of rate limiting nutrients and electron acceptors, such as phosphorus, nitrogen, oxygen, or carbon. Biostimulation can be enhanced by bioaugmentation. Bioaugmentation is the introduction of a group of natural microbial strains or a genetically engineered variant to treat contaminated soil or water. By changing the microbial community to include specific microbes, the characteristics of the microbial community can be improved. Because microorganisms are the heart of any biological wastewater system, it makes sense that by enhancing the microbial community, the overall wastewater system can operate more efficiently. The purpose of bioaugmentation is to supplement the existing microbial community in order to improve its functionality. Bioaugmentation offers many advantages over traditional technology platforms like chemicals, equipments, or other consumables, and has been used in secondary wastewater treatment systems for decades. All wastewater treatment systems incur both capital costs, such as when dealing with expansion or replacement, and operating costs. Employing bioaugmentation technology, to enhance the biomass and ensure the microbial population is operated properly, helps reduce these costs. If a microbial population is not healthy and optimized, operating expenses can be incurred to deal with the effects. For example, polymers can be used to assist settling and powder activated carbon or oxidation chemistries can be used to lower COD. Reducing the use of these consumables saves money. Some examples of costs that can be saved using bioaugmentation include the following:

1. Energy costs are conserved if a plant can recover energy from biogas produced during anaerobic treatment.
2. Reduced consumables like polymers, powdered activated carbon, oxidizers, and others.

3. Reducing sludge volume in lagoons saves dredging or dewatering costs and the final, off-site disposal charges.
4. Improved efficiency of the microbial community reduces the need for capital improvements.

Using bioaugmentation, the microbial community can be enhanced for improved resistance to toxic shocks, better removal of problematic compounds, and increased organic removal. Improving plant efficiency helps remedy these common treatment issues. Moreover, most of the studies have been done on low concentration of melanoidins at laboratory condition. The detail information regarding the nature of metabolic products is yet to be reported.

4.2 ENZYMES CAPABLE FOR DECOLOURISATION OF MELANOIDIN

Although the enzymatic system related with decolourisation of melanoidins is yet to be completely understood, it seems greatly connected with fungal ligninolytic mechanisms. The white-rot fungi have a complex enzymatic system which is extracellular and non-specific, and under nutrient-limiting conditions is capable of degrading lignolytic compounds, melanoidins, and polyaromatic compounds that cannot be degraded by other microorganisms. A large number of enzymes from a variety of different plants and microorganisms have been reported to play an important role in an array of waste treatment applications. Several studies regarding degradation of melanoidins, humic acids and related compounds using basidiomycetes have also suggested a participation of at least one laccase enzyme in fungi belonging to Trametes genus. The role of enzymes other than laccase or peroxidases in the decolourisation of melanoidins by Trametes strain was reported during the 1980s. Several reports claimed that intracellular sugar-oxidase-type enzymes (sorbose-oxidase or glucose-oxidase) had melanoidin decolourizing activities. It was suggested that melanoidins were decolourized by the active oxygen (O_2; H_2O_2) produced by the reaction with sugar oxidases. Decolourisation by microbial methods includes the enzymatic breakdown of melanoidin and flocculation by microbially secreted substances. Ohmomo et al. (1985) used *C. versicolor* Ps4a, which decolourized molasses wastewater 80% in darkness under optimum conditions. Decolourisation activity involved two types of intracellular enzymes, sugar-dependent and sugar-independent. One of these enzymes required no sugar and oxygen for appearance of the activity and could decolourise molasses wastewater (MWW) up to 20% in darkness and 11–17% of synthetic melanoidins. Thus, the participation of these H_2O_2 producing enzymes as a part of the complex enzymatic system for melanoidin degradation by fungi should be taken into account while designing any treatment strategy. One of the more complete enzymatic studies regarding melanoidin decolourisation was reported by Miyata et al. (1998). Colour removal of synthetic melanoidin by *C. hirsutus* involved the participation of peroxidases [manganese peroxidase (MnP) and manganese independent peroxidase (MIP)] and the extracellular H_2O_2 produced by glucose-oxidase, without disregard of a partial participation of fungal laccase. The MnP has also been reported in bacteria as extracellular enzyme for decolourisation of melanoidin (Chandra et al., 2009a). The involvement of MnP and laccase in white rot fungus for degradation of various biopolymers (lignin and tannin) has also been reported (Arora et. al., 2002; Rubia et al., 2002). The detail role of MnP and laccase in bacteria for decolourisation of melanoidin has not been fully investigated. But, recently some bacteria showed ligninolytic enzyme activity has been reported for decolourisation of synthetic as well as melanoidin containing wastewater. For living cells, the major decolourisation mechanism in biodegradation is the production of lignin modifying enzymes (LME), laccase, MnP and lignin peroxidase (LiP) to mineralize complex compounds (Table 4.1).

Table 4.1: Microbes and their enzymes in wastewater decolourisation

Organisms	Wastewater	Percent removal References	Mechanism	References
Proteus mirabilis, Bacillus sp., Raoultella planticola, and Enterobacter sakazakii	Molasses melanoidin	75 (10d)	MnP and Laccase activity	Sangeeta and Chandra (2012)
Alcaligenes Faecalis and Bacillus cereus	Sucrose-glutamic acid Maillard products	73.79%	MnP	Chandra et al. (2009a)
Coriolus sp. No. 20	Melanodin (0.5%)	80 (14d)	Active oxygen	Watanable et al. (1982)
Halosaprheia ratnagiriensis	Paper mill bleach effluent	85 (14d)	Lignin enzymes	Raghukumar et al. (1996)
Merulius tremellosus	Pulp bleach effluent (40 v/v)	50 (14d)	Peroxidase	Lankinen et al. (1991)
Phanerochaete chrysosporium	Olive mill wastewater	70 (10d)	Lignin Peroxidase	Sayadi and Ellouz (1995)
P. chrysosporium	Kraft bleach plant EI Stage effluent	70 (5–8d)	Lignin Peroxidase	Cammarota and Sant Anna (1992)
P. chrysosporium	Pulp bleach effluent (40 v/v)	76 (14d)	Peroxidase	Lankinen et al. (1991)
P. chrysosporium	Alkali extraction stage bleach effluent	90 (3d)	Lignin Peroxidase	Bilgic et al. (1997)
Phlebia radiate	Pulp bleach effluent (40 v/v)	76 (14d)	Peroxidase	Lankinen et al. (1991)
Pycnoporus cinnabarinus	Pigment plant effluent	90 (3d)	Extra Cellular Oxidase	Schliephake et al. (1993)
Schizophyllum commune	Bagasse-based pulp mill effluent	80 (2d)	Lignin enzymes	Belsare and Prasad, (1988)
Sordaria fumicola	Paper mill bleach effluent	55 (14d)	Lignin enzymes	Raghukumar et al. (1996)
Trichoderma sp.	Hardwood extraction effluent	85(3d)	Ligninolytic enzymes	Prasad and Joyee (1991)
Wood rotting fungus (unidentified)	Cotton bleaching effluent (20 ± 50%)	81.5 ± 43.8 (5d)	MnP	Zhang et al. (1998)

However, the relative contributions of LiP, MnP and laccase to the decolourisation of industrial waste may be different for each organism. LME are essential for lignin degradation, however for lignin mineralization they often combine with other processes involving oxidative enzymes. An older concept of ligninolysis reemerges, enzymatic "combustion". By extension, this enzyme-assisted process is applicable to the degradation of many other recalcitrant molecules including distillery effluent. The main LME are oxidoreductases, i.e. two types of peroxidases, LiP and MnP and a phenoloxidase, Laccase.

4.2.1 Lignin Peroxidases (LiP; EC 1.11.1.14)

Lignin Peroxidase (LiP) was detected for the first time in cultures of *Phanerochaete chrysosporium* Burdsall (order Corticiales) in 1983. LiP catalyzes the oxidation of nonphenolic aromatic lignin moieties and similar compounds. LiP has been used to mineralize a variety of recalcitrant aromatic compounds, such as three and four ring polycyclic aromatic hydrocarbons (PAHs), polychlorinated biphenyls and dyes. The extracellular N-glycosylated LiP with molecular masses between 38 and 47 kDa contain heme in the active site and show a classical peroxidase mechanism. The enzyme is basic having an isoelectric point from 3–5 depending on the isoform. Lignin peroxidase requires H_2O_2 as the co-substrate as well as the presence of a mediator like veratryl alcohol to degrade lignin and other phenolic compounds. Here H_2O_2 gets reduced to H_2O by gaining an electron from LiP (which itself gets oxidized). The oxidized LiP then returns to its native reduced state by gaining an electron from veratryl alcohol and oxidizing it to veratryl aldehyde. Veratryl aldehyde then gets reduced back to veratryl alcohol by gaining an electron from lignin or analogous structures such as xenobiotic pollutants. LiP catalyzes several oxidations in the side chains of lignin and related compounds by one-electron abstraction to form reactive radicals. Also the cleavage of aromatic ring structures has been reported. The first step is the reaction of the resting enzyme [Fe (III)PX] with H_2O_2 in a two electron transfer reaction which results in the formation of Compound I. Compound I has one reducing equivalent at the oxyl-ferric iron [Fe(IV) =O] and the other forms a cation radical [R*]. Compound I is then reduced by the substrate (A) in two sequential one-electron steps through Compound II (Fig. 4.1).

The enzyme intermediates, compound I and II, oxidize by one electron two substrates to return to the resting state (PX). Through this catalytic cycle, LiP catalyzes C-C and C-O cleavages in side chains of lignin-like compounds leading in general to the depolymerization of dimers and oligomers. These reactions can involve the oxidation of small molecular weight substrates like veratryl alcohol (VA) (3, 4-dimethoxy phenol) to veratraldehyde where an aryl cation radical intermediate is generated. The latter is highly reactive and can subsequently oxidize lignin subunits. Furthermore, veratryl alcohol can also reverse the deactivation of the enzyme caused by an excess hydrogen peroxide.

1. Enz-heme [Fe(III)] (PX) +H_2O_2→(Enz-heme˙+) [O = Fe (IV)] (Compound I) + H_2O

2. (Enz-heme•+) [O=Fe (IV)] (Compound I) + A → Enz-heme [O = Fe (IV)] (Compound II) + H^+ + R^+

3. Enz-heme [O = Fe (IV)] (compound II) + A → Enz-heme [Fe (III)] (PX) + H^+ + R^+

PX = native or resting enzyme, Enz = enzyme

Fig. 4.1: Catalytic Cycle of Lignin Peroxidase. PX = native or resting enzyme, Enz = enzyme

LiP catalyzes the oxidation of nonphenolic aromatic lignin moieties and similar compounds. LiP has been used to mineralize a variety of recalcitrant aromatic compounds, such as three and four ring PAHs, polychlorinated biphenyls and dyes.

4.2.2 Manganese Peroxidases (MnP; EC 1.11.1.13)

Manganese peroxidases (EC 1.11.1.13) belong to the family of oxidoreductases. Following the discovery of LiP in *Phanerochaete chrysosporium*, MnP secreted from the same fungus was found as another lignin degrading enzyme and subsequent investigations have shown that MnP is distributed in almost all white-rot fungi. MnP seem to be more widespread among white rot fungi than lignin peroxidase. The redox potential of the Mn peroxidase system is lower than that of lignin peroxidase and it has shown capacity for preferable oxidize in vitro phenolic substrates. These are glycosylated glycoproteins with an iron protoporphyrin IX (heme) prosthetic group, molecular weights between 32 and 62.5 kDa and are secreted in multiple isoforms. Its isoelectric point varies from 2.9 to 7.0 depending on the source species of the enzyme and iso-form. MnP preferentially oxidize Mn^{2+} into Mn^{3+}, which is stabilized by chelators such as oxalic acid, itself also excreted by the fungi. Chelated Mn^{3+} acts as a highly reactive (up to 1510 mV in H_2O, low molecular weight, diffusible redox-mediator). Thus, MnP are able to oxidize and depolymerize their natural substrate, i.e. lignin as well as recalcitrant xenobiotics such as nitroaminotoluenes and distillery effluent.

In enzymology, a Mnp is an enzyme that catalyzes the chemical reaction

$$2\,Mn^{2+} + 2\,H^+ + H_2O_2 \rightleftharpoons 2Mn^{3+} + 2\,H_2O$$

The 3 substrates of this enzyme are Mn^{2+}, H^+, and H_2O_2, whereas its two products are Mn^{3+} and H_2O. This enzyme belongs to the family of oxidoreductases, to be specific those acting on peroxide as acceptor (peroxidases).

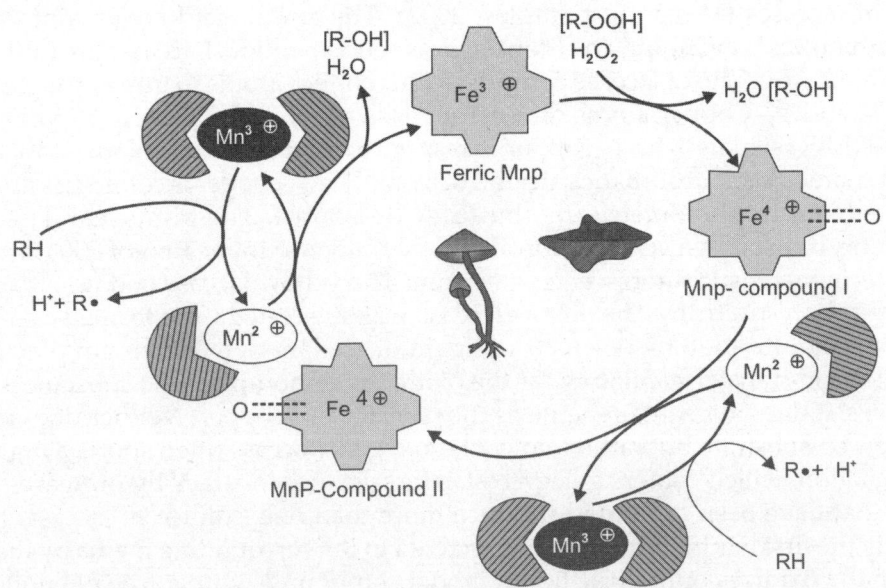

Fig. 4.2: The catalytic cycle of MnP

The systematic name of this enzyme class is Mn^{2+}: hydrogen-peroxide oxidoreductase. Other names in common use include peroxidase-M2, and Mn-dependent (NADH-oxidizing) peroxidase. It employs one cofactor, heme. This enzyme needs Ca^{2+} for activity.

The catalytic cycle of MnP shown in figure 4.2 resembles that of LiP but Mn^{2+} is the preferred electron donor that reduces compound I and II back to the resting state. Mn^{3+} can oxidize the aromatic rings of the lignin subunits or other substrates once it is chelated to organic acids produced by fungi (e.g. oxalate, malonate). Very few phenolic substrates can reduce MnP compound II to the native ferric state because of the steric hindrance at the active. Like LiP, MnP is also deactivated by excess hydrogen peroxide. The effluent decolourisation capability of MnP varies from one species to another and depends on the iso-enzyme and reaction conditions. In general, MnP requires organic acids to decolourize effluent.

4.2.3 Laccases (p-diphenol: oxygen oxidoreductase; EC 1.10.3.2)

Laccases (benzenediol: oxygen oxidoreductase EC 1.10.3.2) belong to multicopper oxidase family. These copper-containing enzymes catalyze the oxidation of various substrates with the simultaneous reduction of molecular oxygen to water. Yoshida first discovered laccases in 1883 after observing that latex from the Japanese lacquer tree (*Rhus vernicifera*) hardened in the presence of air. Since then, laccase activity has been found in other plants, some insects, and few bacteria. However, most laccases were reported from fungal organisms and most biotechnologically useful laccases are also of fungi origin. Probably the first report on the presence of laccase in fungi was from Laborde in 1897 (Mayer and Harel, 1979). Over 60 fungal strains belonging to the phyla Ascomycota, Zygomycota and especially Basidiomycota show laccase activities. The catalytic site of laccase is quite conserved among different species of fungi, but the rest of the enzyme structure shows high diversity. Fungal laccases are mostly inducible, extracellular, monomeric glycoproteins with carbohydrate contents of 10–20% which may contribute to the high stability of laccases (Mayer and Staples, 2002). The amino acid chain contains about 520–550 amino acids including an N-terminal secretion peptide. Laccases are multinuclear enzymes. The active site of laccase comprises four copper atoms in three groups, referred to as T_1, T_2 and T_3. Copper atoms differ from each other in their electron paramagnetic resonance (EPR) signals. The T_1 copper is responsible for the blue colour of the enzyme and has a characteristic absorbance around 610 nm. The T_2 copper is colourless and cannot be detected spectrophotometrically, but EPR detectable. The bi-nuclear T_3 copper is diamagnetic. It displays a spectral absorbance shoulder in the region of 330 nm and also displays a characteristic fluorescence spectrum. The yellow laccase had no blue maxima in the absorption spectrum. The yellow laccase was suggested to be formed as a result of blue laccase modification by products of lignin degradation, which might play a role as natural electron-transfer mediators for the oxidation of non-phenolic substances. Unlike peroxidases, it does not contain heme as the cofactor but copper. Neither does it require H_2O_2 as the co-substrate but rather molecular oxygen. Laccase often sports a high degree of glycosylation, which confers a degree of self resistance to attack by proteases. Almost all fungi that have been examined produce more than one isoform of laccase. Laccases are usually the first ligninolytic enzymes secreted to the surrounding media by the fungus that normally oxidizes only those lignin model compounds with a free phenolic group, forming phenoxy radicals as the mediators that are a group of low molecular-weight organic compounds. Many artificial mediators have been studied, being ABTS [2.2-azino-bis-(3-ethylbenzothiazoline-6-sulphonic acid)] the first described laccase mediator (Bourbonnais and Paice, 1990; Call and Mucke, 1997). There are natural compounds acting as mediator in laccase oxidation such p-hydroxycinnamic acids. The downside however, is that the redox potential although varying between different laccase isozymes, cannot be compared with that of the presence of mediators like veratryl alcohol and Mn^{2+}, their presence increases the effective range of substrates, which can be degraded by laccase. MnP, LiP and laccase are the three major lignin-degrading enzymes with great potential in industrial applications. Production of these enzymes from fungi has been well documented. More than 100 laccase enzymes have been purified and characterized from fungal cultures of basidiomycetes and ascomycetes and the number is still increasing. More than one isoenzyme is produced by most white rot fungi. The catalytic site of laccase is quite conserved among different species of fungi, but the rest of the enzyme structure shows high diversity. At least eight different variants are produced by the well-studied *Polyporus ostreatus* and *Trametes pubescens* MB89, while the ascomycete Podospora anserina also produces several laccase isoenzymes. In some fungi, the addition of copper, sugars and amino acids, ethanol and several phenolic compounds e.g. 2, 5-xylidine increase the production of extracellular laccases or even induce the secretion of additional isoenzymes into the culture medium. Laccase monomers vary in molecular mass from 43 to 383 kDa, but typical fungal laccases have a molecular mass of 60–70 kDa and an acidic isoelectric point around pH 4.0 (range = 2.6–6.9). Most fungal laccases are monomeric proteins, but some also exhibit homodimeric, heterodimeric, or oligomeric structures. Laccases are produced in multiple isoforms, which vary with both fungal species and environmental conditions. Fungal laccases are generally extracellular glycoproteins and often show considerable heterogeneity in molecular weight after purification, possibly attributable to proteolytic and glycosidic activities in the environment. The extent of glycosylation usually ranges between 10 and 25 mol%. However, glycosylation can be much greater and variation is exhibited within the same species. This was evident with *Botrytis cinnerea*, where the laccase saccharide content varied from 50 to 80 mol% with different strains. The glucans consist of arabinose, xylose, mannose, galacose, and glucose units, which are N-linked to the polypeptide. Multiple laccase isoforms in any given strain can thus arise from several laccase genes and from different glycosylation patterns. It has been proposed that in addition to the structural role, glycosylation helps protect of laccase from proteolytic degradation. Fungal laccases typically exhibit pH optima in the acidic pH range, but high activity at an alkaline pH is a desired trait for many industrial applications. In this respect bacterial laccases are advantageous.

4.2.4 Versatile peroxidases (VP; EC 1.11.1.16)

A third group of peroxidases, versatile peroxidases (VP), has been recently recognized, that can be regarded as hybrid between MnP and LiP, since they can oxidize not only Mn^{2+} but also phenolic and nonphenolic aromatic compounds including dyes. Studies indicate that contrary to LiP, MnP may oxidize Mn(II) without H_2O_2 with decomposition of acids, and concomitant production of peroxyl radicals that may affect lignin structure. Due to their Mn-oxidizing activity, the *Pleurotus* VP enzymes were first described as MnP enzymes, but they were later recognized as representing a new peroxidase type. VP is also able to efficiently oxidize phenolic compounds and dyes that are the substrates of generic peroxidases and related peroxidases, or the well-known horse radish peroxidase (HRP). VP oxidizes Mn^{2+}, as MnP does, and also high redox potential aromatic compounds, as LiP do. A novel enzyme which can utilize both veratryl alcohol and Mn^{2+}, versatile peroxidase has been recently described as a new family of ligninolytic peroxidases. The most noteworthy aspect of VP is that it combines the substrate specificity characteristics of LiP, MnP as well as cytochrome c peroxidase. In this way, it is able to oxidize a variety of (high and low redox potential) substrates including Mn^{2+}, phenolic and non-phenolic lignin dimers, veratryl alcohol, dimethoxybenzenes, different types of effluent, substituted phenols and hydroquinones. It has a Mn binding site similar to MnP and an exposed tryptophan residue homologous to that involved in veratryl alcohol oxidation by LiP. It is suggested that the catalytic properties of the new peroxidase is due to a hybrid molecular architecture combining different substrate-binding and oxidation sites. The interest on VP has increased during the last years, both as a model enzyme and as a source of industrial/environmental biocatalysts.

Besides, the recent findings have shown that some bacteria have potential for secretion of ligninolytic enzyme. Bacteria having MnP activity could be more potential for decolourisation and detoxification of industrial waste. Since, sucrose and aspartic acid are predominantly present in sugarcane juice which plays a major role in synthesis of melanoidin during sugar manufacturing process and it remains in molasses. Hence, the model synthetic melanoidins solution could be prepared by refluxing the equimolar (1M) solution of sucrose (Loba chemie, India), aspartic acid (Loba chemie, India) and 0.5M sodium carbonate at 100°C for 7 h for screening of MnP enzyme producing bacteria in presence of model melanoidin (Sucrose-aspartic acid Maillard product, SAA-MP). The potential strains were screened on the basis of growth and MnP activity on modified GPYM agar plates amended with different concentrations of SAA-MP (800, 1600, 2400, 3200 and 3600 mg/L). For bacterial isolation, 5 g of distillery sludge collected from M/s Unnao distillery and brewery Ltd., Unnao (UP), India were transferred to a conical flask having capacity 250 mL containing sterile SAA-MP (2400 mg/L) in modified GPYM medium (75 mL) at pH 7.0. The modified GPYM medium contained (in %): D-glucose, 1.0; peptone, 0.1; K_2HPO_4, 0.1 and $MgSO_4$.$7H_2O$, 0.05. The flasks were incubated at 35 ± 2°C in a rotary shaking incubator (Innova 4230, New Brunswick Scientific, UK) at 120 rpm for 6 days. When decolourisation was observed in samples, an aliquot (100 L) was spread on SAA-MP (2400 mg/L) amended modified GPYM agar plates and incubated at 35±2°C. The morphologically different colonies growing on plates were further purified on SAA-MP (2400 mg/L) amended modified GPYM agar plates by streaking method and the purity of each bacterial culture was checked under the microscope. Hence, twenty different bacterial strains named IITRM1, IITRM2, ...IITRM20 were isolated from effluent contaminated site. Further, to screen the potential bacterial strains, MnP activity was done on modified GPYM in broth amended with different concentration of SAA-MP (800–3600 mg/L) and 0.1% phenol red (w/v). The MnP activity was determined spectrophotometrically using phenol red as a substrate at 610 nm. Five milliliter of reaction mixture contained 1.0 mL sodium succinate buffer (50 mM, pH 4.5), 1.0 mL sodium lactate (50mM, pH 5.0), 0.4 mL manganese sulphate (0.1 mM), 0.7 mL phenol red (0.1mM), 0.4 mL H_2O_2 (50 L), gelatin 1 mg/mL and 0.5 mL of enzyme extract as previously described by Arora et al. (2002). The reaction was initiated by adding H_2O_2 and conducted at 30°C. One millilitre of reaction mixture was taken and 40 µL of 5 N NaOH was added to it to stop the reaction. Absorbance was taken at 610 nm. After every minute the same steps were repeated with 1 mL of the reaction mixture up to 4 min. One unit of enzyme activity is equivalent to an absorbance increase of 0.1 units/min/mL. Result revealed that among twenty isolated aerobic bacterial strains IITRM7, IITRM15, IITRM16 and IITRM17 were found tolerant up to 3200 mg/L melanoidins amended modified GPYM agar plate on the basis of growth and MnP activity. The aerobic bacterial strains IITRM1, IITRM4, IITRM5, IITRM7, IITRM15, IITRM16 and IITRM17 showed MnP activity by changing the colour from deep orange to light yellow at melanoidins (800 mg/L) and phenol red amended modified GPYM agar plates, while this activity was mimicked with increase of SAA-MP concentration. Phenol red changed from deep orange to light yellow during screening of peroxidase activity in bacteria. This change in colour of phenol red amended GPYM agar medium is mainly due to the oxidation of glucose by sugar oxidase enzymes, resulting in the generation of H_2O_2 and media acidification which is required for the melanoidins degradation. The growth of all bacteria was not observed at 3600 mg/L SAA-MP amended GPYM medium. While, the optimum growth and decolourisation by IITRM7, IITRM15 and IITRM16 was noted in 2400 mg/L SAA-MP amended modified GPYM. Out of the twenty bacterial strains, only three IITRM7, IITRM15 and IITRM16 showed optimum growth and exhibited peroxidase activity by changing the deep orange colour of dye to light yellow even at higher concentration (2400 mg/L) of SAA-MP. The PCR-amplified 16S rRNA gene sequence of strain IITRM7 have shown closest relatedness with *Bacillus* species (FJ581030), while strain IITRM15 with *Raoultella planticola* (GU329705) and strain IITRM16 with *Enterobacter sakazakii* (FJ581031). Hence, based on the 16S rDNA sequence similarity, strain IITRM7, IITRM15 and IITRM16 were identified as *B. species*, *R. planticola* and *E. sakazakii*, respectively.

4.3 EFFECT OF DIFFERENT ENVIRONMENTAL AND NUTRITIONAL FACTORS ON THE DECOLOURISATION OF MAILLARD PRODUCT

4.3.1 Effect of Various Carbon Sources on Decolourisation of Maillard Product

Every organism must find in its environment all of the substances required for energy generation and cellular biosynthesis. The chemicals and elements of this environment that are utilized for bacterial growth are referred to as nutrients or nutritional requirements. Many bacteria can be grown the laboratory in culture media which are designed to provide all the essential nutrients in solution for bacterial growth. At an elementary level, the nutritional requirements of a bacterium such as *E. coli* are revealed by the cell's elemental composition, which consists of C, H, O, N, S. P, K, Mg, Fe, Ca, Mn, and traces of Zn, Co, Cu and Mo. These elements are found in the form of water, inorganic ions, small molecules, and macromolecules which serve either a structural or functional role in the cells. The general physiological functions of the elements are outlined in table 4.2. In addition, trace elements are metal ions required by certain cells in such small amounts that it is difficult to detect (measure) them, and it is not necessary to add them to culture media as nutrients. Trace elements are required in such small amounts that they are present as "contaminants" of the water or other media components. As metal ions, the trace elements usually act as cofactors for essential enzymatic reactions in the cell.

Table 4.2: Major elements, their sources and functions in bacterial cells

Element	% of dry weight	Source	Function
Carbon	50	organic compounds or CO_2	Main constituent of cellular material
Oxygen	20	H_2O, organic compounds, CO_2, and O_2	Constituent of cell material and cell water; O_2 is electron acceptor in aerobic respiration
Nitrogen	14	NH_3, NO_3, organic compounds, N_2	Constituent of amino acids, nucleic acids nucleotides, and coenzymes
Hydrogen	8	H_2O, organic compounds, H_2	Main constituent of organic compounds and cell water
Phosphorus	3	inorganic phosphates (PO_4)	Constituent of nucleic acids, nucleotides, phospholipids, LPS, teichoic acids
Sulfur	1	SO_4, H_2S, $S°$, organic sulfur compounds	Constituent of cysteine, methionine, glutathione, several coenzymes
Potassium	1	Potassium salts	Main cellular inorganic cation and cofactor for certain enzymes
Magnesium	0.5	Magnesium salts	Inorganic cellular cation, cofactor for certain enzymatic reactions
Calcium	0.5	Calcium salts	Inorganic cellular cation, cofactor for certain enzymes and a component of endospores
Iron	0.2	Iron salts	Component of cytochromes and certain nonheme iron-proteins and a cofactor for some enzymatic reactions

One organism's trace element may be another's required element and vice-versa, but the usual cations that qualify as trace elements in bacterial nutrition are Mn, Co, Zn, Cu, and Mo.

In order to grow in nature or in the laboratory, a bacterium must have an energy source, a source of carbon and other required nutrients, and a permissive range of physical conditions such as O_2 concentration, temperature, and pH. Hence, in order to optimize the SAA-MP decolourisation process, the effect of various carbon and nitrogen sources was studied in presence of different bacterial combinations as shown in 1st row of table 4.3. The optimum glucose and peptone concentration for decolourisation of melanoidins was at 1 and 0.1%, respectively, probably due to generation of more redox mediators at this concentration that acting as electron donors for the cleavage of ethylinic (C = C) and azomethine (C = N) linkage of melanoidins. These linkage present in conjugated form in melanoidins structure and import colour to these polymer. The decolourisation was increased with increase of glucose concentration, but it decreased when concentration exceeded 2.0%. Higher glucose concentrations might form more acid end-products, thereby inhibiting growth and reducing the decolourisation by consortium due to low pH.

Table 4.3: Effect of composition of bacterial consortium, carbon and nitrogen sources on SAA-MP decolourisation within 144 h

Different Nutrient	Composition of bacterial consortium						
	IITRM7 (1)	IITRM15 (2)	IITRM16 (3)	1 + 2	1 + 3	2 + 3	1 + 2 + 3*
Carbon Sources (1%)							
Glucose (G)	30 ± 1.40	33 ± 1.21	30 ± 2.00	34 ± 1.43	38 ± 0.93	39 ± 1.11	45 ± 0.90
Sucrose (S)	31 ± 1.05	33 ± 1.76	34 ± 1.53	35 ± 1.20	36 ± 0.35	38 ± 1.00	40 ± 0.82
Fructose (F)	20 ± 0.82	21 ± 0.99	22 ± 0.95	23 ± 0.98	24 ± 0.76	25 ± 0.53	28 ± 0.40
Maltose (M)	22 ± 0.76	21 ± 0.75	24 ± 2.31	26 ± 1.11	20 ± 0.12	27 ± 0.34	28 ± 0.50
Lactose (L)	19 ± 0.45	17 ± 0.86	19 ± 0.84	23 ± 0.34	19 ± 0.19	23 ± 0.46	28 ± 0.43
Xylose (X)	21 ± 0.33	23 ± 1.65	28 ± 1.20	29 ± 0.78	30 ± 0.26	32 ± 0.52	33 ± 0.33
Arabinose (A)	17 ± 0.28	17 ± 0.23	20 ± 1.98	21 ± 0.98	23 ± 0.48	22 ± 0.40	24 ± 0.45
Ribose (R)	14 ± 0.15	16 ± 0.18	19 ± 0.85	21 ± 1.56	20 ± 0.45	21 ± 0.38	22 ± 0.40
Carbon (1%) + Nitrogen Sources (0.1%)							
G + Peptone	31 ± 1.23	34 ± 1.18	33 ± 1.56	38 ± 1.23	37 ± 1.65	38 ± 0.78	46 ± 0.58
G + Yeast Ext	27 ± 0.54	30 ± 2.87	31 ± 1.78	33 ± 0.99	34 ± 1.45	35 ± 1.45	36 ± 0.83
G + Beef Ext	16 ± 0.21	17 ± 0.54	19 ± 0.75	21 ± 0.87	23 ± 0.56	25 ± 0.75	28 ± 0.32
G + P + Y*	40 ± 1.11	43 ± 2.10	45 ± 1.34	48 ± 1.24	49 ± 0.51	48 ± 1.23	50 ± 0.47
G + NaNO$_3$	10 ± 0.01	12 ± 0.78	14 ± 0.82	16 ± 0.45	17 ± 0.34	18 ± 0.54	20 ± 0.40
G + NH$_4$NO$_3$	13 ± 0.21	14 ± 0.74	15 ± 0.11	17 ± 0.37	17 ± 0.30	19 ± 0.43	22 ± 0.48
S + Peptone	36 ± 1.23	38 ± 1.78	40 ± 1.78	41 ± 0.67	43 ± 0.78	45 ± 0.43	48 ± 1.45
S + Yeast Ext	24 ± 0.88	26 ± 0.87	29 ± 0.41	30 ± 1.67	31 ± 0.60	34 ± 0.70	37 ± 0.86
S + Beef Ext	13 ± 0.35	14 ± 0.82	16 ± 0.33	19 ± 0.34	21 ± 0.43	22 ± 0.28	26 ± 0.74
S + P + Y	37 ± 1.87	39 ± 1.96	42 ± 1.64	44 ± 1.23	46 ± 0.84	47 ± 1.25	44 ± 0.88
S + NaNO$_3$	11 ± 0.56	13 ± 1.00	15 ± 0.31	16 ± 0.84	18 ± 0.72	19 ± 0.48	21 ± 0.77
S + NH$_4$NO$_3$	09 ± 0.33	11 ± 0.47	13 ± 0.22	17 ± 0.33	19 ± 0.48	21 ± 0.46	23 ± 0.72

All values are means (n = 3) ± SD, P - peptone, Y = yeast extract, * = Optimized conditions

4.3.2 Effect of Nitrogen Sources on Decolourisation of Maillard Product

The maximum SAA-MP decolourisation was observed in presence of peptone (0.1%) and yeast extract (0.1%) as nitrogen source supplemented in modified GPYM medium (Table 4.3). Besides, results also revealed that sucrose (1%) with peptone (0.1%) and yeast extract (0.1%) decolourized SAA-MP only up to 44% (Table 4.3). However, decolourisation was mimicked in presence of inorganic sources i.e. NaNO$_3$ and NH$_4$NO$_3$ with glucose and sucrose both (Table 4.3). Kumar and Chandra (2006) and Miyata et al. (2000) have also studied the inhibitory effect of inorganic nitrogen sources for melanoidins decolourisation. Result also revealed that yeast extract (0.1%) suppress the SAA-MP decolourisation along with sucrose (1.0%) and peptone (0.1%). While, the yeast extract enhanced the decolourisation with peptone and glucose. The differential effect of organic and inorganic sources on decolourisation of SAA-MP might be due to their compositional difference. The organic nitrogen is a complex nitrogen source composed of a spectrum of peptides and free amino acids. During bacterial growth, these are taken up from the media by the cell and directly incorporated into proteins synthesis or transformed into other cellular nitrogenous constituents, which help in generation of extracellular enzyme. While, bacterial cell has spends more energy and time in synthesizing amino acids for protein synthesis from inorganic nitrogen sources. However, the inhibitory role of organic nitrogen at higher concentration possibly generated nitrate or nitrite which may compete with amino-carbonyl complex (melanoidins) and resulted into less decolourisation.

4.3.3 Effect of Temperature, pH and Oxygen on Decolourisation of Maillard Product

Microbial species requires temperature growth range that is determined by the heat sensitivity of its particular enzymes, membranes, ribosomes, and other components. Further, pH dramatically affects bacterial growth, this is because that pH affects the activity of enzymes those that are involved in biosynthesis and growth. Each microbial species also possesses a definite pH growth range and a distinct pH growth optimum. Effects of variable rpm, pH and temperature on SAA-MP decolourisation is shown in Fig. 4.3. The pH 7.0 noted favorable for SAA-MP decolourisation after 144 h incubation, this could be as result of maximum melanoidins solubility at neutral pH. The reduced bioavailability of melanoidins in acidic or alkaline pH has been noted. Similarly, it was also observed that increase in temperature (20–35°C) enhanced decolourisation from 26 to 56% (Fig. 4.3). While further, the increase in temperature up to 45°C adversely affected the growth and decolourisation ability of the consortium. This might be due to the loss of MnP enzyme activity at a higher temperature.

The shaking rate from 100–180 rpm increased the decolourisation (60%), whereas further increase in shaking speed (>200 rpm) decreased the degradation capability of consortium possibly due to the mechanical injury of bacterial cells at a higher shaking speed (Fig. 4.3). Thus, it can be concluded that the environmental factors like pH, temperature, aeration and nutrients play a vital role in microbial degradation process of industrial wastes because the activity of enzymes is greatly influenced by the various environmental factors.

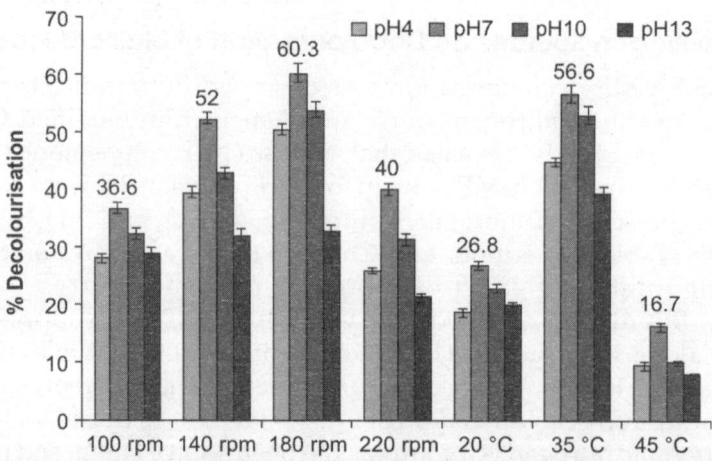

Fig. 4.3: Optimization of rpm, pH and temperature for SAA-MP decolourisation

4.4 BACTERIAL DECOLOURISATION OF SAA-MP MAILLARD PRODUCT AND META-BOLITES CHARACTERIZATION

For decolourisation of SAA-MP, the active bacterial consortia may be constructed by using a loopful of each bacterium (IITRM7, IITRM15 and IITRM16) from modified GPYM agar plate, precultured in 50 mL modified GPYM broth at 35°C with shaking (140 rpm). After 24 h, the bacterial cells of each strain were harvested by centrifugation at 10000 rpm and 4°C for 10 min and then washed with normal saline solution (0.9% NaCl). The washed bacterial cells were subsequently inoculated into fresh SAA-MP containing modified GPYM media to obtain an initial OD_{620} of 0.2. Various bacterial consortia comprising of different bacterial compositions were constructed at the same initial cell density. The combinations of different bacteria were incubated at 35 ± 2°C for 144 h in shaking flask condition (140 rpm). The combination of three showed maximum decolourisation in synergistic manner. The degradation and decolourisation of SAA-MP was assessed with physico-chemical changes.

The maximum decolourisation of SAA-MP (60%) was shown by consortium within 5 days incubation period under optimized conditions (0.1% KH_2PO_4, 0.05% $MgSO_4 \cdot 7H_2O$, 1.0% glucose, 0.1% peptone and 0.1% yeast extract, pH (7.0 ± 0.2), shaking speed (180 rpm) and temperature (35 ± 2°C). Bacterial growth, measured by absorbance at 620 nm showed the stationary phase after 5 days incubation period during SAA-MP degradation (Fig. 4.4).

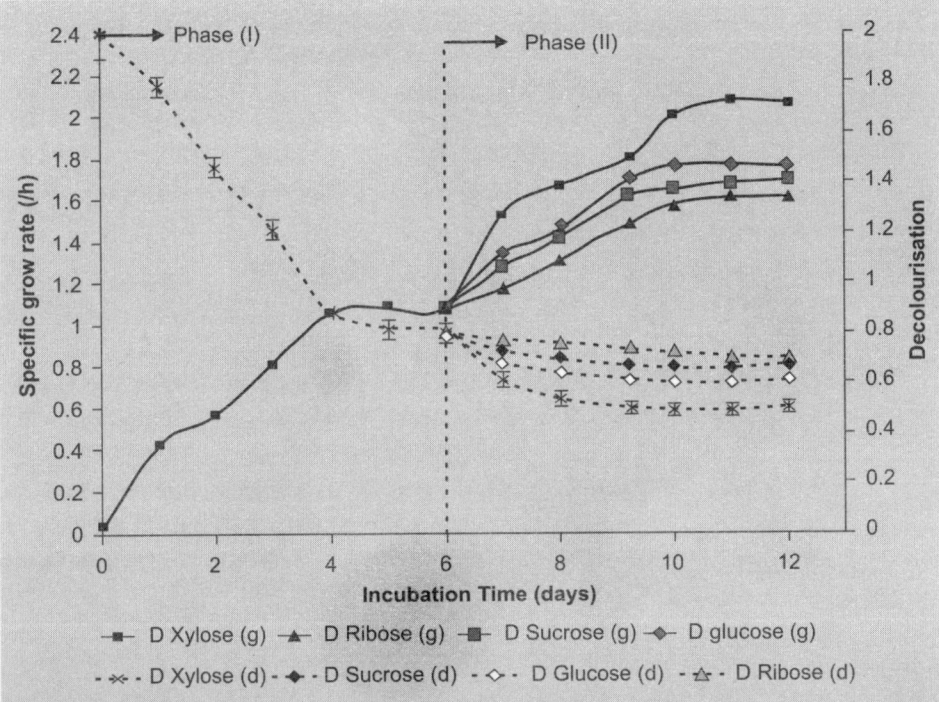

Fig. 4.4: Biostimulation on specific growth rate of bacteria and decolourisation by different sugars. Phase I: modified GPYM media, phase II: addition of different sugars in modified GPYM media when growth rate reached to stationary, g = growth, d = decolourisation.

Further, one percent (1.0%) of different carbon sources (glucose, sucrose, fructose, maltose, lactose, D-xylose, arabinose and ribose) was added in pre optimized flasks to investigate the biostimulation effect when bacterial growth attained stationary phase (Fig. 4.4). To investigate the biostimulation effect of various carbon sources, hexoses, pentoses and disaccharides were added at 6 days of bacterial growth.

The result revealed that the D-xylose enhanced maximum SAA-MP decolourisation process. The order of various carbon sources for decolourisation enhancement were noted as D-Xylose > glucose > sucrose > ribose, this attained stationary phase at 12 days of bacterial growth (Fig. 4.4). This resulted in 75% decolourisation as shown in Fig. 4.4. This result revealed that the bacterial consortium could utilize the biotransformed products in presence of D-Xylose resulted increase of biomass. Similarly, biostimulation effect on indigenous microbial community of soil by addition of organic and inorganic nutrient has also been reported by Stallwood et al. (2005). The physico-chemical analysis of bacterial degraded SAA-MP at different time interval showed reduction of BOD (76.78%), COD (74.20%), colour (75.00%), phenol (79.22%), phosphate (88.39%), ammonium (9.43%) and sodium (80.00%) after 12 days bacterial growth as shown in Table 4.4. The maximum decrease of above values was noted in between 6 to 12 days bacterial incubation. This coincided with the biostimulation activity of D-Xylose. However, the pH of SAA-MP amended modified GPYM medium was noted acidic during the initial 72 h of bacterial growth which subsequently increased up to 7.2. This

Table 4.4: Periodic monitoring of physico-chemical parameters of SAA-MP before and after bacterial treatment

| Parameters | Control | Before biostimulation | | After biostimulation | | |
	0 day	3 days	6 days	9 days	12 days	% Change
Colour	6250 ± 137	3091 ± 72.19	2500 ± 42.85	2050 ± 11.11	1562 ± 14.91	75.00
pH	7.00 ± 0.04	5.50 ± 0.04	7.00 ± 0.03	7.10 ± 0.04	7.20 ± 0.04	2.86
DO	1.70 ± 0.02	2.10 ± 0.03	2.40 ± 0.02	2.90 ± 0.03	2.10 ± 0.02	23.52
COD	26,000 ± 398	21000 ± 412	18453 ± 654	9163 ± 113	6708 ± 16.98	74.20
BOD	16000 ± 276	1176 ± 189	9543 ± 287	5754 ± 93.46	3715 ± 16.98	76.78
TS	6720 ± 194	1360 ± 47.36	1120 ± 32.67	580 ± 18.45	400 ± 12.43	94.05
TDS	3880 ± 106	720 ± 21.78	590 ± 21.46	360 ± 9.76	260 ± 6.78	93.30
TSS	2840 ± 76.45	640 ± 18.56	530 ± 15.78	220 ± 3.75	140 ± 3.33	95.07
Sulphate	4967 ± 142	691 ± 18.47	380 ± 7.98	291 ± 7.98	186 ± 4.65	96.26
Phenol	154 ± 3.87	128 ± 5.65	94.09 ± 3.87	67.00 ± 2.47	32.00 ± 1.24	79.22
Phosphate	5806 ± 102	3907 ± 79.98	1543 ± 47.98	1024 ± 37.98	674 ± 18.94	88.39
Nitrate	157 ± 8.67	168 ± 4.89	130 ± 2.85	107 ± 2.86	192 ± 4.36	22.29
Chloride	48.60 ± 2.45	25.12 ± 1.20	14.12 ± 0.45	65.00 ± 1.86	78.00 ± 4.65	60.49
Ammonium	159 ± 3.65	41.30 ± 5.65	118.23 ± 3.75	135 ± 3.54	144 ± 3.87	9.43
Sodium	820 ± 14.99	658 ± 25.56	324 ± 7.51	284 ± 5.68	156 ± 5.68	80.97

All values are means (n = 3) ± SD in mg/L, except colour (Co-Pt) and pH. DO = dissolve oxygen, TS = total solid, TDS = total dissolved solid, TSS = total suspended solid, arrow shows decreased and increased of parameters after 12 days of the growth.

induced solubility and bioavailability of colourant (melanoidins). Apart from this, there was also increase of chloride and nitrate ions. The release of chloride and nitrate ion might be due to the mineralization of SAA-MP.

The maximum MnP activity was induced at 4th days of bacterial growth in axenic and mixed conditions which subsequently declined and gradually become static in between 5 to 6 days incubation (Fig. 4.5a). But, the addition of D-Xylose at 6th days of bacterial growth increases the MnP activity more than 2 fold and it was noted maximum at 10 days of incubation. This indicated that D-xylose act as stimulator for MnP activity of this bacterial consortium. The stimulatory effect of xylose at stationary growth phase of bacteria might be either due to disappearance of glucose-6-phosphate or production of compounds which induces xyl-operon transcription. Further, the purified MnP showed single band at 43 kDa on denaturing SDS-PAGE during separation (Fig. 4.5b). The bands intensity of IITRM15 and 16 were more intense as compared to IITRM7, indicated high MnP producing capability of IITRM15 and 16.

The bacterial morphological observation under SEM before addition of D-xylose (1st phase) showed reduced cell size. While, after addition of D-xylose (2nd phase) luxuriant growth of each bacterium with bigger cell size were noted (Fig. 4.6). This indicated the biostimulatory effect of D-xylose on SAA-MP decolourisation.

In addition, modern approaches that generate and use metabolite structural information can accelerate the mechanism of effluent degradation. Scientists have been focused on

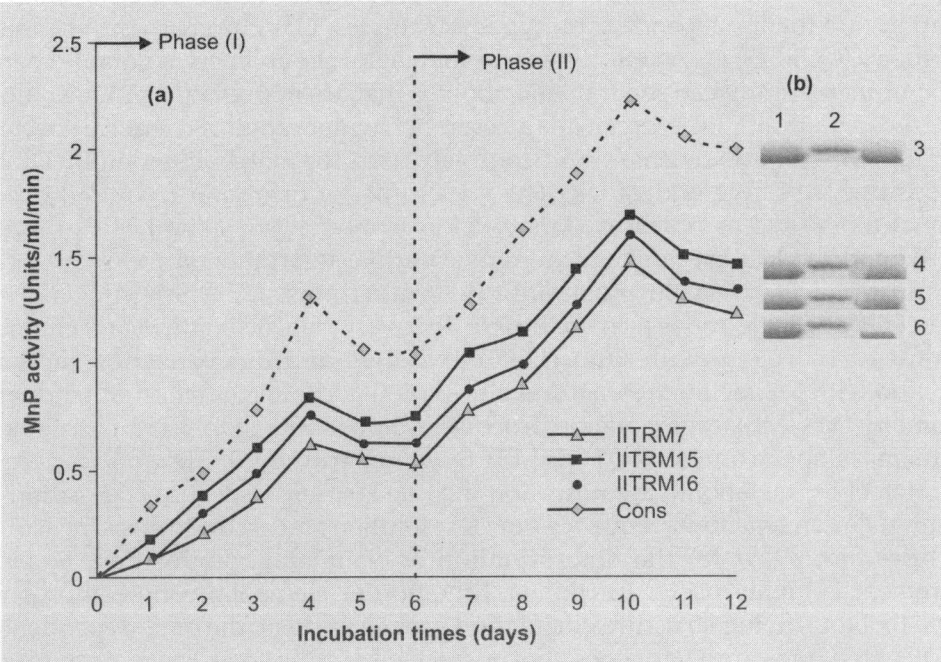

Fig. 4.5: Biostimulation on MnP activity among potential bacterial strains after addition of D-xylose at stationary phase (6th day) during SAA-MP degradation (a), SDS-PAGE of purified MnP, Lane1: standard MnP (43 kDa), 2: protein ladder (45kDa), 3: consortium (43kDa), 4: IITRM15 (43 kDa), 5: IITRM16 (43 kDa), 6:IITRM7(43 kDa), (b).

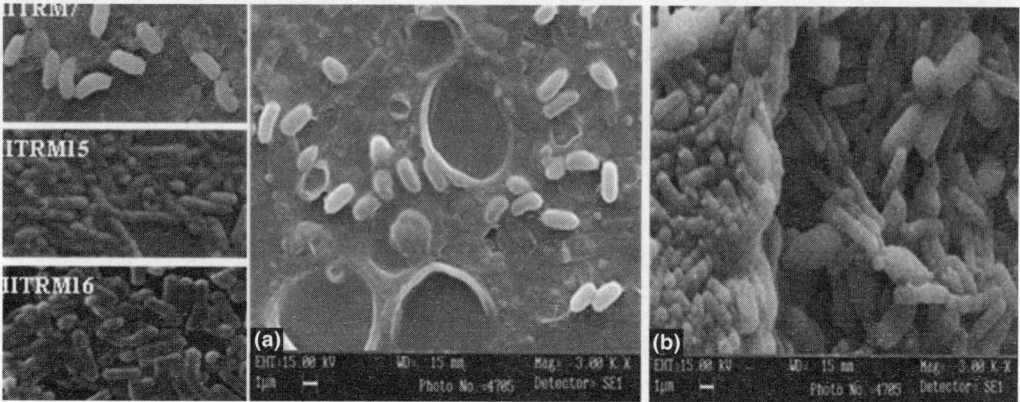

Fig. 4.6: Morphological changes of potential bacterial strains observed under SEM at 6 (a) and 12 days (b) incubation.

developing faster methods for metabolite identification. In order to achieve a more efficient, practical and reliable method for metabolite identification, in vitro microsomal incubations and sample preparation are very important. Chromatographic methods such as Gas Chromatography (GC), High performance liquid chromatography and Liquid Chromatography-Mass Spectrometry (LC-MS), etc. are commonly used in laboratories for the qualitative and quantitative analysis of substances from biological samples. Its applications are totally dependent on types of samples. Gas chromatography and mass spectrometry (GC/MS) are an effective combination for the analysis of volatile chemicals. Liquid chromatography can separate metabolites that are not volatile and have not been derivatized. As a result, LC-MS can analyze a much wider range of chemical species than GC-MS. LC-MS-MS is extensively and routinely used for metabolite identification. It is very sensitive, selective and quick. The structural confirmation frequently requires additional tools such as Nuclear Magnetic Resonance (NMR), synthesis of authentic standards and the lost art of chemical derivatization. Instrumental techniques such as NMR and MS are critical in metabolite determinations. LC-MS-NMR has become commercially available and is used in the late discovery stage to confirm and characterize metabolites. Hydrogen-deuterium (H–D) exchange and derivatisation methods in conjunction with MS facilitate structural elucidation and interpretation of tandem mass spectrometry (MS/MS) fragmentation processes. The advantages of the quadrupole time-of-flight mass spectrometer (QTOF-MS) over ion trap or triple quadrupole mass spectrometers for metabolite identification include fast mass spectral acquisition speed with high full-scan sensitivity, enhanced mass resolution and accurate mass measurement capabilities that allow for the determination of elemental composition. Exact mass measurement is highly useful for the confirmation of elemental composition and is a valuable tool for solving structure elucidation problems. Also, the data-dependent scans on QTOF or ion traps are a very good tool; increasing sample complexity, sample volume restrictions and through put requirements necessitate that the maximum amount of useful information is extracted from a single experiment. Metabolites of melanoidin may be characterized through TLC, HPLC, ESI-MS and LC-MS analysis. The TLC chromatogram showed prominent dark spot in control (Fig. 4.7A). This diminished after bacterial

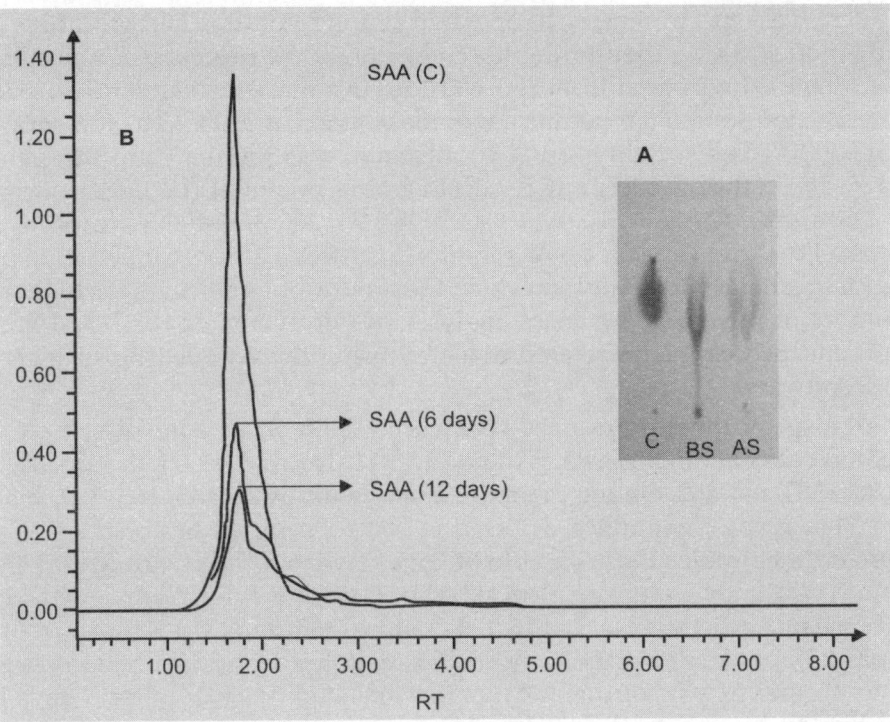

Fig. 4.7: TLC (A) and HPLC chromatogram (B) of SAA-MP after different time bacterial treatment along with control. C = control, BS = before stimulation, AS = after stimulation.

degradation while after biostimulation in presence of D-xylose only very light spot was visible. This apparently indicated decolourisation as well as de-polymerization of colourant present in SAA-MP. Periodic reduction of absorbance peak area by HPLC analysis also confirmed the degradation of SAA-MP at different time interval as compared to control (Fig. 4.7B). The decolourisation was noted 60 and 75% at 6 and 12 days, respectively.

The control and bacterial degraded SAA-MP were centrifuged at 10000 rpm and supernatant was vacuum dried and residue was dissolved in HPLC grade methanol and used for metabolites characterization by electrospray ionization-mass spectrum (ESI-MS) and liquid chromatography-mass spectrometry (LC-MS). The ESI-MS was recorded with Micromass Quattro II triple quadruple mass spectrometer to know the ionization pattern of compounds. For LC-MS, the LC system consist of a SIL-10AD VP auto injector, SCL-10AVP system controller, LC-10ADVP liquid chromatograph and a DGU-12A degasser, all from Shimadzu (Kyoto, Japan). Samples were introduced into the mass spectrometer through a C18 Gemini column (2 × 5 × 200 mm, Phenomenex) eluted at a flow rate of 0.2 mL/min, at ambient temperature. Elution was performed, with 97% solvent A (0.1% formic acid in water) and 3% solvent B (0.1% formic acid, 90% acetonitrile, 5.0% methanol and 5.0% water). Mass spectrometric detection was performed with an Applied Biosystems MDS Sciex (Ontario, Canada) API 2000 triple quadrupole mass spectrometer equipped with an electrospray ionization (ESI) interface in the positive ion mode.

The tandem mass spectrometer was operated at unit resolution in the multiple reactions monitoring mode (MRM), monitoring the transition of the protonated molecular ions to the product ions. Q1 was used from 150–600 amu in a mass-resolving mode to select the parent ion. The ion source temperature was maintained at 350°C. The ion spray voltage was set at 5500 V. The curtain gas (CUR; nitrogen) was set at 15 and the collision gas (CAD) at 7. The collision energy (CE), declustering potential (DP), focusing potential (FP) and entrance potential (EP), were set at 40, 75, 200 and 800 V, respectively. This system was set to the multiple reaction monitoring (MRM) modes, i.e. selecting precursor ions, dissociating them and finally analyzing the product ions reaching the high selectivity and sensitivity of this mode for mass analysis and detection. In the MRM mode, data acquisition and processing were accomplished using the Applied Biosystems Analyst version 1.2 software

The electrospray ionization-mass spectrum (ESI-MS) of control sample showed molecular ion peaks at (m/z) 69, 85, 97, 110, 127, 141, 145, 158, 163, 180, 198, 203, 325, 343, 360, 365, 366, 381 and 382, among them the major peaks were 145, 163, 360, 365 and 381 (Fig. 4.8a). Hence, the compounds detected in control samples on the basis of available literatures and their molecular weight to be 2, 3-dihydro 3, 5-dihydroxy-6-methyl-4(H)-pyran-4-one (145), anhydrohexose form [M – H]$^-$ (163), benzyl-3, 4-ethylenedioxypyrrol-2, 5-dicarboxylate (360), [M + Na$^+$] adduct of maltose (365) and [M + K$^+$] adduct of maltose (381) (Table 4.5). The maltose, maltotriose and maltotetrose adduct have been reported in previous studies as part of the melanoidins skeleton. Further, after 6 days bacterial treatment the ESI-MS spectra showed molecular ion peaks at (m/z) 69, 72, 99, 105, 113, 137, 141, 143, 149, 158, 165, 181, 183, 197, 198, 213, 214, 215, 219, 241, 257, 301, 317 and 339, among them the major peaks were 165, 181, 197 and 213 (Fig. 4.8b).

These compounds were detected as 2'-nitroacetophenone (165), 2-hydroxy-3-methoxyphenyl propanol (181), trans-2-tridecenal (197) and 2-(trichloroacetyl) pyrrole (213) (Table 4.5). Furthermore, ESI-MS spectra showed molecular ion peaks at (m/z) 69, 72, 86, 96, 105, 108, 113, 118, 121, 135, 141, 149, 158, 165, 181, 198, 203, 219, 223, 241, 257, 270, 301, 314, 327, 413 and 459 after 12 days of bacterial SAA-MP degradation (Fig. 4.8c). 3,5-Dihydroxy-2-methyl-4H-pyran-4-one (141), 4-(1H-pyrrol-1-yl) phenol (158) and 1-Hexadecanol (241) were noted as major compounds in 12 days bacterial treated samples (Table 4.5). The fingerprint of LC-MS chromatogram of control and degraded sample has also been shown in Fig. 4.9a–c. The control sample showed major peaks at 4.94, 15.38, 16.67, 23.06 and 27.24 min retention time (RT). Further, the fragmentation pattern of mass ion spectra from RT 4.68–5.14, 15.12–15.45, 16.42–16.87, 22.80–23.26 and 26.72–27.63 min, contained major peaks at m/z 278, 296, 315 and 656 (Fig. 4.9a). Moreover, the LC-MS spectra of bacterial degraded SAA-MP at 6 and 12 days incubation showed reduction of peak at 4.94 min RT, which shifted after 12 days bacterial degradation (Fig. 4.9b–c). However, the peaks noted in range of 16.42–16.87, 22.80–23.26 and 26.72–27.63 min did not change even after 12 days of bacterial treatment (Fig. 4.9a–c). This indicated that compounds available at lower RT were easily degraded by bacterial consortium, but unable to depolymerise the compounds of higher RT. The similar compounds by GC-MS and LC-MS analysis have also been reported by some workers from fungal and bacterial treated melanoidins and distillery effluent (Gonzalez et al., 2000; Chandra et al. 2009a). Similarly, the similar phenolics compounds have also been reported from wine, beer and dried plum by ESI-MS and LC-MS-MS analysis.

Table 4.5: Metabolites identified by ESI-MS in control and bacterial decolourised samples of SAA-MP. All values in brackets are the molecular weight of compounds. C = control, BS = before stimulation, AS = after stimulation

S.No.	m/z	Compounds name	C	BS	AS
1.	69 (69)	3-Pyrroline	+	+	+
2.	72 (74)	3-Hydroxypropanal	−	+	+
3.	81 (81)	1-Methyl-1H-pyrrole	−	+	+
4.	85 (86)	2, 3-Butanedione	+	−	−
5.	86 (86)	Butenoic acid	−	−	+
6.	96(96)	2-Furancarboxaldehyde	−	−	+
7.	99 (98)	Furfuryl alcohol	−	+	−
8.	105 (104)	Vinyl benzene	−	+	+
9.	108 (108)	2, 3-Dimethyl pyrrazine	−	−	+
10.	110 (110)	5-Methyl-2-furan carboxaldehyde	+	−	−
11.	113 (112)	Furan-2-carboxylic acid	−	+	+
12.	118 (117)	Indole	−	−	+
13.	127(126)	5-(Hydroxymethyl)-2-furfural	+	−	−
14.	135 (134)	N-methyl indane	−	−	+
15.	137 (138)	4-Methyl guaiacol	−	+	−
16.	141(142)	3, 5-Dihydroxy-2-methyl-4H-pyran-4-one	+	+	+
17.	143 (142)	p-chloroanisole	−	+	−
18.	145 (144)	2, 3-Dihydro3, 5-dihydroxy-6-methyl-4(H)-pyran-4-one	+	−	−
19.	149 (150)	4-Vinyl-2-methoxyphenol	−	+	+
20.	158 (159)	4-(1H-pyrrol-1-yl) phenol	+	+	+
21.	163 (162)	Anhydrohexose form [M-H]⁻	+	−	−
22.	165 (165)	2'-Nitroacetophenone	−	+	+
23.	181 (181)	2-Hydroxy-3-methoxyphenyl propanol	−	+	+
24.	197 (196)	trans-2-tridecenal	−	+	−
25.	198 (198)	1-(3-Methoxy-phenyl)-1H-pyrrole-2-carbaldehyde	+	+	+
26.	203 (203)	2-Hydroxy-5-(1H-pyrrol-1-yl) benzoic acid	+	−	+
27.	213 (212)	2-(Trichloroacetyl) pyrrole	−	+	−
28.	214 (213)	3, 4-Ethylenedioxypyrrole-2, 5-dicarboxylic acid	−	+	−
29.	198 (199)	1-(4-Chlorophenyl)-1H-pyrrole	−	−	+
30.	219 (221)	1-(p-Tolysulfonyl)pyrrole	−	−	+
31.	223 (225)	1-(tert-Butyl) 2-methylpyrrole-1, 2-dicarboxylate	−	−	+
32.	241 (242)	1-Hexadecanol	−	−	+
33.	257 (256)	Palmitic acid	−	−	+
34.	270 (269)	Diethyl-3, 4-ethylenedioxypyrrole-2, 5-dicarboxylate	−	−	+
35.	301 (301)	Paeonidine-(6acetyl)-3-glucoside	−	+	+
36.	314 (315)	Methyl-5-(benzyloxycarbonyl)-2, 4-dimethyl-3-pyrrolepropionate	−	−	+
37.	317 (317)	Petunidine-(6-coumaryl)-3-glucoside	−	+	−
38.	327 (328)	Hydroxydihydrocannabinol	−	−	+
39.	360 (359)	Benzyl-3, 4-ethylenedioxypyrrol-2, 5-dicarboxylate	+	−	−
40.	365 (365)	[M + Na⁺] adduct of maltose	+	+	+
41.	381 (381)	[M + K⁺] adduct of maltose	+	−	−

100
%
0
365
381
145
163
360
85
127
110
366
325
158
97
382
198
141
180
203
69
343
75 100 125 150 175 200 225 250 275 300 325 350 375 400 425 450 475
(a)

100
%
0
165
158
213
113
141
143
149
167
105
198
214
69
99
137
317
339
72
97
115
215
175
183
263
279
301
355
365
77
321
219
309
413
81
241
257
285
341
377
383
393
421
437
459
80 100 120 140 160 180 200 220 240 260 280 300 320 340 360 380 400 420 440 460
(b)

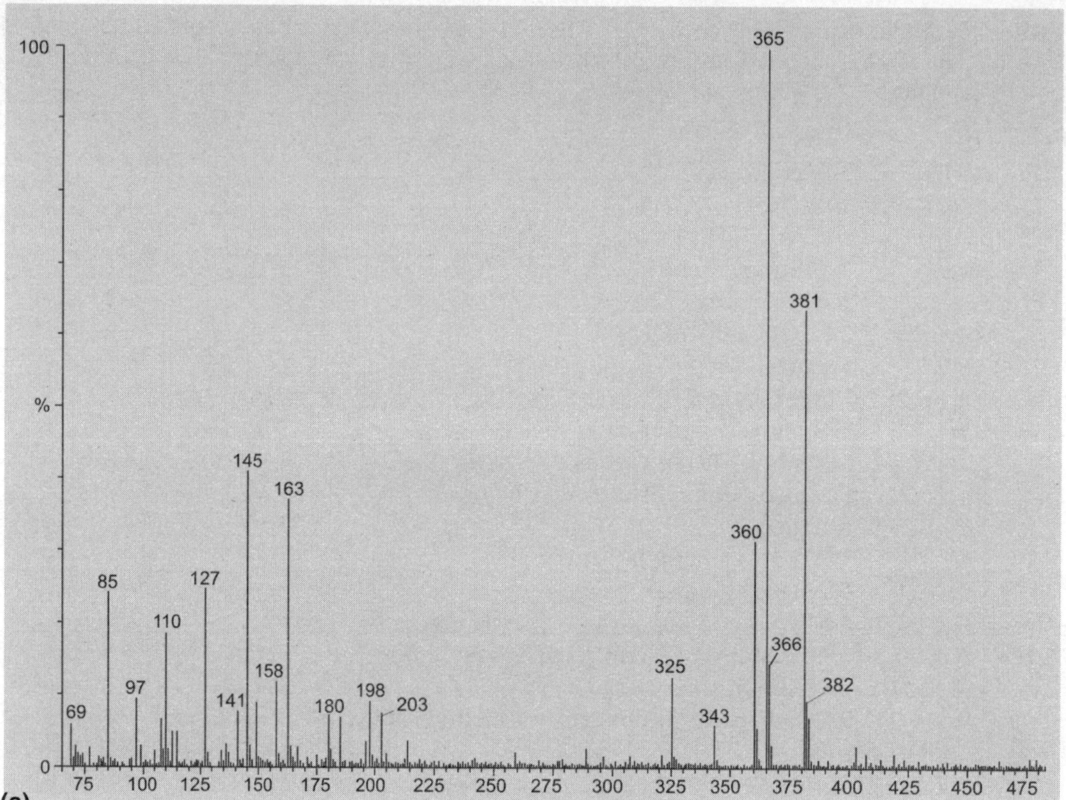

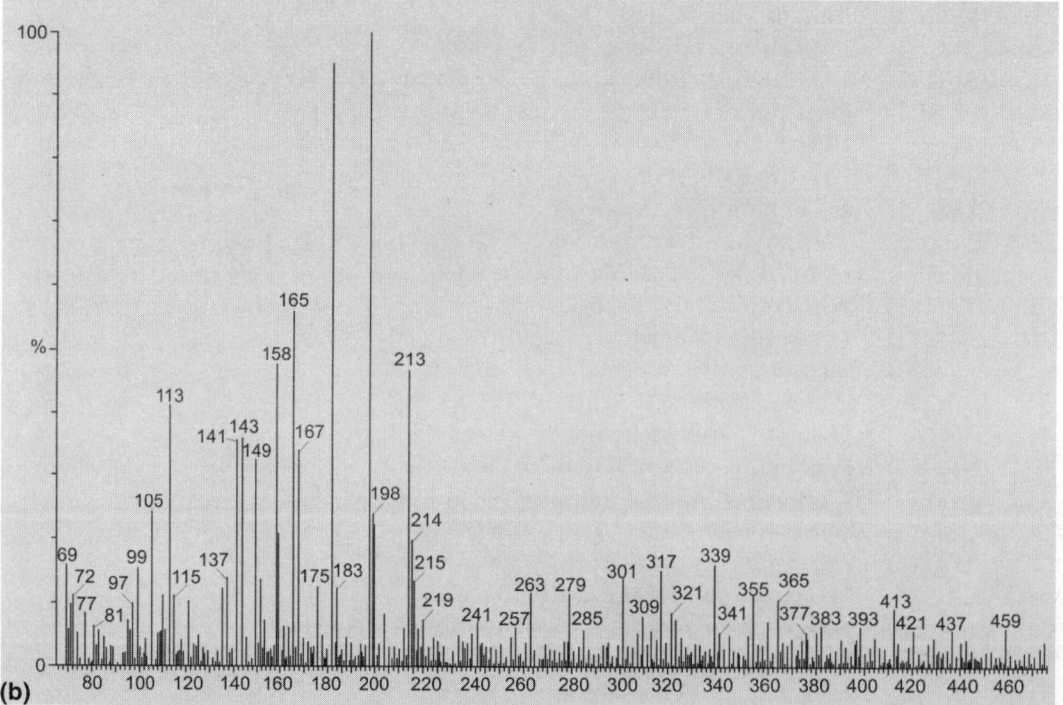

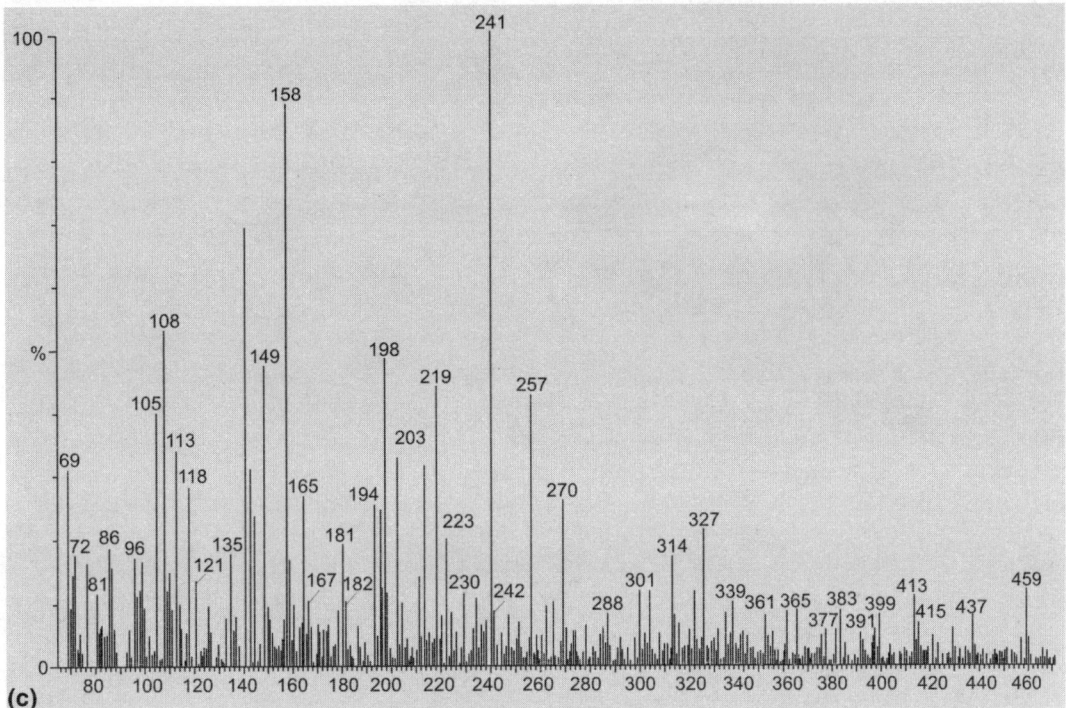

Fig. 4.8: ESI-MS of SAA-MP before bacterial treatment (a), after 6 (b) and 12 days (c) bacterial treatment

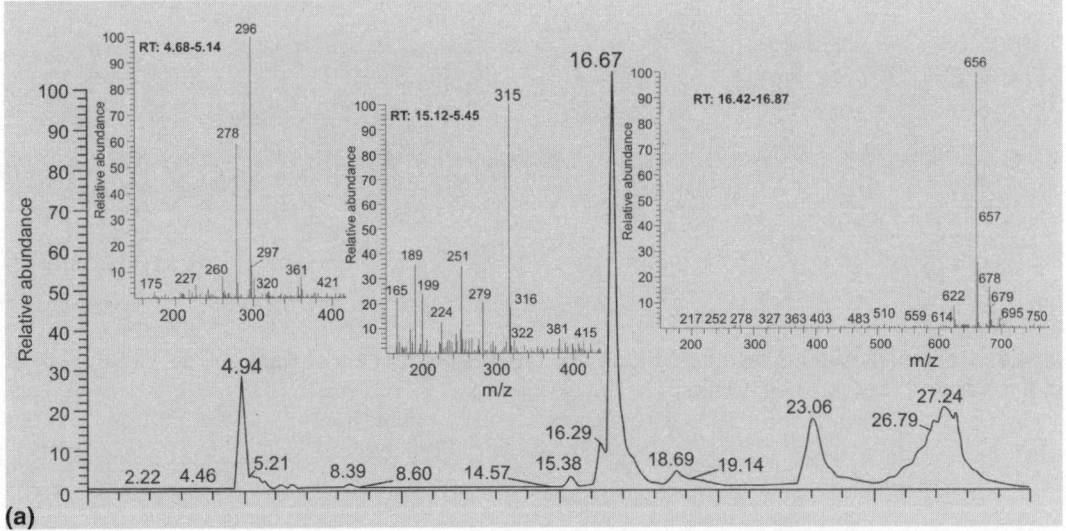

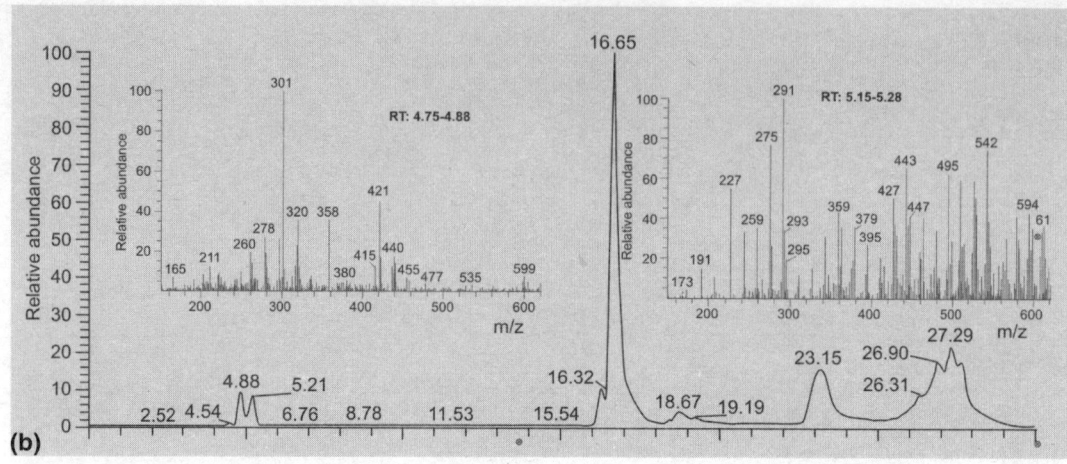

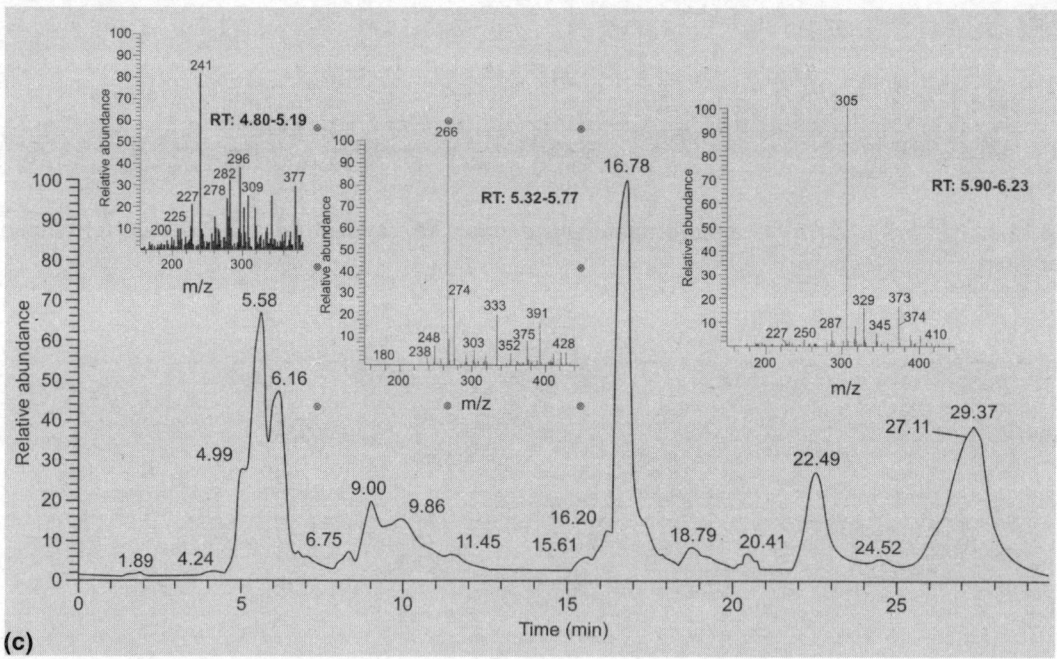

Fig. 4.9: Representative LC chromatograms and MS-MS product ion spectra (inside) of SAA-MP at 0 (a), 6 (b) and 12 days (c) degradation

5

Effect of Phenol and Heavy Metals on Biological Decolourization of Melanoidin

5.1 INTRODUCTION

Melanoidins are found in natural waters, river waters, estuarine, coastal and open waters represent a key link in the transformation of labile organic matter (polysaccharides, amino acids) into more recalcitrant humic material in nature/environment and their similarities with humic acid (HA) make them important as possible buffer compounds for metallic ions. The extent of complexation depends on the amino acids used to make the melanoidin and the highest complexing capacity is obtained for melanoidin prepared from glucose and the basic amino acid lysine. However, the presence of calcium and magnesium ions and other macro and micro constituents further influences the complexing properties of melanoidins towards metal ions. Melanoidins prepared using condensation times longer than two days exhibit complexation properties towards copper ions that appear to depend on the basicity of the amino acid precursor and the molecular mass of the product (Plavsic et al., 2006). The adsorption at lower pH is more pronounced, but is weak at pH 6.0. It is because of the additional binding sites for copper ions available as more groups are dissociated at higher pH values. The higher pH value could cause the ligand configuration to change and make more available binding sites. The highest copper ion complexing capacity values were obtained for melanoidins obtained from Glucose-lysine with molecular mass >10 kDa as lysine contains two amino groups while glutamic acid and valine contains only one (Plavsic et al., 2006). Basic amino acids (i.e. those containing more amino groups than carboxylic groups) preferentially condense with sugars to form nitrogen-rich polymers, which are good complexing agents for copper ions. Metal cations can be classified as A or B-type cations according to the number of electrons in the outer shell. Type A, or a (Hard sphere) cations have the inert gas type (d) electron configuration and are difficult to polarize (e.g. Na^+, K^+, Ca^{2+}, Mg^{2+}) and they preferentially form complexes with fluoride ion and ligands having oxygen as a donor whereas Type B, or b (Soft sphere) cations are of higher polarizability (e.g. Hg^{2+}, Cd^{2+}, Au^+, Cu^+). Type B cations coordinate preferentially with bases containing I, S, or N as the donor atom. Cu^{2+} ions forms very stable organic complexes and are preferentially bound to ligands containing N, S or O as donor atoms (Morales et al., 2005; Painter, 1998). Melanoidins behave as anionic hydrophilic polymers, which can form stable complexes with metal cations and reported that ketone or hydroxyl groups of pyranone or pyridone residues act as donor groups in melanoidins and participate in the chelation with metals as melanoidins have net negative charge and therefore, different heavy metals (Cu^{2+}, Cr^{3+}, Fe^{3+}, Zn^{2+}, Pb^{2+}, etc.) form large complex molecules with melanoidins, amino acids, proteins and sugars in acidic medium and get precipitated (Migo et al., 1997). It is also possible that other metals, e.g. trivalent metal ions like Fe^{3+} are bound to the melanoidin. Trivalent metal ions form stronger complexes with HA than monovalent and divalent metal ions. There were no significant differences between the different melanoidins from the Maillard model systems, but widely different behavior was observed in the ability to bind iron among melanoidins isolated from commercial coffee, sweet wine and beer due to bilinear behavior means the presence of at least two different types of binding sites. Due to antioxidant and antimicrobial properties, low molecular weight compounds bounded to melanoidins exerts antioxidant and antimicrobial activities usually higher than those of the pure melanoidins to whom they are linked. In case of antihypertensive activity, it has been found that the main activity is related to the melanoidin core. However, Melanoidins obtained from Glucose-Histidine reaction mixture are elucidated as furan ring and nitrogen containing brown compounds having peroxyl radical scavenging activity, an indicator of highest antioxidative activity determined by conjugated diene formation from peroxidation of linoleic acid (Yilmaz and Toledo, 2005). Brands et al. (2000) demonstrated that heated sugar-casein model melanoidins consisting variable sugars exhibit different mutagenic activity. For example, ketose sugars (fructose and tagatose) showed a remarkably higher mutagenicity compared with their aldose isomers (glucose and galactose) and generated active oxygen species resulting in DNA strand breaking and mutagenesis. Some other Maillard Reaction Products (MRPs) were also reported to induce chromosome aberrations in Chinese hamster ovary cells and gene conversion in yeast. Mutagenicity and DNA strand breaking activity of melanoidins from a glucose-glycine model was demonstrated by Hiramoto et al. (1997) who reported that the low molecular weight (LMW) fractions act as lipid sink and induced DNA damage, where the effect increased with the concentration added. High concentrations (1%) seems not only to be cytotoxic for the cells, but also lower concentrations between 0.05% and 0.2% reduced cell proliferation and cell viability.

As per previous chapter it is clear that melanoidins are major colourants present in sugarcane molasses based distillery effluent. It contributes high biological oxygen demand (BOD, 5000 mg/L) and chemical oxygen demand (COD, 25000 mg/L). Besides, distillery effluent after biomethanation contains high amount of phenolics (510 mg/L) and heavy metals (Pant and Adholeya, 2007; Bharagava et al., 2008). The higher amount of phenolics and heavy metals are noted due to condensation behavior of residual phenol and heavy metals in anaerobic digestion process, because the simple sugar get converted into methane. Sugarcane molasses based distillery effluent content nearly 2% melanoidin along with mixture of heavy metals and phenolics as major source of environmental pollution due to its complex nature. Because, melanoidin have binding tendency with heavy metals, hence this makes effluent more complexes (Migo et al., 1997). When it is discharged to the surface water, it reduces the sunlight penetration in the river, lakes or lagoons due to dark colour of contribution to mixing sources. This leads to decreased photosynthesis and dissolved oxygen of aquatic resources. Similarly, its disposal on soil is equally detrimental causing inhibition of seed germination and depletion of vegetation by addition of recalcitrant pollutants (Chandra et al., 2004a; Chandra et al., 2008a). Moreover, the assessment based on single pollutant exposure enable us to acquire fundamental knowledge about individual pollutant under carefully controlled condition, which do not reflect real-world exposures. Though, the individual effect of heavy metals and phenol have been reported on bacterial growth and degradation of some environmental pollutants (Yamaoka et al., 2002; Nweke et al., 2007), but the combined effect of phenol and heavy metals on bioremediation of any polymer present in industrial waste is unknown. The biological decolourisation of molasses based distillery effluent is still major challenge. The knowledge on the effect of heavy metals and phenolics on bacterial decolourisation of melanoidin is not reported so far. Therefore, the knowledge of this would be helpful to understand the mechanism of distillery effluent decolourisation. Hence, present chapter has been focused on effect of Fe^{3+}, Zn^{2+}, Mn^{2+} and phenol on sucrose-aspartic acid Maillard product (SAA) decolourisation by bacterial consortium comprising *Bacillus sp.* (IITRM7; FJ581030), *Raoultella planticola* (IITRM15; GU329705) and *Enterobacter sakazakii* (IITRM16, FJ581031) with ligninolytic activity capable for decolourisation of melanoidins (Sangeeta et al., 2011). The changes in physico-chemical properties of SAA due to binding of phenol and different heavy metals have been presented which are common environmental pollutants.

5.2 CHANGES IN PHYSICO-CHEMICAL PROPERTIES OF SUCROSE-ASPARTIC ACID (SAA) MAILLARD PRODUCTS IN PRESENCE OF PHENOL AND HEAVY METALS

It is very essential and important to test the water before it is used for drinking, domestic, agricultural or industrial purpose. Water must be tested with different physico-chemical parameters. Selection of parameters for testing of water is solely depends upon for what purpose we going to use that water and what extent we need its quality and purity. Water does content different types of floating, dissolved, suspended and microbiological as well as bacteriological impurities. Some physical test should be performed for testing of its physical appearance such as temperature, color, odour, pH, turbidity, TDS, etc. while chemical tests should be perform for its BOD, COD, dissolved oxygen, alkalinity, hardness and other characters. For obtaining more and more quality and purity water, it should be tested for its trace metal, heavy metal contents and organic, i.e. pesticide residue. It is obvious that drinking water should pass these entire tests and it should content required amount of mineral level. Only in the developed countries all these criteria's are strictly monitored. Due to very low concentration of heavy metal and organic pesticide impurities present in water it need highly sophisticated analytical instruments and well trained manpower.

In this study the synthetic melanoidin was prepared as described in previous chapter (chapter four). The degradation and decolourisation of SAA in the presence of different heavy metals and phenol has been assessed with physico-chemical changes. To study the effect of phenol and heavy metals on SAA decolourisation, two set of experiments were conducted as detail given in Table 5.1 to generate the simulated conditions of effluent. The stock solution of Zn^{2+}, Fe^{3+} and Mn^{2+} were prepared by using their respective salts $ZnCl_2$, $FeCl_3$ and $MnCl_2$. In addition, the bacterial inoculum was also added in the flasks as mentioned in the above section and incubated it for decolourisation study. The permissible limit of each metal was chosen according to the prescribed standard of EPA (2002). These metals were selected due to their occurrence in post methanated distillery effluent. All the experiments were done without maintaining pH to obtain simulated condition of effluent. The SAA containing GPYM media showed variation in physico-chemical properties due to addition of heavy metals and phenol (Table 5.2a-b). The addition of Fe^{3+} contributed more colour in aqueous solution of SAA as compared to Zn^{2+} and Mn^{2+} even at permissible limit of its disposal. This was due to trivalent of iron.

Table 5.1: Experimental details and growth pattern of potential bacterial consortium in SAA amended GPYM media containing different concentration of Zn^{2+}, Fe^{3+} and Mn^{2+} with and without phenol

Sample	Contents	Growth	Sample	Contents	Growth
C	SAA (1200 mg/L)	++++			
1st set of experiment			**2nd set of experiment**		
SAA + Zn^{2+} (mg/L)			**SAA + Phenol (100 mg/L) + Zn^{2+} (mg/L)**		
S1	SAA + Zn (2.00)	++++	S21	SAA + Phenol + Zn (2.00)	+++
S2	SAA + Zn (10.00)	++++	S22	SAA + Phenol + Zn (10.00)	++
S3	SAA + Zn (20.00)	+++	S23	SAA + Phenol + Zn (20.00)	++
S4	SAA + Zn (30.00)	+	S24	SAA + Phenol + Zn (30.00)	– –
S5	SAA + Zn (40.00)	– –	S25	SAA + Phenol + Zn (40.00)	– –
SAA + Fe^{3+} (mg/L)			**SAA + Phenol (100 mg/L) + Fe^{3+} (mg/L)**		
S6	SAA + Fe (2.00)	++	S26	SAA + Phenol + Fe (2.00)	++
S7	SAA + Fe (10.00)	++	S27	SAA + Phenol + Fe (10.00)	– –
S8	SAA + Fe (20.00)	+	S28	SAA + Phenol + Fe (20.00)	– –
S9	SAA + Fe (30.00)	– –	S29	SAA + Phenol + Fe (30.00)	– –
S10	SAA + Fe (40.00)	– –	S30	SAA + Phenol + Fe (40.00)	– –
SAA + Mn^{2+} (mg/L)			**SAA + Phenol (100 mg/L) + Mn^{2+} (mg/L)**		
S11	SAA + Mn (0.20)	+++	S31	SAA + Phenol + Mn (0.20)	+++
S12	SAA + Mn (1.00)	+++	S32	SAA + Phenol + Mn (1.00)	+++
S13	SAA + Mn (2.00)	+++	S33	SAA + Phenol + Mn (2.00)	++
S14	SAA + Mn (3.00)	++	S34	SAA + Phenol + Mn (3.00)	+
S15	SAA + Mn (4.00)	++	S35	SAA + Phenol + Mn (4.00)	+
SAA + Mix metals			**SAA + Phenol (100mg/L) + Mix metals**		
S16	SAA + Zn (2.00) + Fe (2.00) + Mn (0.20)	+++	S36	SAA+ Phenol +Zn(2.00) + Fe (2.00) + Mn (0.20)	+ +
S17	SAA + Zn (10.00) + Fe (10.0) + Mn (1.00)	+++	S37	SAA+Phenol+Zn (10.00) + Fe (10.0) + Mn (1.00)	++
S18	SAA + Zn (20.00) + Fe (20.0) + Mn (2.00)	++	S38	SAA+Phenol+ Zn (20.00) + Fe (20.0) + Mn (2.00)	+
S19	SAA + Zn (30.00) + Fe (30.0) + Mn (3.00)	– –	S39	SAA+ Phenol + Zn (30.00) + Fe (30.0)+Mn (3.00)	– –
S20	SAA + Zn (40.00) + Fe (40.0) + Mn (4.00)	– –	S40	SAA+Phenol + Zn (40.00) + Fe (40.0) + Mn (4.00)	– –

++++: luxuriant growth, +++: moderate growth, ++: slow growth, +: very slow growth, – – no growth.

The SAA amended GPYM media containing permissible limit of Zn^{2+} (S1), Fe^{3+} (S6) and Mn^{2+} (S11) showed significant decolourisation after bacterial treatment. However, Fe^{3+} concentration (>15 times of permissible limit) started precipitation. Similar observations were also noted with Mn^{2+} and Zn^{2+} but less precipitation. Sample S16 showed significant decolourisation and COD reduction after bacterial treatment. This study showed the comparative binding tendency of heavy metals with melanoidins produced by sucrose-aspartic acid reaction. This also showed that the developed bacteria having potential to tolerate selected metals even in mixed condition. The used bacterial consortium was unable to decolourise the SAA solution containing >15 times higher permissible concentration of Zn^{2+} and Fe^{3+} (S5, S10) and COD and BOD reduction was not noted (Table 5.2a).

Table 5.2a: Effect of heavy metals on physico-chemical parameters of SAA, before and after bacterial treatment

Sample	Before bacterial treatment			After bacterial treatment		
	Colour	COD	BOD	Colour	COD	BOD
Conc.	6250 ± 250	26000 ± 763	11000 ± 278	1875 ± 56	5720 ± 175	3840 ± 143
1st set of experiment						
SAA + Zn²⁺ (mg/L)						
S1	6256 ± 135ns	27000 ± 292ns	12000 ± 276*	1438 ± 22*	5130 ± 143 ns	2520 ± 78*
S2	6458 ± 175 ns	28318 ± 763*	13000 ± 236*	1808 ± 45*	10477 ± 265*	4550 ± 96ns
S3	6876 ± 87*	30138 ± 683*	14653 ± 254*	1994 ± 76*	17178 ± 176*	7619 ± 134*
S4	8450 ± 164*	32132 ± 753*	15932 ± 232 ns	5915 ± 72*	25705 ± 543*	13064 ±260*
S5	8575 ± 213*	36053 ± 413*	16381 ± 234 ns	8575 ± 93*	36000 ± 487*	16300 ± 365*
SAA + Fe³⁺ (mg/L)						
S6	6810 ± 231ns	29786 ± 432*	14321 ± 324*	2941 ± 65*	11318 ± 387*	5585 ± 155*
S7	7590 ± 172*	32321 ± 462*	15861 ± 453 ns	3795 ± 76*	16806 ± 386*	7296 ± 98*
S8	8696 ± 241*	36123 ± 653*	17621 ± 245*	6087 ± 132*	24924 ± 435*	12334 ± 435*
S9	9688 ± 251*	43144 ± 532*	21673 ± 654*	10172 ± 176*	43200 ± 655*	21760 ±675*
S10	9770 ± 231*	47893 ± 671*	23521 ± 547*	10551 ± 198*	47893 ± 677*	23678 ± 643*
SAA + Mn²⁺ (mg/L)						
S11	6340 ± 165 ns	28321 ± 437*	13321 ± 432*	2193 ± 57ns	7646 ± 102*	3863 ± 87ns
S12	6586 ± 186 ns	30143 ± 671*	14732 ± 124*	2733 ± 65*	12961 ± 333*	6776 ± 134*
S13	6950 ± 152*	33543 ± 436*	16321 ± 546ns	3780 ± 56*	19119 ± 432*	10608 ± 156*
S14	8540 ± 173*	36231 ± 487*	19535 ± 243*	6405 ± 87*	27897 ± 542*	15628 ± 264*
S15	8832 ± 231*	41132 ± 586*	21111 ± 254*	7065 ± 145*	33728 ± 765*	17099 ± 276*
SAA + mix metals						
S16	6656 ± 152 ns	30532 ± 432*	14931 ± 343ns	3061 ± 124*	26715 ± 463*	11383 ± 228*
S17	7980 ± 265*	34512 ± 453*	16317 ± 234ns	4628 ± 87*	25884 ± 528*	12440 ± 289*
S18	9050 ± 270*	45532 ± 598*	23498 ± 432*	8552 ± 165*	33347 ± 666*	20969 ±365*
S19	12320 ± 300*	70133 ± 975*	30132 ± 432*	12320 ± 164*	69599 ± 876*	29763 ±376*
S20	14890 ± 362*	98543 ± 908*	42383 ± 654*	14890 ± 253*	97891 ± 954*	41353 ±564*

All the values are mean of three replicates (n = 3) ± SD in mg/L except colour (Co-Pt). Statistical significance of the difference between the values of the control and treated samples in a column was evaluated by means of ANOVA. Significance levels: * p<05; ns p>0.05. COD: chemical oxygen demand; BOD: biological oxygen demand.

The binding of SAA with mixed metal solution of iron, manganese and zinc at permissible limit (Table 5.1). Furthermore, in different concentration of zink, iron and manganese at the constant concentration of phenol (100 mg/L) has shown the increasing tendency of BOD, COD and colour as it has shown in table 5.2b from sample S21–S40. The chromogenic activity of iron with melanoidin has also been reported (Morales et al., 2005). The addition of phenol with heavy metals in SAA solution in separate study has shown the increased colour as compared to metal solution only where phenol was absent except Mn^{2+}. The order of metals for colour contribution with SAA was noted mixed metals > Fe^{3+} > Mn^{2+} > Zn^{2+} (Table 5.2a). It appears that the observed effect followed the schulz-hardy rule which stated that trivalent is more effective than divalent and monovalent.

Table 5.2b: Effect of phenol on physico-chemical parameters of SAA, before and after bacterial treatment

Sample	Before bacterial treatment			After bacterial treatment		
	Colour	COD	BOD	Colour	COD	BOD
Conc.	6250 ± 250	26000 ± 763	11000±278	1875±56	5720±175	3840 ± 143
2nd set of experiment						
SAA + phenol(100 mg/L) + Zn²⁺ (mg/L)						
S21	6678 ± 154 ns	30145 ± 652*	14321 ± 342*	2637 ± 65ns	7837 ± 365ns	4009 ± 87ns
S22	6924 ± 134 ns	32154 ± 476*	15893 ± 432ns	3081 ± 76*	13183 ± 487*	6357 ± 154*
S23	7212 ± 165*	34873 ± 587*	16891 ± 432ns	4543 ± 57*	23016 ± 465*	11485 ± 243*
S24	8721 ± 198*	36238 ± 654*	17243 ± 453ns	8721 ± 265*	36238 ± 685*	17243 ± 265*
S25	8932 ± 203*	38183 ± 415*	19983 ± 342*	8932 ± 276*	38183 ± 489*	19983 ± 354*
SAA + Phenol (100 mg/L) + Fe³⁺ (mg/L)						
S26	7315 ± 214*	33453 ± 563*	16321 ± 187ns	3950 ± 367*	13046 ± 355*	6691 ± 786*
S27	7917 ± 376*	36830 ± 654*	18981 ± 275*	8233 ± 187*	36830 ± 473*	18981 ± 387*
S28	8374 ± 434*	39170 ± 584*	19761 ± 453*	8960 ± 187*	40170 ± 642*	19761 ± 342*
S29	9912 ± 254*	46453 ± 543*	23811 ± 453*	10903 ± 254*	46553 ± 657*	23934 ± 435*
S30	10543 ± 354*	50501 ± 675*	24921 ± 309*	11808 ± 322*	50501 ± 854*	25921 ± 465*
SAA + Phenol (100 mg/L) + Mn²⁺ (mg/L)						
S31	6874 ± 187 ns	32000 ± 682*	15721 ± 423ns	3100 ± 243*	2130 ± 45*	4087 ± 87*
S32	7289 ± 234*	34321 ± 653*	16883 ± 325ns	3879 ± 176*	2915 ± 55*	8441 ± 183ns
S33	7474 ± 265*	36123 ± 542*	17932 ± 324*	4529 ± 165*	4484 ± 64ns	12373 ± 276*
S34	9232 ± 278*	42762 ± 543*	19981 ± 435*	7754 ± 186*	7754 ± 87ns	17383 ± 365*
S35	9412 ± 185*	48672 ± 587*	20123 ± 654*	9412 ± 187*	8094 ± 176ns	17909 ± 276*
SAA + Phenol (100 mg/L) + Mix metals						
S36	6815 ± 563ns	34325 ± 875*	16561 ± 345ns	4388 ± 154*	3475 ± 78ns	13070 ± 298*
S37	8232 ± 275*	58000 ± 865*	21678 ± 654*	6132 ± 186*	5614 ± 165ns	18671 ± 354*
S38	9443 ± 287*	70000 ± 867*	32516 ± 764*	9443 ± 254*	68000 ± 875*	31516 ± 576*
S39	16312 ± 345*	90321 ± 980*	41220 ± 783*	16312 ± 376*	90321 ± 999*	41220 ± 875*
S40	19532 ± 333*	120000 ± 3212*	49010 ± 876*	19532 ± 453*	120000 ± 3212*	49010 ± 876*

All the values are mean of three replicates (n = 3) ± SD in mg/L except colour (Co-Pt). Statistical significance of the difference between the values of the control and treated samples in a column was evaluated by means of ANOVA. Significance levels: * p<05; ns p>0.05. COD: chemical oxygen demand; BOD: biological oxygen demand.

5.3 SCREENING OF POTENTIAL BACTERIAL CONSORTIUM FOR DECOLOURISATION OF SAA

The study at variable concentration of phenol (50, 100, 200 and 300 mg/L) conducted with addition in glucose-peptone-yeast extract-mineral salt (GPYM) broth that to investigate the general bacterial growth and their tolerance. The phenol concentration showing optimum bacterial growth at 100 mg/L was chosen for further study. In the bacterial degradation of melanoidin study has shown that SAA was degradable in presence of GPYM up to 1200 ppm. The flasks were inoculated with 3% (V/V) of bacterial consortium consisting mixture of three pre-isolated bacterial strains, i.e., *Bacillus* sp., *R. planticola* and *E. sakazakii* in equal volume (1.00%, V/V) having cell density 3.30×10^4, 3.50×10^4 and 3.80×10^4 cell/mL, respectively. Further, this was incubated at 180 rpm and $(37 \pm 1°C)$ temperature in shaking flask condition incubated at temp controlled shaker (Kuhner, Switzerland) for twelve consecutive days. The bacterial growth and SAA decolourisation was monitored spectrophotometrically (UV-2300 Spectrophotometer, Techcomp, Korea) by taking absorbance at 620 and 475 nm, respectively against the initial absorbance at the same wavelength. Subsequently, the viable cells of *Bacillus* sp., *R. planticola* and *E. sakazakii* from day 6 and 12 of bacterial degraded samples were counted using plate spreading method on nutrient agar after 24 and 48 h incubation.

5.4 BACTERIAL DECOLOURISATION OF SAA IN PRESENCE OF HEAVY METALS AND PHENOL

The Zn^{2+} concentration up to 10 times permissible limit (S1 – – S3) was found stimulatory for the bacterial growth and decolourisation as compared to control (only SAA amended GPYM medium) (Figs. 5.1 and 5.2). Further, higher concentration of Zn^{2+} at 20 times of permissible limit (S5) showed inhibitory effect. The stimulatory effect might be attributed due to the use of zinc as trace element by bacterial consortium. In contrast, even permissible concentration of Fe^{3+} (S6) in SAA solution inhibited the bacterial growth and decolourisation process as compared to control (Figs. 5.1 and 5.2). The bacterial cell count was also inhibited by Fe^{3+} >16 times as compared to Zn^{2+} at same concentration (Fig. 5.1). Further, increases of Fe^{3+} concentration up to 5 times of the permissible limit (S7) suppressed the bacterial growth. The addition of Fe^{3+} prolonged the lag phase of bacterial cell growth, but this gradually attained the growth cycle, which indicated that the potentiality of bacterial consortium. However, only few bacterial cells could appear at 10 time's higher concentration of permissible limit of Fe^{3+} with SAA (S8). Furthermore, increased Fe^{3+} concentrations (S9 and S10) ceased the bacterial growth and increased the colour as compared to control even after bacterial treatment (Figs. 5.1 and 5.2). Similarly, the presence of Mn in SAA solution (S11 – – S15) also inhibited the bacterial growth and decolourisation process (Figs. 5.1 and 5.2). But, bacterial consortium showed growth as well as decolourisation up to 20 fold higher concentration of Mn. Besides, all three metals in mixed condition (S16 – – S20) with SAA solution contributed more inhibition for bacterial growth with prolonged lag phase as compared to samples S1 – – S15 (Fig. 5.1).

While, Zn^{2+} with phenol (S21 – – S25) inhibited the bacterial growth as well as decolourisation process as compared to Zn^{2+} without phenol (S1 – – S5) even at permissible concentration (Figs. 5.1 and 5.2). The bacterial consortium showed 30% decolourisation in presence of Zn^{2+} even up to 15 time's higher concentration of permissible level.

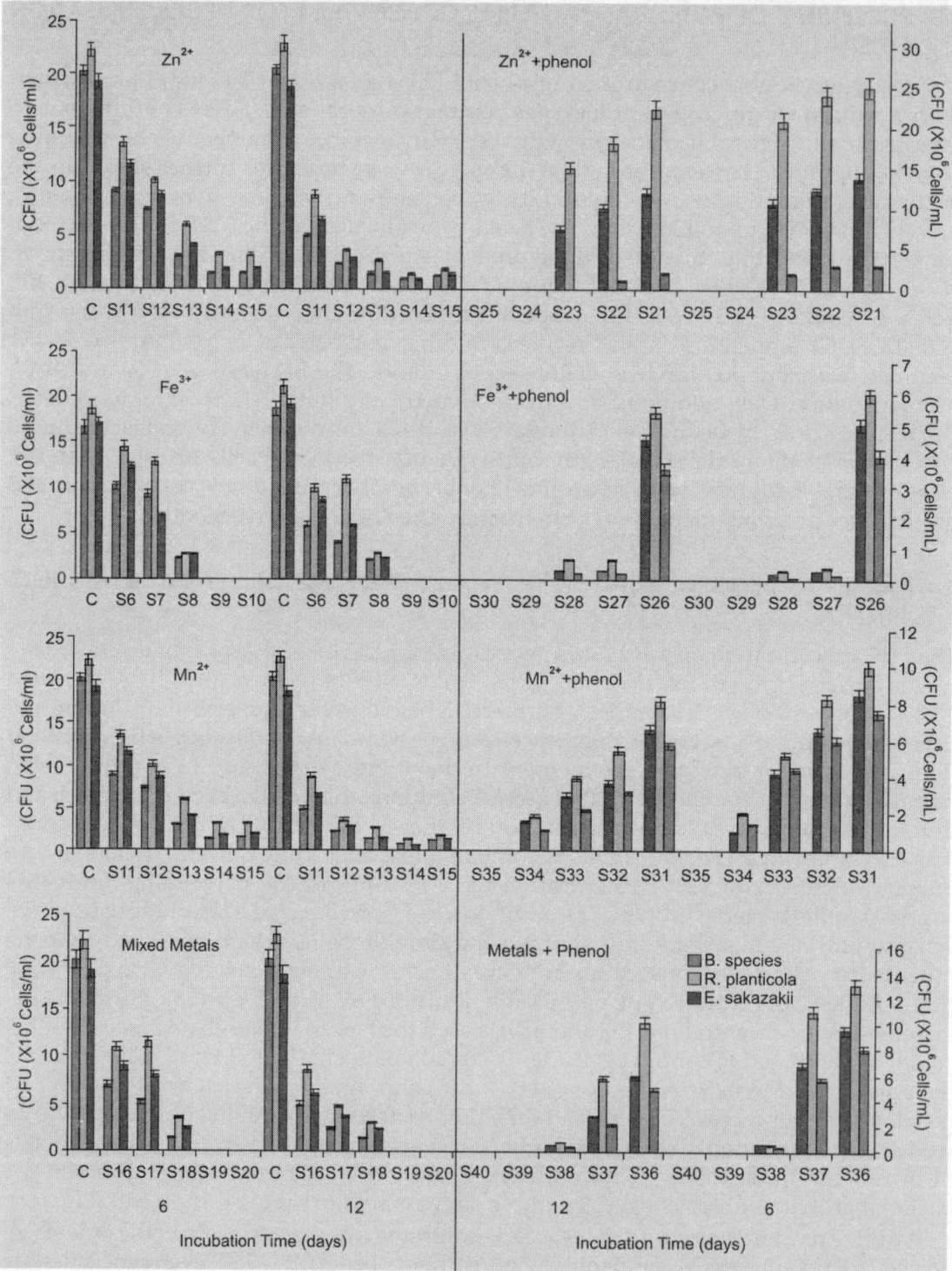

Fig. 5.1: Bacterial growth pattern and CFU of bacterial consortium in presence of different metals and phenol in SAA solution

However, in presence of phenol (100 mg/L) bacterial consortium could decolourise SAA only up to 37% at 10 times permissible concentration of Zn^{2+} (Fig. 5.2). But, bacteria could not grow at further higher concentration of Zn^{2+} with phenol. Phenol (100 mg/L) along with Fe^{3+} even at low concentration (S26) drastically suppressed the bacterial growth and decolourisation (46%) (Figs. 5.1 and 5.2). Result also revealed that the Fe^{3+} along with phenol above than 5 times of permissible limit (S27–S30) completely ceased the bacterial growth and increased the colour of media (Fig. 5.2). Bacterial consortium at 20

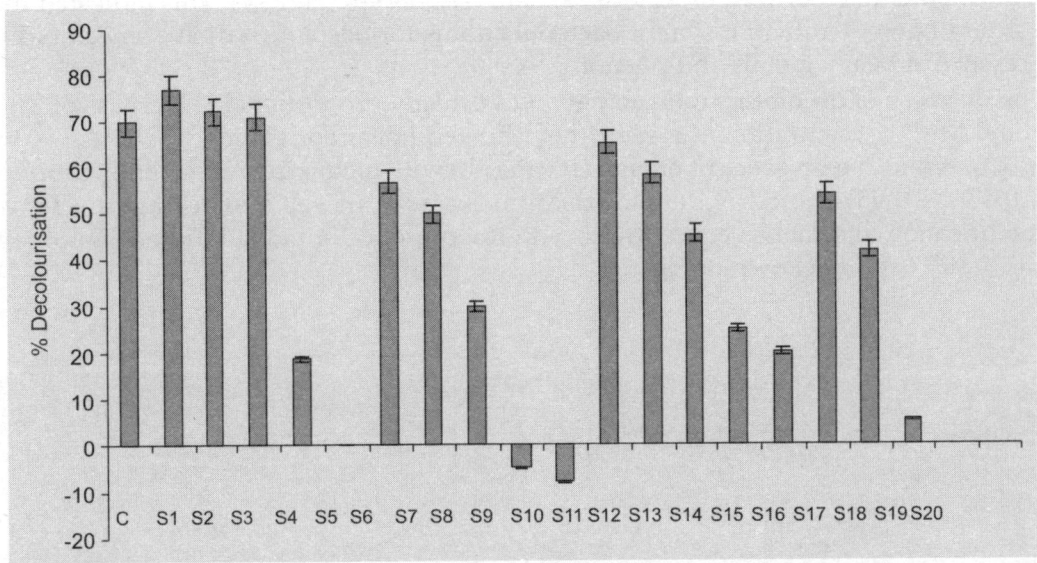

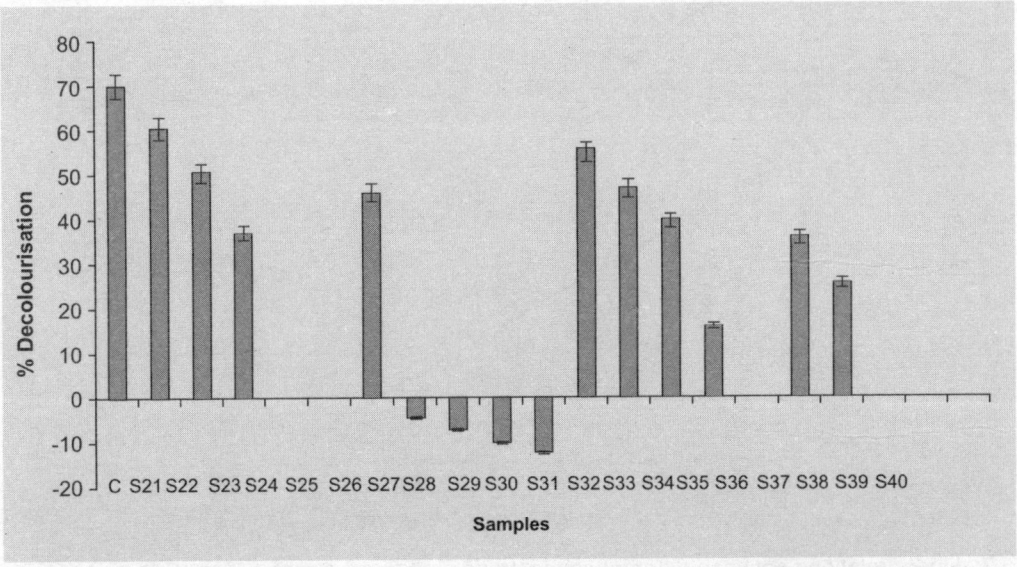

Fig. 5.2: Effect of heavy metals and phenol on bacterial SAA decolourisation

time's of permissible limit for Mn^{2+}, with and without phenol (S15 and S35) showed decolourisation 20% and 14%, respectively (Fig. 5.2). However, the presence of phenol with mixed metal solution (S36 – – S40) increased the complexity of melanoidin and reduced the pH (pH 4.00); this further suppressed the bacterial growth and concomitantly increased the BOD and COD (Fig. 5.1 and Table 5.2). Consequently, heavy metals with phenol (set 2) reduced the bacterial cell count and SAA decolourisation as compared to solution containing only heavy metals (set 1). This indicated that the multi-metals makes more complex environment with melanoidin in presence of phenol for bacterial growth and degradation. The comparative Colony Forming Unit (CFU) revealed that *R. planticola* was most potential followed by *E. sakazakii* and *Bacillus* sp. (Fig. 5.1). This indicated that the order of potentiality of different bacterial strains for decolourisation of melanoidins in presence of heavy metals and phenol.

The decrease of decolourisation activity was directly proportional to increase of Zn^{2+}, Fe^{3+} and Mn^{2+} concentration, therefore they showed linear correlation (R^2: 0.94 – – 0.98) (Fig. 5.3). While, the presence of phenol (100 mg/L) with metals in SAA severely dropped (R^2: 0.45 – – 0.95) the SAA decolourisation activity in set 2 experiment. Hence, decolourisation and metals concentrations in the presence of phenol did not follow the linearity (R^2: 0.45) as shown in Fig. 5.3.

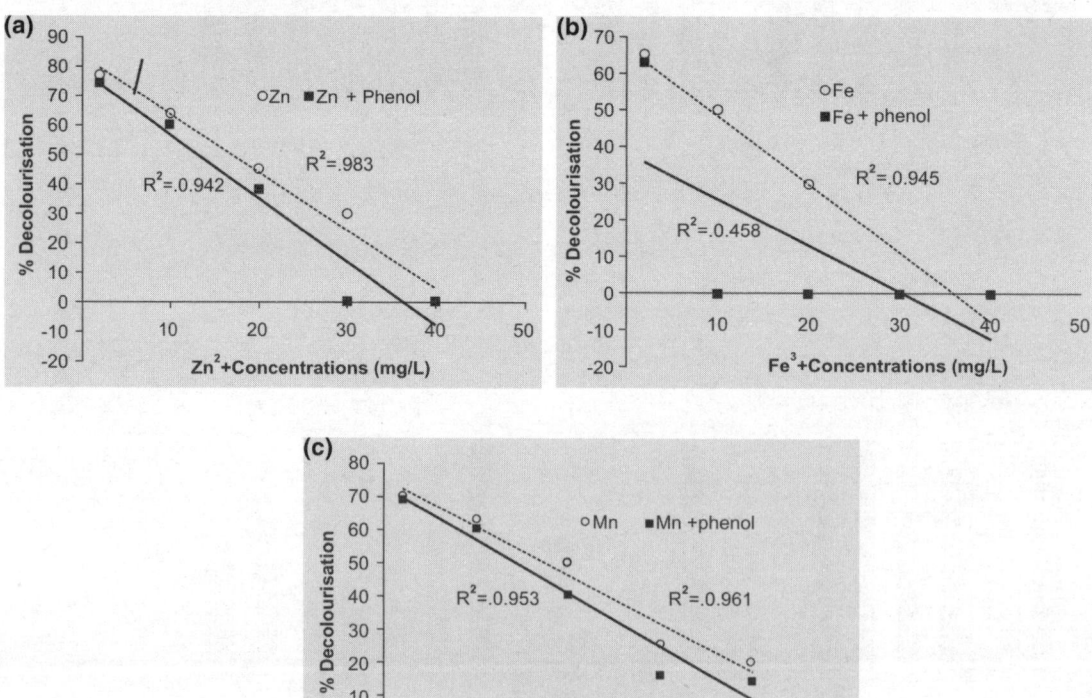

Fig. 5.3: Linear relations between different concentrations of various metal (Zn^{2+}, Fe^{3+} and Mn^{2+}) and mean decolourization value

The degradation and decolourisation of SAA in presence of phenol and heavy metals were also assessed with HPLC. The samples were analyzed by using a Waters, 515 HPLC systems equipped with reverse phase (RP) column C-18 (150 mm × 4.6 mm, particle size 0.5 μm) at 27°C and 2487 UV/Vis detector via millennium software. Samples (20 μL) were injected followed by implementation of HPLC grade water (100%) at the flow rate of 1.00 mL/min.

The mobile phase consisted HPLC grade acetonitrile: water in 70:30 (V/V) ratio and the detection wavelength was set at 290 nm to monitor the degradation of SAA.

The melanoidin peak in HPLC analysis also showed maximum reduction in presence of Zn^{2+} (S1) followed by SAA alone than S21 after bacterial treatment, this supported the above observation (Fig. 5.4a). However, the higher concentration of Zn^{2+} (>15.00 mg/L)

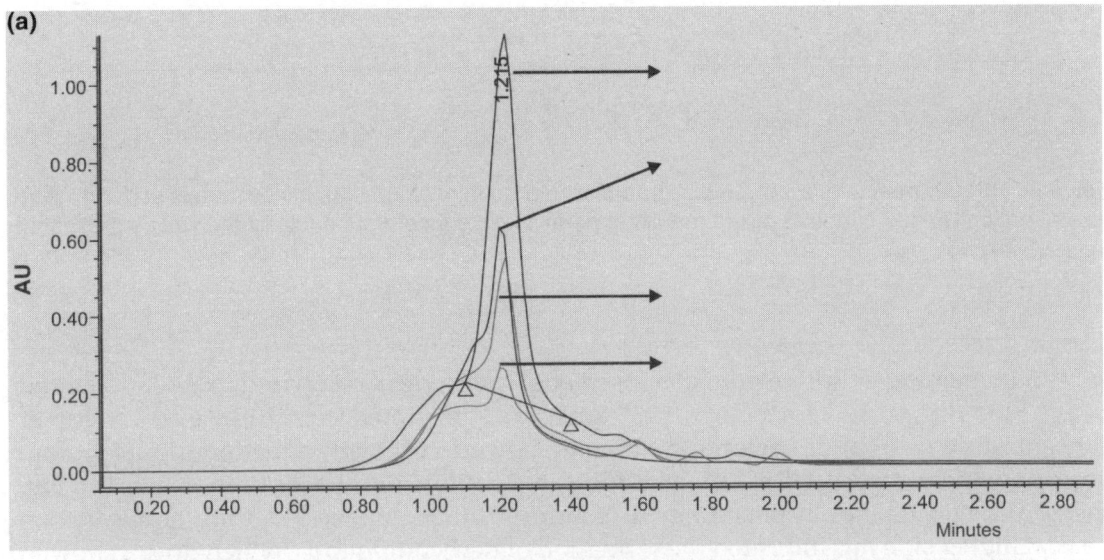

(a)

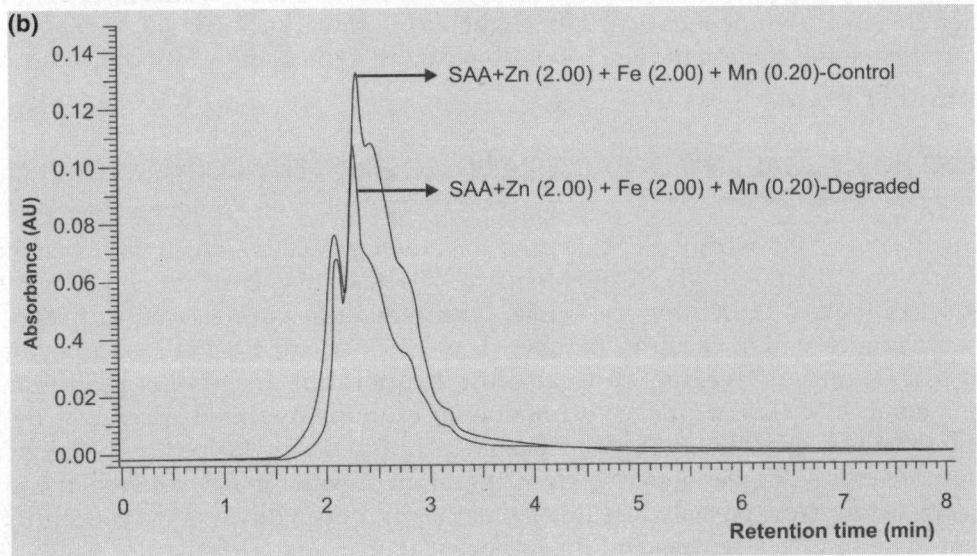

(b)

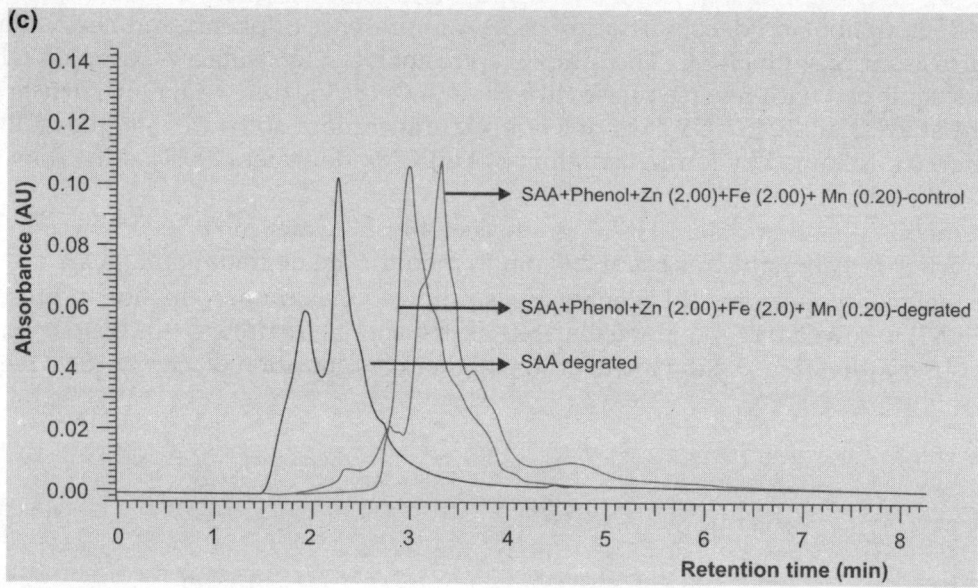

Fig. 5.4: HPLC chromatogram of SAA degradation in presence of Zn^{2+} (a), mixed metals at permissible concentration (b) and phenol + mixed metals at permissible concentration (c). All the values presented in bracket are in mg/L.

in SAA containing culture media increased absorption peak compared to SAA degraded peak. But, the presence of Zn^{2+}, Fe^{3+} and Mn^{2+} in mixed condition even at lower concentration (S16) after bacterial treatment showed very less reduction in melanoidin absorption peak during HPLC analysis indicated the low decolourisation (Fig. 5.4b). But, the addition of phenol in mixed metals conditions (S36) showed shifting of absorption peak at higher side this indicated the formation of new compounds, which do not showed reduction in absorption peak rather than slight sifting towards lower side as compared to S16 after bacterial treatment (Fig. 5.4c). This explains the complexation of SAA with metals and phenol.

5.5 MORPHOLOGICAL VIEW OF BACTERIA UNDER STRESS ENVIRONMENT OF PHENOL AND HEAVY METALS

Scanning Electron Microscopy (SEM) is used to visualize bacteria which had been fixed, dehydrated, and dried. Rapidly frozen (cryofixed) bacteria may be examined at very low temperatures (below-120°C) by cryo-SEM. Both techniques are carried out with the specimen placed in a high vacuum chamber. The specimen must not release any volatile substances. Drying or freezing at a very low temperature meets this requirement. Environmental SEM (ESEM) makes it possible to examine hydrated specimens by not exposing them to high vacuum. A sophisticated technical design keeps the specimen at a temperature several degrees above the freezing point of water in a small area inside the microscope where a low partial pressure of water vapour provides ions at a concentration sufficient to neutralize electrons and thus to prevent charging artifacts.

Every specimen destined for SEM must be dry. As living microorganisms, bacteria contain proteins and a high proportion of water in their cells. It is essential to fix them first in order to preserve their structure while they are being further prepared for SEM. To accomplish such steps in reasonable time, the specimens should be relatively thin (<2 mm) and small (only a few mm). Large solid materials which contain bacteria on their surfaces such as contaminated meat, skin, vegetables, composted materials, or agar gel plates with bacterial colonies are first excised and trimmed to approximately 10 mm × 10 mm specimens as thin as possible (1–2 mm), and fixed before they are further reduced into smaller (approx. 5 mm × 5 mm) particles. If the bacteria are inside a specimen such as set-style yogurt, cheese, kefir grains and other materials which may easily be cut, it is advantageous to trim them into prisms 10 to 15 mm long and less than 1.5 mm × 1.5 mm in cross section. Viscous specimens such as stirred yogurt and cream are best encapsulated in agar gel tubes in the form of 10 to 15 mm long columns, whereas heat-sensitive foods such as raw egg yolk may be encapsulated in calcium alginate gel tubes. Bacterial colonies on agar gel plates are excised with up to 1 mm of a clear agar gel rim. If possible, the bacteria destined for SEM should be grown on thin plates (<2 mm) to make the excision easier. Solid specimens are fixed in a buffered (0.1 M, pH 6.5–7.0) fixative such as 2–3% glutaraldehyde for periods ranging from 5 min to 24 h. If specimen constituents not based on protein such as lipids (fat globules in food products) or polysaccharides (mucus in the intestines) are also present, post fixation using osmium tetroxide (OsO_4) or Ruthenium Red, respectively, is used. Although double bonds (–C=C–) in unsaturated fatty acids react with osmium tetroxide, osmium is easily removed by subsequent spontaneous hydrolysis. It results in the formation of a diol and the release of free osmium trioxide (OsO_3). No hydrolysis takes places, however, if a heterocyclic base such as pyridine or imidazole is present in the fixative. Imidazole-buffered OsO_4 preserves the fat and oil particles for SEM in specimens such as cheese, yogurt, hard-boiled eggs, and comminuted meat products.

Fixed specimens are washed with the corresponding buffer and dehydrated in a graded ethanol series. The fixative and dehydrating solutions must penetrate the entire specimens to make the SEM examination relevant. The fragments acquire electrically conductive surfaces after they are sputter-coated with a 20 nm layer of gold. They are then ready to be examined by SEM.

SEM can achieve resolution better than 1 nanometer. Specimens can be observed in high vacuum, in low vacuum, and in wet conditions (in environmental SEM). SEM allows the visualization of images at high magnification (50x – 10.000x and above). In this technique, an electron beam scans the surface of the sample to produce a variety of signals, the characteristics of which depend on many factors, including the energy of an electron beam and the nature of the sample, since a beam of electrons hit the sample and the response is collected by a detector. There is no usage of light and the color of the sample does not influence on the image. The most common detectors used in SEM are secondary electron detectors (ETD – Everhart-Thornley Detector or SE1/SE2 in high-vacuum or LFD – Large-Field Detector, in low-vacuum) and back-scattered electron detector (BSED). The difference between them is that secondary electrons have been ejected from the outer electron shell of an atom as a result of impact from a high energy electron, having relatively low energies (up to about 50 eV, compared to the 1–30 keV of the beam electrons) and highlight the surface features (topography) of the sample. On the other hand, back-scattered electrons are beam electrons that have undergone sufficient elastic 'collisions' with atomic nuclei and consequent changes in direction to exit the surface of the sample. BSED will provide an image with difference of phases, based on the difference of the atomic number (Z) of the surfaces. Regions of lower atomic number (such as Al, Si, C) in a sample will appear darker than areas of higher atomic number (such as Fe, Cu), which will appear brighter (phenomenon known as Z contrast).

For SEM analysis, degraded sample was centrifuged ($6500 \times g$) to separate the bacterial cells for 20 min. The pellets were washed thrice with distilled water to remove the culture medium contents. Subsequently, the bacterial cells were fixed, dehydrated according to the method described previously for SEM analysis (Sangeeta et al., 2011) Then dried cells were mounted on the metal stubs with colloidal sliver paste and coated with a thin conductive film of gold in a sputtering coater and examined under scanning electron microscope (LEO 435 VP, LEO Electron Microscopy Ltd., USA). The bacterial consortium effectively showed SAA decolourisation in GPYM media with normal cell morphology as shown in Fig. 5.5a.

However, an adverse morphological change on bacterial cells was observed under SEM at higher concentration of heavy metals with and without phenol (Fig. 5.5b–e). The bacterial morphological shrinkage at adverse conditions has also been previously noted (Kaletunc et al., 2004). The ten fold higher concentration of mix metals in media pronounced the shrinkage of bacterial cell morphology (S18), (Fig. 5.5d). This indicated that higher concentration of heavy metals in mixed condition is more toxic to bacterial growth. Furthermore, it was also observed that phenol (100 mg/L) along with 10 fold higher mixed metals concentration (S38) showed more shrinkage and clumped bacterial

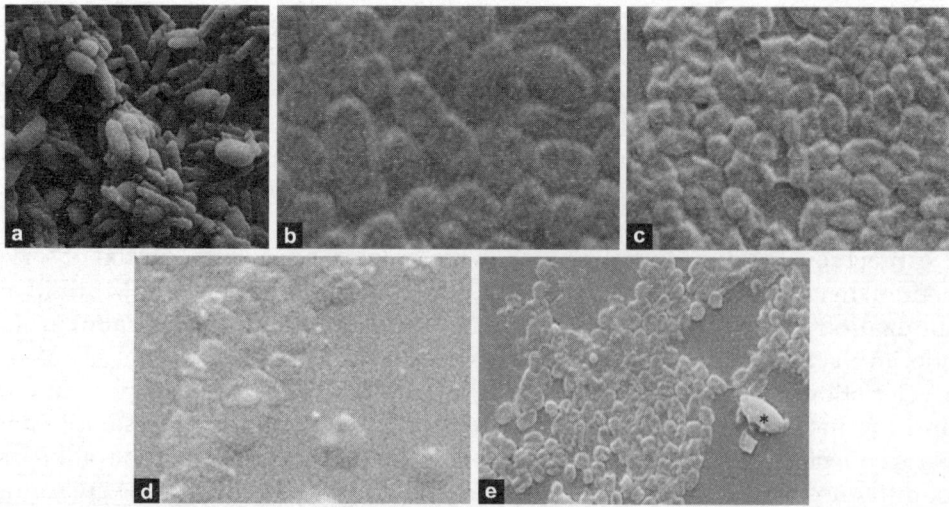

Fig. 5.5: Scanning Electron Micrograph (SEM) depicting the effect of heavy metals with and without phenol (100 mg/L) during SAA decolourisation at a magnification 5000X. a: bacterial strains under untreated condition, b: SAA + Zn^{2+} (2.00 mg/L) + Fe^{3+} (2.00 mg/L) + Mn^{2+} (0.20 mg/L) (S16), c: SAA + Phenol + Zn^{2+} (2.00 mg/L) + Fe^{3+} (2.00 mg/L) + Mn^{2+} (0.20 mg/L) (S36), d: SAA + Zn^{2+} (20.00 mg/L) + Fe^{3+} (20.00 mg/L) + Mn^{2+} (2.00 mg/L) (S18), e: SAA + Phenol + Zn^{2+} (20.00 mg/L) + Fe^{3+} (20.00 mg/L) + Mn^{2+} (2.00 mg/L) (S38), * shows *Bacillus sp.* with thick cell wall.

cells of *R. planticola*, while *Bacillus sp.* was observed very few with thick cell wall (Fig. 5.5e). This might be due to polysaccharides covering around the cell wall which is adoptive feature of bacteria in adverse conditions. The shrinkage of bacterial cell size, secretion of polysaccharides and clumping in adverse conditions has also been reported by Kaletunc et al. (2004) and Xie et al. (2010).

5.6 EFFECTS OF MnP ENZYME ACTIVITY IN PRESENCE OF HEAVY METALS AND PHENOL

The supernatant received after the biomass separation from bacterial degraded melanoidin were used for the assessment of MnP activity. The MnP activity was measured by oxidation of phenol red as described in previous chapter four. The effect on MnP activity was noted at permissible concentration of metals with and without phenol in SAA amended GPYM solution as shown in Fig. 5.6a. The permissible concentration of zinc (S1) showed stimulatory effect on MnP activity as compared to control. The maximum enzyme activity was noted at 4th day of bacterial growth and enzyme activities remains constant up to day 9. However, the enzyme activity was declined during subsequent incubation. The presence of Fe^{3+} (S6) suppressed the MnP activity for melanoidin decolourisation and it was lower than the presence of Mn^{2+} in melanoidin solution (S11). The order of enzyme induction by different metals was $Zn^{2+} > Mn^{2+} > Fe^{3+} >$ metals in

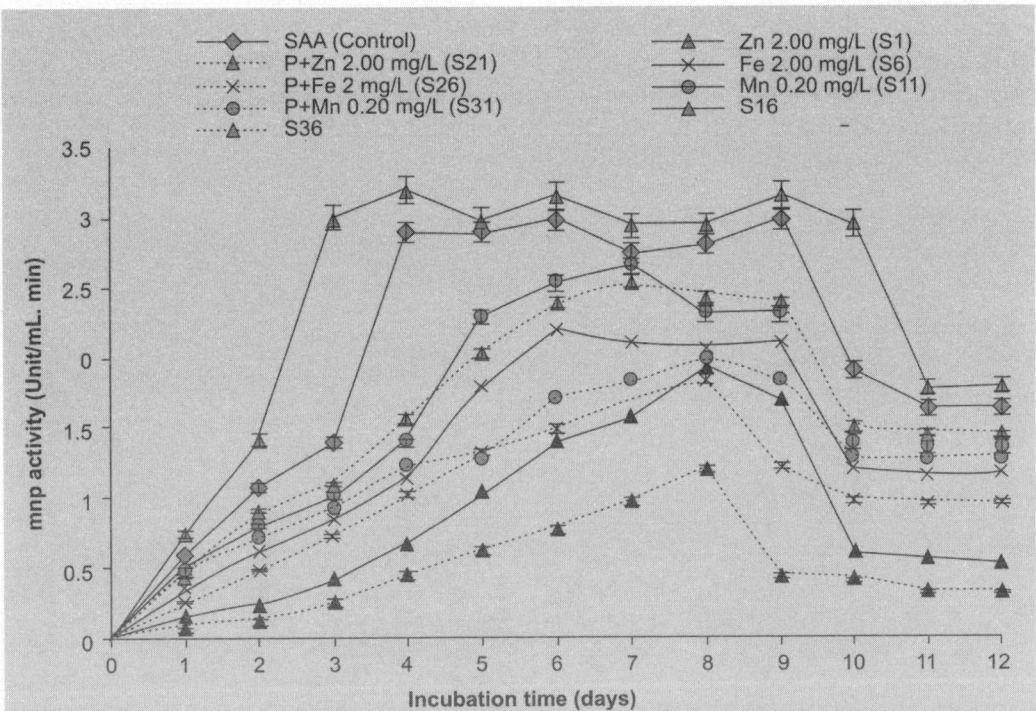

Fig. 5.6: MnP activity of potential bacterial consortium in SAA containing GPYM media amended with metals, with and without phenol. Insert figure shows purified MnP, Lane 1: Ladder, 2: standard, 3: SAA, 4: SAA + Zn^{2+} (2.00 mg/L), 5: SAA + Fe^{3+} (2.00 mg/L), 6: SAA + Mn^{2+} (0.20 mg/L). P = phenol, S16 = SAA + Zn^{2+} (2.00 mg/L) + Fe^{3+} (2.00 mg/L) + Mn^{2+} (0.20 mg/L), S36 = SAA + Phenol + Zn^{2+} (2.00 mg/L) + Fe^{3+} (2.00 mg/L) + Mn^{2+} (0.20 mg/L).

mix condition. However, the presence of phenol along with these metals (set 2) showed inhibitory effect on enzyme activity as compared to control and set 1 (Fig. 5.6a). The mechanism for the bacterial enzyme inhibition might be due to the changes in the molecular structure of the enzyme caused by the inhibitors, i.e. phenol and heavy metals (Shen et al., 2006; Nweke et al., 2007). For the heavy metals, it is assumed that either they bind to sulfhydrile group of the active site or organic chemicals may denature the entire protein structure (Silva et al., 2007).

In addition, the intensity of purified MnP band in SDS-PAGE coincided to the pattern of enzyme activity (Fig. 5.6b). The MnP band intensity was found in order of S1 > C > S11 > S21 > S6 > S31 > S26 > S16 > S36.

The findings concluded that lower concentration of zinc stimulated bacterial growth and SAA decolourisation. However, Fe^{3+} showed inhibitory effect on bacterial growth even at permissible limit. Further, it was also observed that presence of these metals along with SAA contributed the colour, COD, BOD and TS. While, the combination of phenol along with Fe^{3+} in SAA solution, further increased the colour intensity and pollution parameters. But, the bacterial decolourisation of SAA in presence of metals alone showed reduction in melanoidin absorption peak indicated depolymerisation along with slight decrease in pH. While, presence of phenol in these solution inhibited the bacterial growth and enzyme activity and do not show direct reduction of peak also. The order of bacterial dominancy was noted as *R. planticola* > *E. sakazakii* > *Bacillus* sp. However, the shrinkage and reduced number of bacterial cells at higher concentration of heavy metals along with phenol was noted under SEM. Hence, the study concluded that the presence of heavy metals and phenol in melanoidin containing industrial wastewater contributed colour and toxicity, which inhibited the microbial degradation process.

6

Health Hazards of PMDE on Aquatic and Terrestrial Ecosystem

6.1 AQUATIC SYSTEM

Nowadays green revolution has led to the environmental pollution, especially water pollution due to the over growing urbanization and industrialization. Untreated and partially treated effluents from the industrial wastewater systems and treatment processes in unsewered areas contribute significant quantities of nutrients, suspended solids, dissolved solids, oil, metals (arsenic, mercury, chromium, lead, iron, and manganese) and biodegradable organic carbon to the water environment. Growing industrial establishments can results in hazards on the local environment in the city if proper attention is not paid. Water is life; Water is the foundation of health, hygiene, progress and prosperity. Water is essential nutrient that is involved in every function of the living beings and it helps to transport of nutrients and waste products in and out of the cells. The water is the most needful natural resource for human utilization nowadays; unfortunately it is undergoing extreme contamination by industrialization and urbanization. The pollution of our water resources can have serious and wide-ranging effects on the environment and human health. Drinking water sources become contaminated, causing sickness and disease. The World Health Organization estimated in 1996 that every eight seconds a child died from a water-related disease and that each year more than five million people died from illnesses linked to unsafe drinking water or inadequate sanitation. Pollutants accumulate in food through water, making it dangerous or inedible. The presence of these toxic substances in our food and water can lead to reproductive problems and neurological disorders. The quality requirement varies distinctly with respect to the specific uses. For instance drinking water has specified quality, which is not at all essential for industrial purposes or other domestic uses. Though underground water appears to be less prone to pollutants yet there are a number of potential sources of underground water pollution. Underground water is an important resource in our environment. It replenishes our streams, rivers, habitats and also provides fresh water for irrigation, industry, and communities. The most of great cities depend on the underground water as a primary source of drinking water. However, underground water is highly susceptible to contamination from industrial waste, septic tanks, agricultural runoff, landfills, and pipe leaks. Wastewater from manufacturing or chemical processes in industries contributes to water pollution. Industrial wastewater usually contains specific and readily identifiable chemical compounds. During the last fifty years, the number of industries in India has grown rapidly. But water pollution is concentrated within a few subsectors, mainly in the form of toxic wastes and organic pollutants. Out of this a large portion can be traced to the processing of industrial chemicals and to the food products industry. Distilleries are one of the major polluting industries especially with reference to water pollution with about 80% of the raw materials ending up as waste. In India there are about 330 distilleries, the total installed capacity is about 3500 million liters of alcohol. Distilleries generate a huge amount of wastewater (spentwash) with enormous quantity of organic and inorganic nutrients, thus having high Na, K, Ca, Mg, TKN, BOD and COD load. For production of each liter of alcohol, 12–15 liter of effluent is produced. Approximately 40 billion liters of effluent is generated per annum from 330 distilleries in the country. They discharge their untreated wastes directly into the natural environments which cause various adverse effects on soil, water, air and health. It also affects the farm animals (Fig. 6.1). They drink it and resulted in increased livestock mortality, poor health, and reduced milk yield. Even the human beings lived in distillery effluent polluted Area is affected by skin allergies, headache, vomiting sensation, irritating eyes, fever and stomach pain. This kind of water have dissolved impurities like carbonate, bicarbonate, sulphate, chloride of calcium, magnesium, iron, sodium participant and potassium and colloidal impurities like colouring matter, organic waste, finely divided, silica and clay.

It is analyzed that distillery industry produces a huge amount of wastewater which is highly polluted and having very high Chemical and Biological Oxygen Demand (COD and BOD), dark brown reddish colour and have high load of organic matter, when discharge into natural water bodies, causes severe environmental pollution. They also alter the physico-chemical characteristics of the receiving aquatic bodies and affect aquatic flora and fauna. The effluent in the environment, poses serious health hazard for the rural and semi-urban population that uses the stream and river water for agriculture and domestic purposes. The distillery effluent has obnoxious odour and unpleasant colour, when it is released into the environment without proper treatment. Farmers have been using these effluents for irrigation and found that the growth, yield and soil health were

Fig. 6.1: PMDE showing aquatic pollution at Unnao region (UP) and health hazards (b), PMDE discharge from M/S Kedia distillery Ltd. (c) and PMDE prior to mixing in the river

reduced. The contaminants like chloride, sulphate, phosphate, magnesium and nitrate are discharged with the effluent by various industries, which create nuisance by the way of physical appearance, odour and taste. Such harmful water is injurious to plants, animals and human beings. Some of the contaminants, such as certain level of minerals or compounds are not only harmful to health, but also create a long term effects.

6.1.1 Impact of Distillery Effluent on Environment

The effluent of distillery is red brown in colour with unpleasant odour of indol, sketol and other sulphur compounds. Distillery effluent is a complex, multicomponent stream that is known to cause considerable fouling. Discharge of distillery effluent with high TDS would also have adverse impact on aquatic life and to make unsuitable water for drinking purpose, if used for irrigation reduce the crop yield, corrosion in water system and pipe line. High amount of BOD in the wastewater leads to the decomposition of organic matter under the anaerobic condition that produces highly objectionable products including methane (CH_4), ammonia (NH_3) and hydrogen sulphide (H_2S) gas. Fall in DO levels causes undesirable odours, tastes and reduce the acceptability of water for domestic purpose. In steam generation, DO is one of the most important factors causing corrosion of the boiler material. Generally, industrial wastewater changes pH level of the receiving water body. Such changes can affect ecological aquatic system; excessive acidity particularly can result in release of hydrogen sulphide (H_2S) to air. Alkaline nature of wastewater causes declination in plant growth and crop growth. In addition, spentwash contains low molecular weight compounds such as lactic acid, glycerol, ethanol and acetic acid and also contain small amount of heavy metals in water bodies causes several health problems. Heavy metals, e.g. Hg, Fe, Cu, Mn, Ni Cd and Cr, etc. can accumulate and they enter in food chain and biomagnifies to toxic level.

6.1.2 Impact on Water Quality and Aquatic Life

The effluent discharged from breweries, textile industry, fertilizer factory and antibiotic factories were hazardous to flora and fauna of island streams and other natural water resources. The discharge of effluent into river increased BOD and COD of river water. The high BOD and COD deplete dissolved oxygen content of river water, thereby creating an anaerobic condition in the river bed which, in turn, affected the aquatic life. Aquatic toxicity of distillery effluent was due to its high BOD (40,000 mg/L), COD (90,000 mg/L) and other toxic components. The distillery effluent is a potential water pollutant in two ways. Firstly, it is highly coloured and the receiving waters would become turbid affecting the penetration of light, preventing oxygenation by photosynthesis and, thus, would be detrimental to aquatic life. Secondly, distillery effluent itself has a high pollution load having very high BOD and COD, requiring more oxygen for biodegradation, leading to decreased DO levels, thus, causing eutrophication of contaminated water courses (Fig. 6.2, eutrophication is the process by which a body of water acquires a high concentration of nutrients, especially phosphates and nitrates. These typically promote excessive growth of algae. As the algae die and decompose, high levels of organic matter and the decomposing organisms deplete the water of available oxygen, causing the death of other organisms, such as fish. Eutrophication is a natural, slow-aging process for a water body, but human activity greatly speeds up the process.).

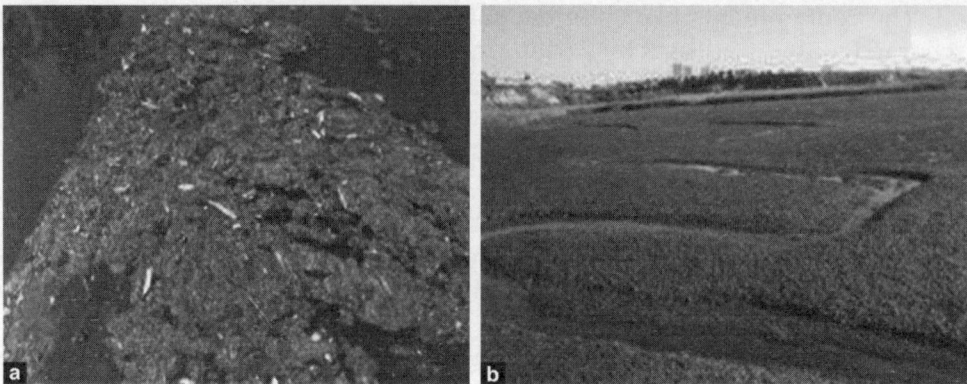

Fig. 6.2: Showing eutrophication at different distillery effluent contaminated site

The colour causing compounds reacted with metal ions to form substances which were very toxic to fish and other aquatic life. Similarly, the suspended solids in wastewater reduce the light penetration and plant production as a result in receiving water by increasing turbidity it can also clog the fish gills. However, Low DO in water bodies affect the aquatic life as DO rops fish (*Erpetoichthys calabaricus*) and other species are threatened and may get killed. Due to heavy organic loading, only tolerant species of organisms survived.

The pollution standards stipulate that BOD of effluent should be less than 30 mg/L for disposal into inland surface waters and less than 100 mg/L for disposal on land. BOD can be 500 mg/L, in case land application effluent is envisaged as a secondary treatment system for further removal of BOD.

6.1.3 Impact on Groundwater

The term "leachate" refers to liquids that migrate from the waste carrying dissolved or suspended contaminants. Leachate results from precipitation entering the land and from moisture that exists in the waste when it is disposed. If uncontrolled, leachate can be responsible for contaminating groundwater and surface water. Leachate can be a major source of contamination to groundwater, surface water and soil. The risks from leachate are due to its high organic contaminant concentrations. Toxic substances may however be present in variable concentration and their presence is related to the nature of waste deposited. The movement of water along with industrial effluents by diffusion and convection in the soil involves a series of complex processes. These processes are very important for monitoring the contamination of soil, groundwater and for environmental geochemical studies. Many metals and other contaminants migrate into the soil horizontally as well as vertically and cause groundwater pollution that depends on factors like mobility of chemicals and soil conditions. Application of spentwash has resulted in leaching of high amounts of organic and inorganic apart from river water, pollution of groundwater due to disposal of effluent on open land has been reported. The distillery effluents have an adverse effect on the soil groundwater ecosystem, and unless managed properly, it may lead to severe land and groundwater pollution affecting various components of the soil ecosystem. The effluent is rich in organic and inorganic ions, which may leach down and pollute the groundwater. An on-farm experiment was conducted to assess the impact of long-term irrigation with PMDE on nitrate, sulphate, chloride, sodium, potassium, and magnesium contents in the groundwater of two sites in northwest India. Electrical conductivity (EC), pH, TDS, sodium adsorption ratio (SAR) and colour were also determined to assess the chemical load in the groundwater. Nitrate content in the groundwater samples ranged from 16.95 mg/L in the unamended fields to 59.81 mg/L in the PMDE-amended fields during the 2-year study (2001–2002). Concentrations of TDS in water samples from tube well of the amended field was higher by 40.4% over the tube well water of the unamended field. Colour of the water samples of the amended fields was also darker than that of the unamended fields. The study indicated that the organic and inorganic ions added through the effluent could pose a serious threat to the groundwater quality if applied without proper monitoring. Similarly, irrigation with distillery effluent impaired the groundwater quality of Gajraula region especially of agricultural zone making it unsuitable for drinking purpose. This change in groundwater quality resulted because of irrigation done by improperly treated effluent whose physico-chemical nature deviated very much from the required standard characteristics for irrigation. Hence, it is primary requirement to treat the effluent properly as prescribed by Bureau of Standard (BIS) or World Health Organization (WHO). As the present study reveals that shallow aquifers (8 and 10 m depth) are more prone for contamination, therefore, a caution must be taken while performing ferti-irrigation especially in area of low water table depth and it would be safe to use water from higher depth well, especially for drinking purpose in the region. Further, every crop has specific requirement of nutrients. Hence, effluent treatment before application should match well with nutrient requirement of crops. Otherwise, excess nutrients may become toxic. The study also indicated that period of ferti-irrigation should also be decided on the basis of prevailing weather conditions, as maximum groundwater contamination was noted during monsoon and summer, therefore, winter could be chosen as best time for practicing ferti-irrigation. Further, to reduce environmental impact of ferti-irrigation certain strategies as dilution of untreated or improperly treated industrial effluent, intermittent and discontinuous irrigation, crop-specific and area-specific irrigation are recommended.

6.1.4 Health Impacts

The majority of waterborne microorganisms that cause human disease come from animal and human fecal wastes. But, industrial waste also play a role to stimulated pathogenic organisms. These contain a wide variety of viruses, bacteria, and protozoa that may get washed into drinking water supplies or receiving water bodies. Microbial pathogens are considered to be critical factors contributing to numerous waterborne outbreaks. Diseases caused by bacteria, viruses and protozoa are the most common health hazards associated with untreated drinking and recreational waters Contaminated water is a vehicle for several waterborne diseases, such as cholera, typhoid fever, shigellosis, salmonellosis, campylobacteriosis, giardiasis, cryptosporidiosis and Hepatitis A (WHO, 2004). Also, many microbial pathogens in wastewater can also cause chronic diseases with costly long-term effects, such as degenerative heart disease and stomach ulcer. The density and diversity of these pollutants can vary depending on the intensity and prevalence of infection in the sewered community. The detection, isolation and identification of the different types of microbial pollutants in wastewater are always difficult, expensive and time consuming. To avoid this, indicator organisms are always used to determine the relative risk of the possible presence of a particular pathogen in wastewater. Viruses are among the most important and potentially most hazardous pollutants in wastewater. They are generally more resistant to treatment, more infectious, more difficult to detect and require smaller doses to cause infections. Because of the difficulty in detecting viruses, due to their low numbers, bacterial viruses (bacteriophages) have been examined for use in faecal pollution and the effectiveness of treatment processes to remove enteric viruses. Bacteria are the most common microbial pollutants in wastewater. They cause a wide range of infections, such as diarrhea, dysentery, skin and tissue infections, etc. Disease-causing bacteria found in water include different types of bacteria, such as *E. coli* O157:H7; Listeria, Salmonella, Leptosporosis, etc. The major pathogenic protozoans associated with wastewater are Giardia and Cryptosporidium. They are more prevalent in wastewater than in any other environmental source. Drinking water with high nitrate content often causes methemoglobinemia (blue-baby disease) in infants. When this happens, nitrate is reduced to nitrite in the digestive system, which in turn attacks the hemoglobin in infants resulting in methemoglobinea. Methemoglobinemia is associated with nitrates in drinking water above the maximum contaminant level (10 mg/L) as set by the US Environmental protection Agency (EPA, 2002). Nitrite can also interact with amine chemically or enzymatically to form nitrosoamines which are carcinogens. Also, the presence of nitrogen and phosphorus in fresh water can also create environmental conditions that favour the growth of toxin-producing cyanobacteria and algae. The resulting toxins can cause gastroenteritis, liver damage, nervous system impairment and skin irritation. In some cases, liver cancer in humans is thought to be associated with exposure to cyanobacterial toxins through drinking water line and exposure to these toxins has usually been through contaminated drinking-water or recreational water contact (WHO, 2006). In addition, eutrophication also leads to the production of algal blooms and plant growth in streams, ponds, lakes, reservoirs and estuaries and along shoreline. Algal blooms are responsible for depletion of dissolved oxygen and contribute to serious water quality problems. Eutrophication of water sources may also create environmental conditions that favour the growth of toxin-producing cyanobacteria. Chronic exposure to such toxins produced by these organisms can cause gastroenteritis, liver damage, nervous system impairment, skin irritation and liver cancer in animals (WHO, 2006).

6.2 PMDE EFFECT ON TERRESTRIAL ECOSYSTEM

Wastewater reclamation and reuse is one of the best alternatives for compensating water shortages. In arid and semi-arid regions, wastewater reclamation and reuse has become an important element in water resources planning. In the agriculture, the irrigation water quality is believed to have effects on the soils and agricultural crops. Applications of industrial wastes as fertilizer and soil amendment have become popular in agriculture. Moreover, agricultural irrigation with wastewater effluents became a common practice in arid and semiarid regions, where it was used as a readily available and inexpensive option to fresh water. Some characteristics of effluents material are favourable for agriculture since the effluent is rich in organic matter, nitrogen (N), phosphorous (P), potassium (K) and magnesium (Mg). The disposal of wastewater is a major problem faced by industries, due to generation of high volume of effluent and with limited space for land based treatment and disposal. On the other hand, wastewater is also a resource that can be applied for productive uses since wastewater contains nutrients that have the potential for use in agriculture, aquaculture, and other activities.

Fig. 6.3: Wheat plants (a) and mustard plants (b) growing near the PMDE and tannery effluent contaminated sites

Wastewater generation results of increasing fresh water scarcity, their nutrients enrichment required advanced treatment for other applications including application in the agricultural lands. The use of soil as a medium for the treatment and disposal of industrial wastewater is becoming common practice. The increasing application of wastewater in agricultural fields may serve as a viable method of disposing the wastewater; improve soil fertility and sustaining agriculture production in non irrigated areas having shortage of fresh water for irrigation. The effluent causes concern of environmental pollution owing to its very high organic content. The effluent contains considerable amount of organic matter and plant nutrients, particularly potassium and sulphur, this can be applied to arable land as irrigation water and as an amendment. It may act as a source of plant nutrients and has been reported to increase the yield of the crops. Thus, application of distillery effluent to arable land as irrigation water and as a source of plant nutrients offers a promising alternative for its safe disposal. The application of distillery effluent improved the water retention characteristics of the soil; whereas, non-judicious use of distillery effluent might adversely affect the crop growth and soil properties by increasing soil salinity. The utilization of industrial waste as soil amendment has generated interest in recent times. Wastewater irrigation; reduce the need for chemical fertilizers, resulting in net cost savings to farmers. Hence, the impact of distillery effluent on various parameters of soil and plants are described in this chapter.

6.2.1 Effect on Physical Properties of Soil

Application of distillery effluent on soils is one of the most economical resources for the soil fertility amelioration through improvement in soil water-holding capacity, texture, structure, nutrients retention, roots penetration, and reduction in soil acidity. Nowadays in our country due to the increasing number of sugar mills and distillery units, application of distillery effluent on soil nearly become mandatory. However, its application in soil also results in environmental problems, because apart from organic content and nutrients, sludge also includes heavy metals, coloured compounds, dissolved inorganic salts, chlorinated lignin, and phenolic derivatives. The distillery effluent contains a significant quantity of salt (EC, 25.3 dS/m) its indiscriminate use may affect the physico-chemical properties of soil in the long run (Table 6.1). These compounds may change soil physico-chemical properties and soil enzyme activities. Soil enzymes activities play an essential role in catalyzing reactions which are necessary for the decomposition of organic matter and nutrient cycling in ecosystems, involving a range of plants, microorganisms, animals and their debris. Therefore, changes in enzymes activity could alter the availability of nutrients for plant uptake and these changes are potentially sensitive indicators of soil quality. Cellulase and Urease are the two important enzymes which play a significant role in soil environment. Cellulase is a core enzyme which contains exo, endo and β-glucosidases. This enzyme synergistically acts on cellulose, the most abundant polysaccharide of plant cell walls and representing significant input to soils.

Urease catalyzes the hydrolysis of urea and amides to carbon dioxide and ammonia. It acts on carbon-nitrogen (C–N) bonds other than peptide linkage. Urease is a constitutive intracellular enzyme with three subunits of α, β and γ and two nickel ions. Furthermore, liberation of these enzymes by microbes during litter decomposition may be influenced by too many factors like temperature, pH and substrate concentration in the soil environment. The discharge of effluents from distillery has altered the physico-chemical

Table 6.1: Physico-chemical properties of experimental unamended soil and soil amended with different dosages of distillery effluent (values represent mean n = 3 ± SE)

Parameter	Soil	10%	50%	100%
pH	6.8 ± 0.27	7.3 ± 0.17	8.47 ± 0.35	8.2 ± 0.31
Sand (%)	31.6 ± 2.56	18.35 ± 0.54	13.00 ± 1.05	8.5 ± 1.9
Silt (%)	56.4 ± 3.56	69.06 ± 2.67	71.50 ± 4.76	73.87 ± 4.22
Clay (%)	12.0 ± 1.05	12.75 ± 0.81	15.46 ± 1.54	18.20 ± 2.50
Moisture (%)	70 ± 1.67	71 ± 1.62	72 ± 3.43	74 ± 3.22
Organic matter (%)	1.09 ± 0.03	1.89 ± 0.056	2.86 ± 0.86	3.72 ± 0.75
EC (Ms/cm)	1.55 ± 0.024	1.85 ± 0.12	1.93 ± 0.083	2.02 ± 0.12
Phosphate (mg/kg)	265 ± 2.88	316 ± 3.27	383 ± 6.92	432 ± 8.43
Sulphate (mg/kg)	55.68 ± 1.53	188.67 ± 1.82	278.50 ± 6.45	321.36 ± 8.77
Phenol (mg/kg)	3.09 ± 0.21	34.78 ± 1.47	85.02 ± 8.82	102.45 ± 11.23
Total Nitrogen (mg/kg)	56.76 ± 2.34	73.34 ± 2.66	92.32 ± 7.33	116.56 ± 13.08
Sodium (mg/kg)	13.67 ± 0.82	29.14 ± 0.85	48.45 ± 2.18	69.44 ± 2.07
Chloride (mg/kg))	96.72 ± 1.51	109.28 ± 3.57	168.45 ± 8.89	211.14 ± 9.19
Magnesium (mg/kg)	9.48 ± 0.32	9.88 ± 0.48	14.42 ± 0.77	17.22 ± 0.63
Calcium(mg/kg)	11.55 ± 0.21	19.33 ± 0.64	45.3 ± 2.21	62.12 ± 3.46
Aluminium (mg/kg)	3.94 ± 0.27	4.34 ± 0.47	8.12 ± 0.34	10.88 ±0.64
Potassium (mg/kg)	44.39 ± 0.017	72.82 ±0.13	108.55 ± 0.41	183.33 ± 0.88
Cadmium (mg/kg)	0.06 ± 0.0021	0.32 ± 0.016	1.02 ± 0.042	1.38 ± 0.04
Chromium (mg/kg)	0.89 ± 0.014	2.88 ±0.065	2.92 ± 0.12	3.08 ± 0.07
Copper (mg/kg)	2.25 ± 0.043	18.22 ± 0.78	62.88 ± 3.27	88.17 ± 4.77
Iron (mg/kg)	3.73 ± 0.069	79.12 ± 0.67	114.34 ± 2.11	173.74 ± 3.38
Maganese (mg/kg)	1.86 ± 0.047	32.66 ± 0.48	78.55 ± 1.94	112.55 ± 2.7
Nickel (mg/kg)	1.11 ± 0.021	23.66 ± 0.63	39.33 ± 0.87	75.6 ± 1.34
Zinc (mg/kg)	12.18 ± 0.15	38.65 ± 0.86	76.55 ± 2.24	91.61 ± 2.45
Lead (mg/kg)	4.31 ± 0.062	8.86 ± 0.32	23.72 ± 1.28	22.44 ± 1.55

All values presented in mg/kg except electrical conductivity (mS/cm) and pH.

properties and enhanced the cellulase and urease activity of the soil, but it was declined with the time. Furthermore, by increasing the effluents concentration, the enzyme activity was improved upto 50% and later decreased. Similarly, distillery effluent discharged soil had relatively higher clay and silt contents than the control soil. The organic materials are stabilized through the activity of microbial flora in the soil. The application of distillery effluent in irrigation increases the saturated hydraulic conductivity and decrease the bulk density of the soil. It was observed earlier that an increase in hydraulic conductivity and infiltration rate and improvement in aggregate stability following addition of distillery slops and molasses to the columns of a saline sodic soil. The significant increase was recorded in infiltration rate (Infiltration rate in soil science is a measure of the rate at which soil is able to absorb rainfall or irrigation. It is measured in inches per hour or millimeters per hour. The rate decreases as the soil becomes saturated. If the precipitation rate exceeds the infiltration rate, runoff will usually occur unless there is some physical barrier. It is related to the saturated hydraulic conductivity of the near-surface soil. The rate of infiltration can be measured using an infiltrometer) and decrease in bulk density of an inceptisol (Inceptisols are a soil order in USDA soil taxonomy, they form quickly through alteration of parent material, they are older than entisols) with application of distillery effluent. The irrigation of distillery effluents reduced the alkalinity of the soil, while increased the organic carbon, available nitrogen, phosphorus, potassium, calcium, magnesium and sodium content in soil. The available nutrient status of the soil was increased due to effluent irrigation. The effluent of the Doon distillery Dehradun (Uttarakhand) decreased the pH and moisture content and increased it, bulk density EC, Cl^-, OC, HCO_3^-, CO_3^{2-}, Na^+, K^+, Ca^{2+}, Mg^{2+}, Fe^{2+}, TKN, NO_3^{2-}, PO_4^{3-}, SO_4^{2-} and Zn, Cd, Cu, Pb and Cr of the soil. The micronutrients such as Fe, Zn, Cd, Cu, Pb and Cr was also recorded higher in the soil irrigated with distillery effluent, which leads to toxicity of soil at higher concentration in comparison to control. The nutrients and trace elements of distillery effluent irrigation contributed some significant changes to the soil quality and affected the natural composition of the soil. Such alterations improved the fertility and enhanced the nutrients status of soil at lower concentration of effluent irrigation. Thus, effluent irrigation improved the soil nutrient status. Soil permeability is an important parameter when planning for liquid waste disposal to agricultural land. The reduction in hydraulic conductivity by effluent irrigation was due to accumulation of solids at the soil surface. The enrichment factor (EF) indicated the order of accumulation of various heavy metals in the soil after distillery effluent irrigation. Among various micronutrients the maximum enrichment factor (EF) was shown by Pb and minimum by Zn and it was in order of Pb > Cr > Cd > Cu > Zn after irrigation with distillery effluent. Thus application of distillery effluent to the agricultural field, as an amendment, might be a viable option for the safe disposal of this industrial waste with improvement in physico-chemical properties of the soil. However, the level of application should be within the prescribed limit to avoid development of soil salinity in the long run. Spentwash leads to significant levels of soil pollution and acidification in the cases of inappropriate land discharge. It is reported to inhibit seed germination, reduce soil alkalinity, and cause soil manganese deficiency. Some researchers identified spentwash as an agent causing manganese deficiency in soil. For instance, distillery effluent in combination with bioamendments such as farmyard manure, rice husk and Brassica residues was also used to improve the properties of sodic soil.

Beside distillery effluent, distillery sludge also contains high concentrations of nitrogen, phenol, chloride and heavy metals as compared to the garden soil. The values of pH (8.5) and EC (3.70 mS/cm) are also higher in sludge then the garden soil where the pH and EC are 7.4 and 2.03 mS/cm, respectively. High values of pH and EC in the sludge may be due to the presence of high concentrations of soluble salt. High concentrations of heavy metals and salts in sludge are due to the condensation process which takes place during sugar manufacturing and alcohol production. Also, cation exchange capacity (CEC) and organic matter (OM) of distillery sludge are significantly higher than garden soil. The pH of the soil increased from 7.4 to 8.0 with rise in sludge amendment rate indicated that sludge could act as a buffer medium for garden soil. Additions of distillery sludge also increased the EC significantly of sludge amendment soil which adversely affected the growth of *P. mungo* L. Simultaneously, phosphate and total nitrogen content in garden soil are also increased with increase in the amendment rate of distillery sludge.

6.2.2 Effect on Seed Germination and Seedling Growth

The effect of different concentrations (0, 5, 10, 15, 20, 25, 50, 75 and 100%) of distillery effluent (raw spentwash) on seed germination (%), speed of germination, peak value and germination value in some vegetable crops viz. tomato, chilli, bottle gourd, cucumber and onion did not showed any inhibitory effect on seed germination at low concentration except in tomato, but in onion the germination was much higher (84%) at 10% concentration as beside 63% in the control. Irrespective of the crop species, at highest concentrations (75 and 100%), complete crash of germination was noted. The impact of distillery effluents on growth and advance of rice seedlings and reported that the germination percent, number of roots, shoot and root length, fresh and dry weight of the seedlings showed an opposite relationship with effluent concentration. The rice seeds treated with different concentration of spentwash (0, 5, 10, 25, 50, 75 and 100%). At higher concentration (25% and above) both the speed of germination and seedling growth were retarded. Effluents were rich in inorganic constituents like ammoniacal nitrogen, chemicals and traces of heavy metals and these markedly suppressed the germination percent and early growth of the seedling as the concentration of the effluent increase. At 5% concentration by and large growth of seedling was better than in control and suggested that by diluting the effluent to 5% the effluent can be used as an alternative for chemical fertilizers. Moreover, the shoot and the root length and the number of on the side roots shaped in the case of sorghum reached maximum values when treated with 2.5% distillery spentwash. The distillery spentwash did not note any inhibitory effect on seed germination at low concentration. But up to 10% concentration the distillery spentwash noticeably better the seed germination and seedling growth in White sorghum. However, effect of distillery effluent on seed germination is governed by its concentration and is crop specific. In a study by Ramana et al. (2002a), the germination percent in, i.e. cucumber, chilli, onion, bottle gourd, tomato decreased with an increase in concentration of the effluent. The germination was inhibited in all the five crops studied with concentration exceeding 50 per cent. At the same time, organic wastes contained in distillery effluent are valuable sources of plant nutrients especially N, P and K.

Similarly, percentage seed germination decreased (except at 10%) with increase in sludge concentrations. After 10 days of sowing, soil amended with 20% (w/w) sludge caused 20% inhibition in seed germination followed by 30, 40 and 50% inhibition at 60,

80 and 100% (w/w) respectively versus the control. However, the seedling died within 15 days of germination at 100% (w/w) sludge. The inhibition of seed germination at higher concentrations of sludge-amended soils may be due to the high salt concentrations in distillery sludge creating high osmotic pressure. The high osmotic pressure caused inhibition of seed germination. Furthermore, the vegetative growth parameters of *P. mungo* (shoot length and root length) at all tested concentrations showed inhibitory effect (except 10% sludge amended soil) versus the control. The root length and number of leaves of *P. mungo* at 10% (w/w) sludge concentration increased by 3.45 and 16.66 % respectively compare to control. At 80% (w/w) sludge amended soil (after 60 days of study), 48.94 and 21.66% reduction in shoot length and root length was observed. In general 10% (w/w) sludge amended soil exhibited vigorous growth along with an increase in the number of nodules/plant but delayed flowering was observed as compared to control (Fig 6.4). However, at higher concentration of the sludge amended garden soil resulted in marked reduction in nodule/plant. The reduction of nodule/plant was 89.65% at 80% (w/w) sludge amended soil versus the control along with drastic decrease in leaves area and nodule/plant (Fig. 6.4) and numbers. This decrease in *P. mungo* L nodulation may be due to the presence of soluble salts and heavy metals, which are toxic to *Rhizobia*. These results supported the induction of plant growth at concentrations ≤ 10%.

This might be due to the induction of plant growth hormone while at higher sludge concentrations reversal in plant growth was observed. It has been reported that high sludge content was suppressive for plant growth hormone(s) (auxin and gibberline) which are responsible for the growth and development of plants. The reduction in plant growth at high concentrations of distillery sludge might be due to the entrance of the metal into the protoplasm resulting in the loss of intermediatory metabolites which are essential for further growth and development of plants. Dry matter of shoot, roots and leaves were inducible up to 20% (w/w) sludge concentration versus the control. However, it decreased drastically at 80% (w/w) sludge amended soil. The maximum biomass (0.94 g) was in shoot of the control sets. Dry matter follows the order leaves > shoot >root at 10% (w/w) sludge amended soil. The chlorophyll content was more in 10% (w/w) sludge amended soil which was gradually decreased with increase in concentration. The control showed low chlorophyll content than amended soil (up to 80% w/w sludge concentration). The effect of the distillery effluent is crop specific and due care should be taken before using the distillery effluent for pre-sowing irrigation purposes.

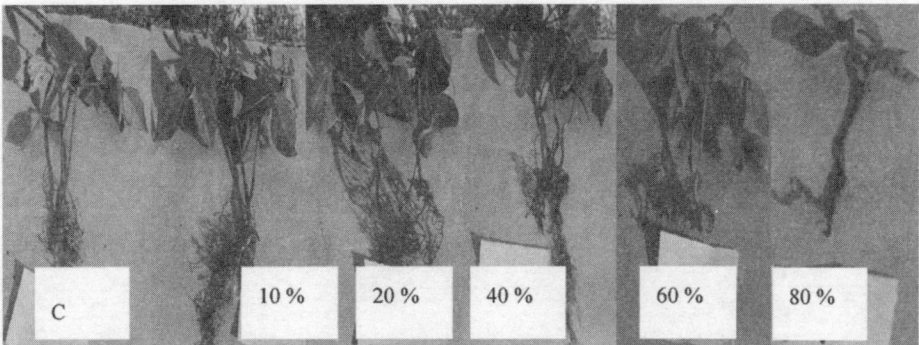

Fig. 6.4: Morphological effect of distillery sludge amendment soils at different concentrations (10–80%) on the growth of root, shoot and leaves of *P. mungo* after 60 days vs. the control

6.2.3   Effect on Amylase Activity

Seed germination is a complex physiological and biochemical process in plants that can be affected severely by several environmental factors. Starch is the major component of most of the world's crop yield and degradation of starch is essential for seed germination. In germinating seeds, starch degradation is initiated by α-amylase producing soluble oligosaccharides from starch. These are then hydrolyzed by α-amylase to liberate maltose and finally, α-glucosidase breaks down maltose into glucose providing energy to germinating seeds.

Bharagava and Chandra (2010b) showed that the optimum amylase activity (0.6 U) was recorded in seeds of *Phaseolus mungo* L treated with 10% (v/v) concentration of untreated PMDE and thereafter a continuous decline in α-amylase activity was observed at higher concentration (>10%) of PMDE (Fig. 6.5). However, the seeds treated with tap water (control) have shown low amylase activity (0.3 U) than the seeds treated with 10 and 20% (v/v) concentration of untreated and treated PMDE, respectively. Moreover, the seeds treated with bacterial degraded PMDE at 20% (v/v) concentration have shown the maximum amylase activity (0.9 U) indicating that the toxicity of PMDE is reduced significantly after bacterial treatment. The reduction in amylase activity at higher (>10 and 20%) concentration of untreated and treated PMDE might be due to the high salt load and metals content affecting various physiological and biochemical process of seed germination. Enzymes are vulnerable to various environmental factors. Their activity may be significantly diminished or destroyed by a variety of physical or chemical agents resulting in a loss of the functions performed by the enzymes.

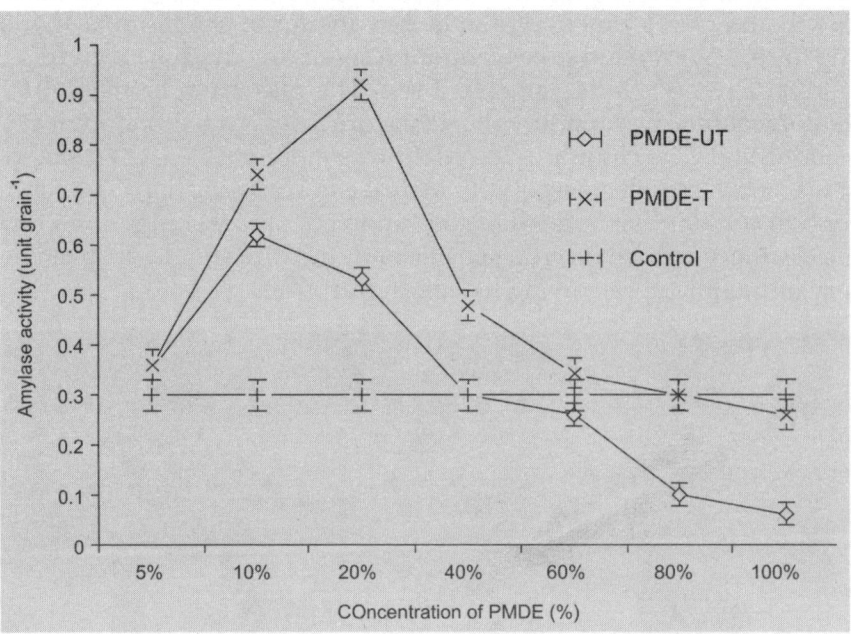

Fig. 6.5: Amylase activity shown by *Phaseolus mungo* L. seeds treated with different concentrations of bacteria treated and untreated PMDE. PMDE: post-methanated distillery effluent; UT: untreated; T: treated.

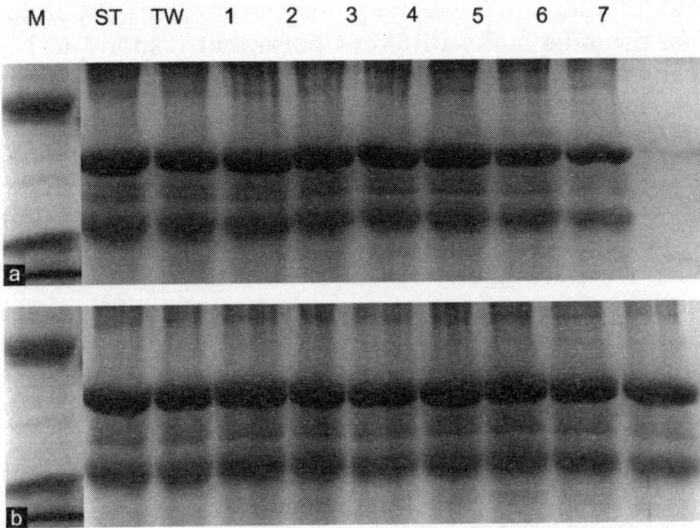

Fig. 6.6: Effect of different concentrations of Post methanated distillery effluent (PMDE) on amylase induction and its content in germinating seeds of *Phaseolus mungo* L. before (A) and after bacterial treatment (B). Lane M: Protein marker; Lane ST: Amylase standard; Lane TW: Seeds treated with tap water/control; Lane 1, 2, 3, 4, 5, 6 and 7: Seeds treated with 5, 10, 20, 40, 60, 80 and 100% (v/v) post methanated distillery effluent (PMDE).

Further, the denaturing SDS-PAGE of α-amylase enzyme extracted from germinating seeds treated with different concentration of untreated and treated PMDE has yielded three bands of different molecular weight and intensity/concentration (Fig. 6.6). Band intensity has indicated the concentration of α-amylase enzyme produced in germinating seeds treated with different concentration of untreated and treated PMDE. Results indicated that the concentration of α-amylase enzyme (i.e. band intensity) decreases gradually as the concentration of untreated PMDE increases. Phaseolus seeds treated with 60 and 80% (v/v) concentration of untreated PMDE have shown reduced α-amylase activity clearly and no α-amylase activity or enzyme production was observed in seeds treated with 100% (v/v) concentration of untreated PMDE.

6.2.4 Effect on Photosynthetic Pigments

In plants, metals present in effluent exert their toxic action mostly by damaging chloroplast and disturbing photosynthesis. An essential component of photosynthesis present in chloroplast with porphyrin (tetrapyrrole) nucleus with a chelated magnesium atom at the centre and a long chain hydrocarbon (phytol) side chain attached through a carboxylic acid group. Regarding the green pigment content, heavy metal-treated plants showed a remarkable decrease of chlorophyll that causes photosynthesis rate enormously decrease in response to elevated heavy metal concentration. Reduction in chlorophyll content as a consequences of many metabolic reactions like inhibition of enzymes activity such as α-aminolevulinic acid dehydratase (ALAdehydratase) and protochlorophyllide reductase, replacement of Mg with heavy metals in chlorophyll structure, decrease in the source of essential metals that involved in chlorophyll synthesis such as Fe^{2+} and Zn^{2+}, destruction of chloroplast membrane by lipid peroxidation due to increase in peroxidase activity and lack of antioxidants such as carotenoids, decrease in density, size and the synthesis of chlorophyll and inhibition in the activity of some enzymes of Calvin cycle. In another word, chloroplast contains many different parts that respond to heavy metal stress therefore any changes in chlorophyll synthesis and activity used as the index of direct toxic effects of heavy metals. The inhibition of photosynthesis is the consequence of interference of metal ions with photosynthetic enzymes and chloroplast membranes. Photosynthesis is reduced indirectly by heavy metal accumulation in leaves which influences the functioning of stomata. The increase in the chlorophyll content at low sludge concentration was due to the presence of essential nutrients and metal ions. However, these metals which were acting as nutrient crossed the threshold limit at high sludge concentrations and work as toxic agent through direct inhibition of photosynthesis. In contrary to above findings an increase in chlorophyll ratio (a/b) in Co and Ag stress shows that chlorophyll 'b' is more sensitive to Co and Ag that disrupt the balance between energy trapping in photosystem II and cause a decrease in electron transport. While decrease in chlorophyll ratio (a/b) in response to Cd and Pb treatments suggest that chlorophyll 'a' is more sensitive to Cd and Pb. (Within the thylakoid membranes of the chloroplast, are two photosystems. Photosystem I optimally absorbs photons of a wavelength of 700 nm. Photosystem II optimally absorbs photons of a wavelength of 680 nm. The numbers indicate the order in which the photosystems were discovered, not the order of electron transfer. Under normal conditions electrons flow from PSII through cytochrome bf (a membrane bound protein analogous to Complex III of the mitochondrial electron transport chain) to PSI. Photosystem II uses light energy to oxidize two molecules of water into one molecule of molecular oxygen. The 4 electrons removed from the water molecules are transferred by an electron transport chain to ultimately reduce $2NADP^+$ to 2NADPH. During the electron transport process a proton gradient is generated across the thylakoid membrane. This proton motive force is then used to drive the synthesis of ATP. This process requires PSI, PSII, cytochrome bf, ferredoxin-$NADP^+$ reductase and chloroplast ATP synthase).

In addition, caratenoids are tetraterpeniod (C40) compounds widely distributed in plants, function as accessory pigments in photosynthesis and as colouring matters in flowers and fruits. Some of the commonly occurring carateniods are simple unsaturated hydrocarbons based on lycopene and their oxygenated derivatives known as xanthophylls. β-carotene is the most common pigment in this group found in higher plants. Carotenoid is a non-enzymatic antioxidant pigment that protects chlorophyll, membrane and cell genetic composition against reactive oxygen species (ROS) under heavy metals stress. In plant cell protective role of this pigment might be due to quenching triplet chlorophyll, replacing peroxidation and destruction of chloroplast membrane. Decrease in carotenoid content is a common response to metal toxicity (Rout et al., 2001), but increase is due to important role of this pigment in detoxifying ROS (Chandra et al., 2009b). The carotenoid content decreased in response to heavy metals indicate a severe effect on cell and its component parts at first carotenoid content increased to protect the cell against these heavy metals, but in high concentration (100 μM) these heavy metals activate some mechanisms and degrade carotenoid pigments. The metal accumulation potential and its physiological effects in Indian mustard plants (*Brassica nigra* L.) grown in soil irrigated with post methanated distillery effluent (25%, 50%, 75%, 100%, v/v) were studied after 30, 60 and 90 days after sowing by Baragava et al. (2008). An increase in the chlorophyll contents was recorded at the lower concentrations of PMDE at initial exposure periods followed by a decrease at higher concentrations of PMDE compared to their respective controls (Fig. 6.7).

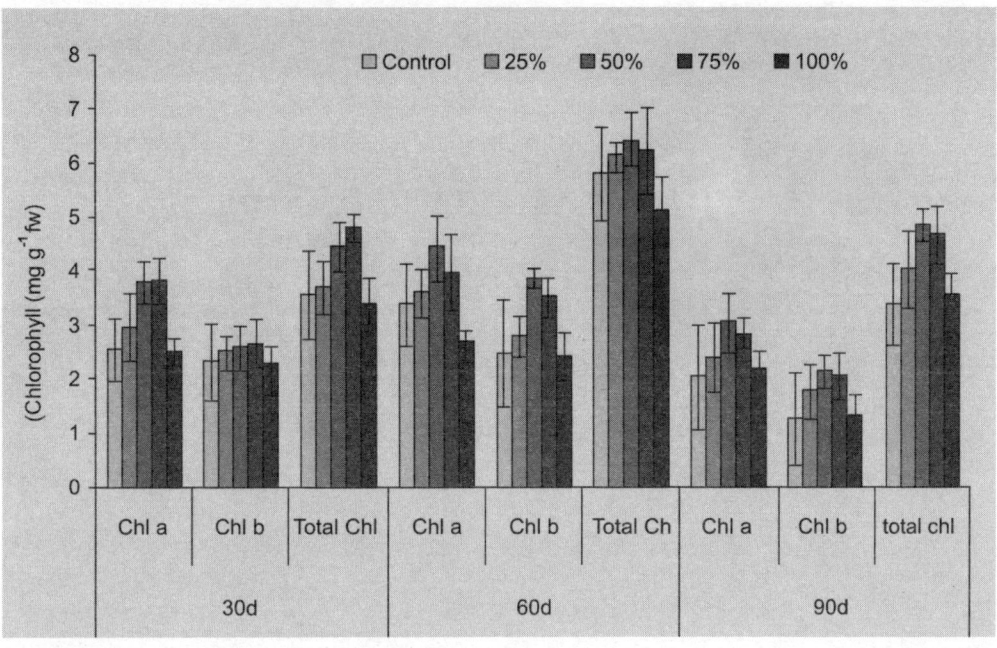

Fig. 6.7: Effect of PMDE on Chlorophyll-a, b and total chlorophyll content in mustard plants

6.2.6 Effect on Protein Content

Lipids and proteins are important constituents of the cell that easily damage in environmental tress condition. Hence, any change in these compounds can be considered as an important indicator of oxidative stress in plants it is thought that decrease in total soluble protein content under heavy metals stress may be due to increase in protease activity, various structural and functional modifications by the denaturation and fragmentation of proteins, DNA-protein cross-links, interaction with thiol residues of proteins and replacement them with heavy metals in metalloproteins. The protein content was increased with 20% (w/w) sludge-amended soil. However, >20% sludge there was a slight decrease in root, shoots and leaves protein. The higher content of protein may be due to the induction of other stress protein in plant at higher sludge concentrations reflecting the manurial properties of distillery sludge at lower concentrations (Chandra et al., 2008a). Similarly, compared to control plants, the protein content in root, shoot and leaves also increased in mustard plants irrigated with 50% (v/v) PMDE at 60 days of growth period (Fig. 6.8). But at higher concentrations (> 50%, v/v) of PMDE, there was a decrease in protein content of root, shoot and leaves at 60 and 90 days of growth period. It has also been reported that cadmium is able to decrease protein content by inhibiting the uptake of Mg and K and promote post translational modification, decrease in synthesis or increase in protein degradation and the prevention of Rubisco activity. The increase in total soluble protein content under heavy metal stress may be related to induce the synthesis of stress proteins such as enzymes involved in Krebs cycle, glutathione and phytochelatin biosynthesis and some heat shock proteins (Mishra et. al., 2006).

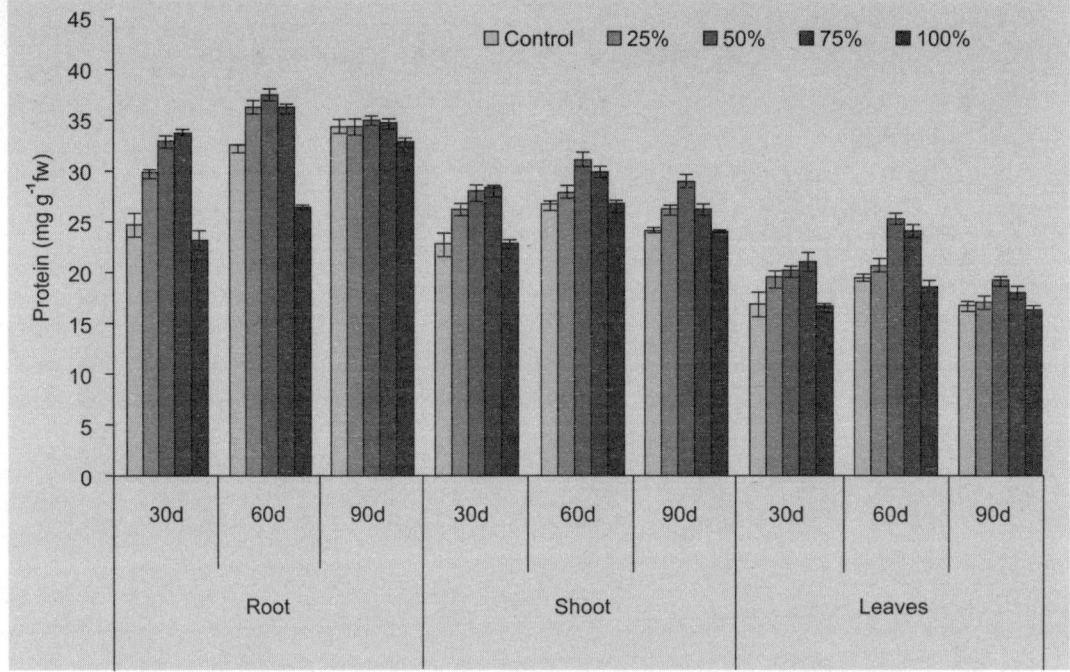

Fig. 6.8: Effect of PMDE on protein content in mustard plants

6.2.6 Effect on Lipid Peroxidase

Malondialdehyde (MDA), a major cytotoxic product of lipid peroxidation acts as an indicator for free radical production. Therefore, MDA formation can be considered as a measure of lipid peroxidation. MDA content in roots and leaves increased with rise in sludge amendment ratio versus the control at all exposure periods, indicating the enhanced lipid peroxidation in sludge grown *P. mungo*. To protect from oxidative stress conditions induced by free radicals, plants adopted cellular entities consisting of non enzymatic cellular entities. These included ascorbic acid, cysteine, non-protein thiol etc. Therefore, after 20, 40 and 60 days growth of *P. mungo* at all sludge concentrations, the non enzymatic cellular entities in both root and leaves increased versus the control. After a period of 40 days, cysteine, non protein thiol and ascorbic acid content in leaves of *P. mungo* increased with rise in sludge concentrations. At 80% distillery sludge, the values of cysteine (300%), non-protein thiol (541) and (ascorbic acid) 229% were maximum. The increase in cysteine, non protein thiol and ascorbic acid contents in root and leave were upto a period of 60 days and at 10 and 20% sludge amendment soil grown plants. At higher sludge concentrations an inhibitory trend of these parameters were observed. Maximum increase in ascorbic acid (21%) was in root followed by non-protein thiol (20%) in root and cysteine (11%) in leaves of *P. mungo* versus the 40 days content ascorbic acid, cysteine and non-protein thiol at 60 days of growth.

6.2.7 Effect on Antioxidants

Antioxidants are considered to play an important role in the detoxification of toxic oxygen species generated in presence of metal ions. However, to protect from oxidative stress conditions induced by free radicals, plants have adopted several non-enzymatic cellular entities as ascorbic acid, cysteine, and nonprotein thiol, etc. Cysteine, a–SH containing amino acid is a key constituent of phytochelatins and plays an important role in metal detoxification. An increase in cysteine content was recorded in root and leaves of wheat and mustard plants irrigated with effluents. The results are well corroborated with those of Nussbaum et al. (1988), who reported an increase in cysteine content in root of Zea mays at high Cd level followed by decrease with increase in metal concentration. The increase in cysteine content corresponds to the level of tolerance exhibited by metal-treated plants whereas the decrease in cysteine content in plants at higher exposure periods may be due to the decreased activities of sulfate reduction enzymes, ATP sulfurylase, and adenosine-5-phosphosulfate sulfo transferase under metal stress. Ascorbic acid together with glutathione also plays a prominent role in scavenging of free radicals. The maximum induction in ascorbic acid content was observed in root of wheat and mustard plants irrigated with effluents after 90 days of growth period.

6.2.8 Impact on Microorganisms

Soil microbial population was studied, as they are an important entity of soil ecosystem helping in nutrient recycling processes and regulating the soil productivity. Distillery effluent treatments at lower and up to 30% concentrations increased the soil microbial population over control, that may be attributed to the carbon and nutrients present in the distillery effluent. An increase in bacterial population over control was found and the highest count was recorded with 20% distillery effluent treatment at third irrigation which was significant among different distillery effluent treatments. However, 30% distillery effluent treatment at third irrigation had higher fungi and actinomycetes (0.425 $\times 10^6$ and 6.6×10^6 cfu/g of soil respectively) population in soil. The fungal population was in lesser number compared to bacteria or actinomycetes population. Impact of distillery effluent on soil microflora of *Phaseolus aureus* L was also investigated and found that the irrigation of the pots by 1–10% distillery effluent stimulated the growth of the soil microflora (increase number of bacteria, fungi and actinomycetes). Further, 15–20% distillery effluent had toxic effect on soil micro flora. Higher concentration of raw distillery effluent reduced the bacterial population. However, the bacterial treated distillery effluent <10% had stimulatory effect on fungal and actinomycetes population. Melanoidins which impart colour to spentwash have antioxidant properties and are often toxic to microorganisms involved in waste treatment. The growth rates of Rhizobium and Azotobacter were also reduced after distillery effluent application. The toxic effect of raw wastewater was minimized when it was mixed with stabilization pond effluent (1:1), this was demonstrated by an increase in the populations of all the microorganisms studied. This study indicated that higher distillery effluent concentrations had a negative impact on soil microflora. Therefore, monitoring and integrated approaches are needed to effectively utilize PMDE as valuable resource in agriculture and reduce its negative effects on the environment.

6.2.9 Accumulation and Distribution of Toxic Metals in Wheat and Mustard Plants

The process of metal uptake, accumulation and distribution by different plants is strongly influenced by the soil characteristics such as pH, CEC, organic matter content, and concentration of available metals in soil, solubility sequences and plant species. Soil pH is an important factor controlling the availability of metals to plants. The bioavailability of Cu and Zn is mainly regulated by pH and organic carbon content whereas low solubility of Mn was observed at neutral to alkaline pH. The unchanged or enhanced transpiration rate, which is a major force that drives metal uptake and translocation to shoots, could be responsible for the increased accumulation of metals in the above ground parts and preventing their precipitation in root and vascular system. It is well documented that lowering of pH increases soluble metal concentrations in the soil and enhances stability of humic acid metal complex. Moreover, different crop plants exercise differently in accumulation and distribution process of metals in their different parts (root, shoot, leaves, and fruits/seeds) and the efficiency of different crop plants in absorption and distribution of metals are judged either by plant metal uptake or by transfer factor of metals from soils to plants. The translocation process of metals from root to shoot includes long distance in xylem and storage in vacuoles of leaf cells and it is affected by several factors. The accumulation and distribution pattern of different metals in different parts of wheat and mustard plants irrigated with effluents was recorded as leaves > roots > seed > shoot for Cu, roots > leaves > seeds > shoot for Cd, roots > leaves > shoot > seeds for Cr, seeds > roots > shoot > leaves for Zn, roots > leaves > shoot > seeds for Ni, roots > leaves > shoot > seeds for Fe, roots > shoot > leaves > seeds for Mn, roots > leaves > shoot > seeds for Pb in wheat plants and roots > leaves > seeds > shoot for Cu, leaves > shoot > seeds? roots for Cd, leaves > seeds > roots > shoot for Cr, roots > leaves > seeds > shoot for Zn, roots > seeds > leaves > shoot for Ni, leaves > roots > seeds > shoot for Fe and Mn, leaves > shoot > seeds > roots for Pb in mustard plants. The concentration of all tested metals varied widely in different parts of wheat and mustard plants. The variability in metals content in different parts of wheat and mustard plants may also be due to the compartmentalization and translocation in the vascular system. The concentration of all tested metals in wheat and mustard plants irrigated with effluents were found significantly higher than the permissible limit set by FAO/WHO for Cu, Cd, Cr, Zn, Ni, Fe, Mn and Pb as 3.00, 0.21, 0.02, 27.4, 1.63, 20.0, 2.00 and 0.43 (mg/kg dry weight basis) for consumption. Hence, it may be advisable that wheat and mustard plants irrigated with effluents should not be taken as food by human beings and cattles because these metal rich plants may cause several clinical problems. The ratio of metals between soil and plants is an important criterion for the selection of crop plants for the cultivation on soils contaminated with high level of toxic metals. If the ratio > 1, means higher accumulation/concentration in the plant parts than soil.

Metal accumulation in the plants grown at different sludge concentrations of amended soil showed different magnitude and relative distribution. The trend of metal accumulation follows the order Zn > Fe > Mn > Cu > Cr > Ni > Cd. This trend varied from one part to another depending on distillery sludge amendment ratio. The results revealed that the accumulation of heavy metals at 40% w/w concentrations was highest in the roots except iron and manganese, which accumulated maximum in the shoots. Overall, Cu and Zn were accumulated maximum while Cd and Ni were accumulated least in *P. mungo* L. The accumulation of all heavy metals was minimum in fruit parts at 10% sludge amended soil grown *P. mungo*. However, accumulation pattern of N, P, K and Mg were in the order shoot > leaf > root. Accumulation of Ca was maximum in upper part of the plants including shoot and leaves. High accumulation of Zn, Fe and Cu, in various parts of *P. mungo* indicated the fast mobility of these metals in the plant while lead absence may be due to its poor availability in soil for plant uptake.

6.3 GUIDELINES FOR INDUSTRIAL EFFLUENT REUSE IN AGRICULTURE

Reuse of industrial effluent induces risk to infect soil, water and crops though imminent in agricultural perspectives. Detail guidelines are available for safe and secure use of effluent in different perspectives. However in effluent reuse practice, to abate adverse impacts of effluents, protecting soil health and crop quality the steps need to follow are:

6.3.1 Characterization of Effluent

Prior to put into practice the effluent needs to thoroughly analyzed and estimated for its constituents, which reveals the type and quality of effluent.

6.3.2 Evaluation of Effluent Potential by Comparing its Characteristics with Different Irrigation Water Quality Standards

After characterization, prioritization of effluent constituents as per their potential to cause damage is required in this process. This prioritization will help to eliminate the effluent if contain hazardous element/compounds, such as heavy metals beyond their corresponding standards at the first instance. Heavy metal causes irreversible damage on living beings. Thus the step will help to disclose effluent's ability to be used or not-used in irrigation purpose.

6.3.3 Match the effluent characteristics with relevant properties of dominant soil types

Soils are differed by their contents and characteristics. Different soils are dominated in different regions. This could be used as an yardstick for effluent selection and improve its prospect of utilization in cropping, such as soil with acidic pH and low salinity can be used for receiving effluent of high pH and high salt content. In this way the effluent's liabilities could be conditioned and its potential can utilize for cropping. The soil based effluent reuse option could be employed if the information of soil as par availability of fresh water resources is known in various locations/regions.

6.3.4 Identification of Effluent Quality Indicators

For surveillance of effluent quality, testing of number of properties in regular interval is a major burden in effluent reuse program. Identification of salient properties of effluent will help to relieve the load while monitor its quality in an effective manner. Choosing of appropriate mechanism to screen effluent properties and select the properties as indicators is vital in this process

6.3.5 Check the Quality of Crops, Soil and Adjacent Water Resources

All steps are mandatory for effluent reuse practice. But for continued use of same kind of effluent, monitoring of indicating properties (effluent quality indicators) and a strict vigil on soil/water/crop quality are highly required to promote its safe use in crop production.

7

Phytoremediation of Heavy Metals from Post Methanated Distillery Effluent

7.1 INTRODUCTION

Phytoremediation is an emerging technology that uses various plants to degrade, extract and immobilize contaminants from soil and water. This technology has been receiving attention recently as an innovative, cost effective alternative to the more established treatment methods used at hazardous waste sites. Many hazardous waste sites are contaminated with salts, organics, heavy metals, trace elements, and radioactive compounds. The simultaneous clean-up of multiple, mixed contaminants using conventional chemical and thermal methods are both technically difficult and expensive; these methods also destroy the biotic component of soils. Besides, several organisms, regularly exposed to higher levels of metals have developed some resistance mechanisms. Micro-organisms are especially important in this aspect and their tolerance strategies include intracellular accumulation, precipitation, biotransformation, etc. Conventional methods for removing heavy metals from industrial effluents (e.g. precipitation and sludge separation, chemical oxido-reduction, ion exchange, reverse osmosis, electrochemical treatment and evaporation) are often ineffective and comparatively costly when applied to dilute and very dilute effluents.

Recently, phytoremediation processes has been attracting enormous attention for removing heavy metals from aqueous wastes has been promising so far. Phytoremediation of heavy metals offers the advantages of low operating cost, minimizing secondary problems with metal-bearing sludge and high efficiency in detoxifying very dilute effluents. Phytoremediation of a site contaminated with heavy metals or radionuclides involves "farming" the soil with selected plants to "biomine" the inorganic contaminants, which are concentrated in the plant biomass. For soils contaminated with toxic organics, the approach is similar, but the plant may take up or assist in the degradation of the organic compounds. Phytoremediation actually benefits the soil, leaving an improved, functional, soil ecosystem at costs estimated at approximately one-tenth of those currently adopted technologies.

Heavy metals comprise a broad group of metals of atomic weight higher than that of sodium and having a specific gravity of more than 5.0. When these are present in high amounts it becomes highly toxic and harmful for living organisms. Heavy metal pollution is a wide spread problem in all parts of the world. The release of heavy metals from industries into the environment has resulted in severe problems for both human health and aquatic ecosystems. Toxic heavy metal contamination of distillery effluent is a significant problem. Heavy metals accumulate in living tissues throughout the food chain which has humans at its top. These toxic metals can cause accumulative poisoning, cancer and brain damage when found above the tolerance levels. High levels of lead in drinking water may lead to Alzheimer's disease. Alzheimer's disease is a neurological disorder in which the death of brain cells causes memory loss and cognitive decline. A neurodegenerative type of dementia, the disease starts mild and gets progressively worse. Fishes and other water inhabiting animals face several health irregularities due to presence of toxic metals and nuclides in water. Anaemia and blood deformities are reported in fishes grown in metal-contaminated water streams. These metals, in fact, are ubiquitous in natural environment and always present in a certain trace concentration. But when this concentration of metals rises due to anthropogenic causes, it may disrupt normal biogeochemical cycles and lead to environmental problems. Metal polluting industries include electroplating, chrome-tanning, textiles, photographic sensitizer manufacturing, nuclear processing and industries of batteries, paints, dyes, preservatives, insecticides, etc. But the most important is mining and other metallurgical activities. Besides these, distilleries effluent is also source of heavy metals contamination. Heavy metals such as zinc, lead and chromium have a number of applications in basic engineering works, paper and pulp industries, leather tanning, organo-chemicals, petrochemicals, fertilisers, etc. Major lead pollution comes from automobiles and battery manufacturers. For zinc and chromium the major sources are fertiliser industry and leather tanning respectively. These metals coming from different sources may exist in free metallic or compound form, either in solid or liquid form. Different types of morphological and physiological changes were reported in plants growing in mineral and metal deficient and abundant soil (Fig. 7.1a and b). These metals have severe effects like chlorophyll loss, nutrient deficiency, root shortening, etc. in plants.

Fig. 7.1: (a) Morphological changes (chlorophyll loss and shortening of leaves) in plant growing in metals deficient soil

Nitrogen deficiency (N) early stage

Nitrogen deficiency (N) progression

Nitrogen deficiency (N) late stage

Nitrogen abundance (N) early stage

Nitrogen abundance (N) late stage

Phosphorus deficiency (P) early stage

Phosphorus deficiency (P) progression

Phosphorus deficiency (P) late stage

Potassium deficiency (K) early stage

Potassium deficiency (K) progression

Potassium deficiency (K) late stage

Magnesium deficiency (Mg). early stage

Magnesium deficiency (Mg). progression

Sulfer deficiency (S). early stage

Sulfer deficiency (S). progression

Sulfer deficiency (S). late stage

Zine deficiency (Zn). early stage

Zine deficiency (Zn). progression

Zine deficiency (Zn). late stage

Zine deficiency (Mn). early stage

Manganese deficiency (Mn). progression

Manganese deficiency (Mn). late stage

Iron deficiency (fe) early stage

Iron deficiency (fe) progression

Iron deficiency (fe) late stage

Fig. 7.1: (b) Effect of deficiency and abundance of some minerals/metals on plant's leaves

Metal translocation from plants to subsequent trophic levels of ecosystem affects all animals. Microbial growth is also inhibited partially or completely by presence of excessive amounts of toxic metals and this may ultimately lead to imbalance of the total ecosystem. Physiological functions of microorganisms like different metal-dependent enzymes, electron transport system (ETS), nitrogen fixation etc. are often hampered by metal toxicity. Such toxic heavy metals have been associated with different disorders in human including brain and bone structure damage, nervous system disorder, etc. From different sources like canals and rivers heavy-metals often reach coastal and estuarine waters and often found to be a serious threat to the ecologically very significant estuarine biota.

Besides, several researches have been reported that wetland plants are capable for degradation of industrial wastewater of distilleries, pulp paper mill and tanneries. It has tremendous properties to grow near the industrial effluent contaminated site which shows their tolerance against the various pollutants (Fig. 7.2). Both natural and constructed wetlands are used to treat wastewater and storm water runoff from urban, industrial, and agricultural sources. Wetlands pertaining to their characteristically high productivity may represent the low-cost and high-value cleanup systems described by the US EPA (2001). The use of wetland areas as natural filters for the amelioration of pollutants transported in rivers or lakes is considered to be a successful, low-cost, cleanup option to ameliorate the quality of surface waters. Indeed, wetlands have been widely utilized in the last decades to clean polluted water over almost all the world. Vegetation comprising wetland plants is the most important component of a wetland system. The common plants in wetlands are common reed (*Phragmites spp.*), cattail (*Typha spp.*), rush (*Juncus spp.*), and bulrush (*Scirpus spp.*) and some other important wetland plant species are listed in Table 7.1.

Adaptations that the plants make to live in these adverse conditions can take many forms, but are generally grouped into morphological, physiological and reproductive adaptations. Morphological adaptations are changes in the structure or form of the plant

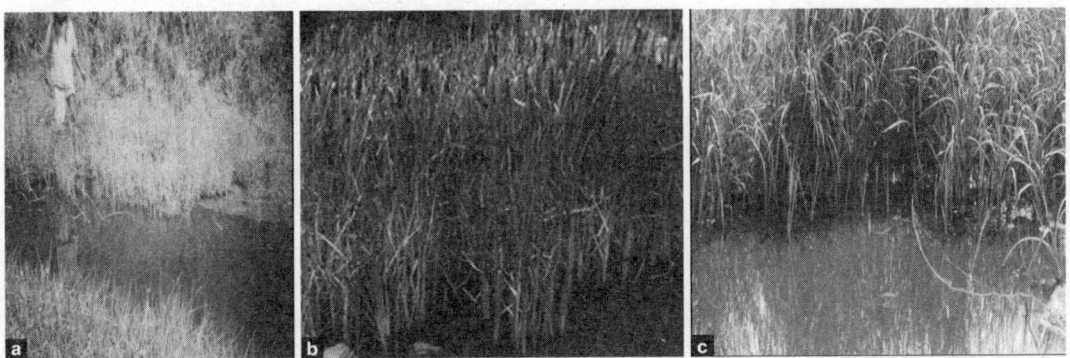

Fig. 7.2: Wetland plants (a: *Phragmites cummunis*, b: *Typha anguistifolia*, c: *Cyperus esculentus*) growing near distillery effluent contaminated sites

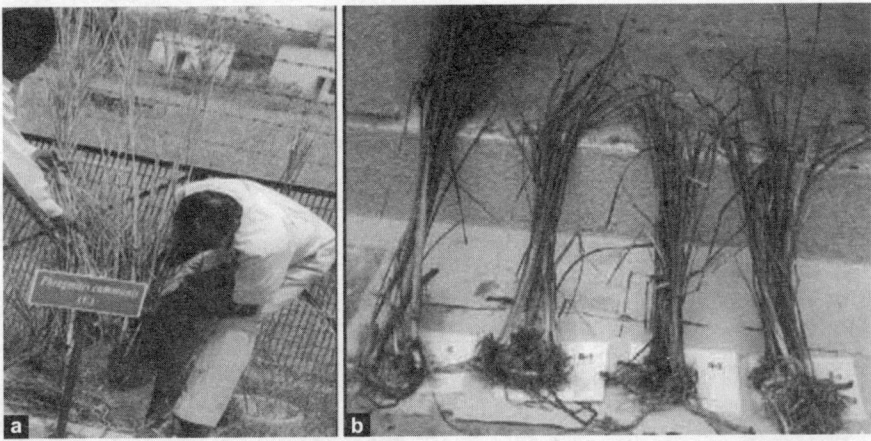

Fig. 7.3: Showing massive root system of (a) *Phragmites cummunis* and (b) *Typha anguistifolia*

which aid them in growing in their particular environment. In wetlands, hydrophytes, especially trees, may have buttressed bases which help provide additional stability in soft wetland soils. They may have adventitious roots, which are multiple root stems growing down from the main trunk (Fig.7.3). Mangroves trees are an example of a plant with this adaptation.

Table 7.1: Important Wetland Plant Species

S.No.	Common Name	Scientific Name	Family
1.	Reed	*Phragmites australis, Phragmites karka , Phragmytes sp.*	Poaceae
2.	Water fern, Water velvet	*Azolla caroliniana*	Salviniaceae
3.	Water bloom/algal bloom	*Mycrocystis sp.*	Microcystaceae
4.	Balrush/Cattail	*Typha latifolia, Typha augustata, Typha domingensis*	Typhaceae
5.	Poplar trees	*Populus deltoids*	Salicaceae
6.	Pond weed/Curly leaf pond weed	*Potamogeton natans Potamogeton crispus*	Potamogetonaceae
7.	Parrot's feather	*Myriophyllum spicatum*	Haloragaceae
8.	Umbrella plant	*Cyperus alternifolius Cyperous papyrous*	Cyperaceae
9.	Duckweed	*Wolffia globosa*	Araceae
10.	Water hyacinth	*Eichhornia crassipes*	Potenderiaceae
11.	Smart weed	*Polygonum hydropiper*	Polygonaceae
12.	Smooth cordgrass	*Spartina alterniflora*	Poaceae
13.	Water zinnia	*Wedelia trilobata*	Asteraceae
14.	Irish leaved rush	*Juncus xihoides*	Juncaceae
15.	Fuzzy water clover	*Marsilea dromondii*	Marsileaceae
16.	Reed canarygrass	*Phalaris arundinacea*	Poaceae
17.	Salt marsh bulrush	*Scirpus robustus, Scirpus lacustris*	Cyperaceae
18.	Rabbitfoot grass	*Polypogon monspeliensis*	Poaceae
19.	Zebra rush	*Scirpus tabernaemontani*	Cyperaceae
20.	Taro	*Colocasia esculenta*	Araceae
21.	Khus khus	*Vetiveria zizaniodes*	Poaceae
22.	Asian rice	*Oryza sativa*	Poaceae
23.	Waterpurslane/Water primrose	*Ludwigia sp.*	Poaceae
24.	Cutgrasses	*Leersia sp., Leersia oryzoides*	Poaceae
25.	Smooth cord grasses	*Spartina alterniflora*	Poaceae
26.	Indian lotus/bean of India	*Nelumbo nucifera*	Nymphacaceae
27.	Arrow head/dark potato/swan potato	*Sagittaria australis Sagittaria latifolia*	Nynphacaceae
28.	Gaint bur reed	*Sparganium eurycarpum*	Spaganiceae
29.	Pickerel weed	*Pontederia cordata*	Pontederiaceae
30.	Wild celery	*Vallisnaria americana*	Araceae
31.	Green arrow arum	*Peltendra virginica*	Araceae
32.	Pond weed	*Elodea Canadensis*	Hydrocharitaceae
33.	Benghal dry flower/tropical spider wood	*Commelina benghalensis*	Commelinaceae

Other plants have aerenchyma tissue, which is spongy, hollow tissue often found in the stems which increase the plant's buoyancy, and the number of air spaces in the plant. This type of tissue may help the leaves of the plant to float and may store air for the plant when it is under water. Wetland plants with floating aquatic plants, such as water lilies, have this feature. Shallow root systems are morphological adaptations to provide additional stability to the plant growing in wetland soils. Finally, some plants have developed specialized cells to enhance the movement of oxygen to the roots from the stems of the plants. Physiological adaptations are methods which plants use to change the metabolic pathways in which they process energy. For example, some plants can store the accumulated chemicals in a non-toxic form in their roots until a dry spell when the chemicals can be released. Another adaptation is the ability to lower the rate in which metabolic activity takes place under stressful conditions. The larch (*Larix laricina*) is thought to possess this ability. Many plants can transfer oxygen from the roots into the pore spaces of the soil surrounding the roots, to minimize root degradation and to maintain nutrient uptake under anaerobic soil conditions.

Wetland plants are morphologically adapted to growing in waterlogged sediment by virtue of large internal air spaces for transportation of oxygen to roots and rhizomes. The extensive internal lacunal system, which normally contains constrictions at intervals to maintain structural integrity and to restrict water invasion into damaged tissues, may occupy up to 60% of the total tissue volume. The internal oxygen movement down the plant serves not only the respiratory demands of the buried tissues, but it also supplies the rhizosphere with oxygen by leakage from the roots. This oxygen leakage from roots creates oxidised conditions in the otherwise anoxic substrate and stimulates both aerobic decomposition of organic matter and growth of nitrifying bacteria (Fig. 7.4).

It is well documented that aquatic macrophytes release oxygen from roots into the rhizosphere and that this release influences the biogeochemical cycles in the sediments through the effects on the redox status of the sediments. Qualitatively this is easily

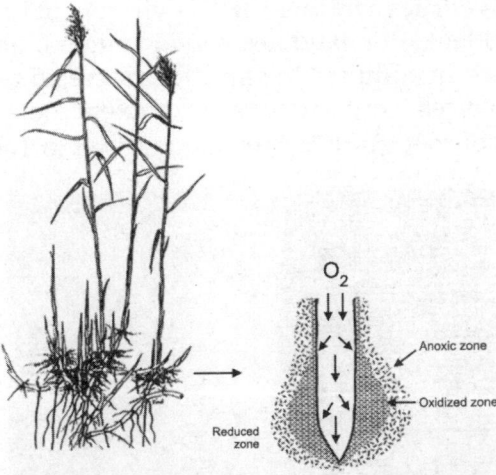

Fig. 7.4: Sketch of the common reed, *Phragmites australis*. The oxygen leakage from roots creates oxidized conditions in the otherwise anoxic substrate and stimulates both aerobic decomposition of organic matter and growth of nitrifying bacteria.

visualised by the reddish colour associated with oxidised forms of iron on the surface of the roots and experimentally by submerging a root system into a reduced solution containing methylen blue (Fig. 7.5). But the quantitative magnitude of the oxygen release under in situ conditions remains a matter of controversy.

Cattails (*Typha spp.*) can maintain root growth under very low oxygen levels in the soil, a condition which would end root growth in many other plants. Reproductive adaptations include prolonged seed viability in wet conditions, and the ability of the seed to be triggered to grow in dry conditions. Many wetland plants have seeds which can germinate under low oxygen, soil conditions, and have seedlings which can survive low oxygen conditions during their early development.

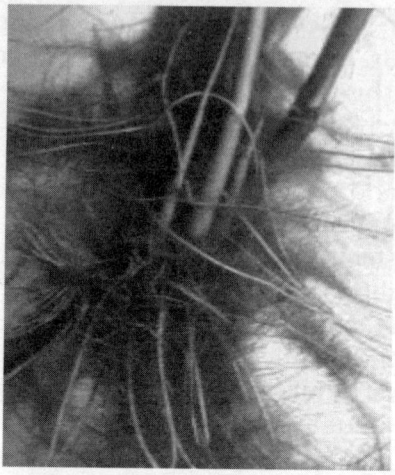

Fig. 7.5: Root release of oxygen by *Phragmites australis*. The blue colour around the roots is formed by radical oxygen release from the roots oxidising the reduced form of methylen blue.

Consequently, root zone or rhizosphere of wetland plants supports a population of aerobic microorganisms known as rhizospheric microbes. Wetland plants not only assimilate pollutants directly into their tissue, but also act as catalysts for bioremediation by increasing environmental diversity in the rhizosphere, thereby promoting various chemical and biochemical reaction that enhanced degradation/remediation process. These rhizospheric microbes plays a major contribution for bioremediation of different pollutants like agrochemicals, heavy metals, toxic organic compounds, polyaromatic hydrocarbon (PAH), industrial and sewage wastes.

7.2 PLANT MICROBES INTERACTION WITH ENVIRONMENTAL POLLUTANTS

The main reason for the improved degradation of pollutants in the phytoremediation is presumably the increase in the number and metabolic activity of microbes. The exudation of nutrients by plant roots creates a nutrient-rich environment in which microbial activity is stimulated (Fig. 7.6). Plant root exudates contain sugars, organic acids, and amino acids as main components. In addition, the mucilage secreted by root cells, lost root cap cells, the starvation of root cells, or the decay of complete roots provides nutrients.

Some of the functions of root exudates are summarised in Table 7.2.

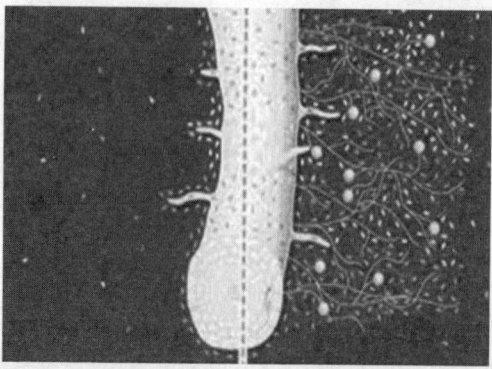

Fig. 7.6: Showing microbes interaction in plant's rhizospheric and non-rhizospheric zone

Table. 7.2: Organic compounds and enzymes released by plants in root exudates and their function in the rhizosphere

Class of compounds	Components	Functions
Carbohydrates	arabinose, glucose, fructose, galactose, maltose, raffinose, rhamnose, ribose, sucrose, xylose	lubrication, protection of plants against toxin, microbial growth stimulation
Amino acids and amides	all 20 proteinogenic amino acids, aminobutyric acid, homoserine, cystathionine, mugineic acid phytosiderophores	inhibit nematodes and root growth of different plant species, microbial growth stimulation, chemoattractants, osmoprotectants, iron scavengers
Aliphatic acids	formic, acetic, butyric, propionic, maleic, malic, citric, isocitric, oxalic, fumaric, malonic, succinic, tartaric, oxaloacetic, pyruvic, oxaloglutaric, glycolic, shikimic, acetonic, valeric, gluconic, quinic	plant growth regulation, chemoattractants, microbial growth stimulation
Aromatic acids	p-hydroxybenzoic, caffeic, pcoumeric, ferulic, gallic, gentisic, protocatechuic, salicylic, sinapic, syringic	plant growth regulation, chemoattractants
Phenolics	flavanol, flavones, acetosyringone, flavanones, anthocyanins, isoflavonoids	plant growth regulation, allelopathic interactions, plant defence, phytoalexins, chemoattractants, initiate legumerhizobia, arbuscular mycorrhizal and actinorhizal interactions, microbial growth stimulation, stimulate bacterial xenobiotic degradation
Fatty acids	linoleic, linolenic, oleic, palmitic, stearic acid	plant growth regulation
Vitamins	p-aminobenzoic acid, biotin, choline, n-methionylnicotinic acid, niacin, panthothenate, pyridoxine, riboflavin, thiamine	microbial growth stimulation
Sterols	campestrol, cholesterol, sitosterol, stigmasterol	plant growth regulation
Enzymes and proteins	amylase, invertase, phosphatase, polygalacturonase, protease, hydrolase, lectin	plant defence, Nod factor degradation
Hormones	auxin, ethylene and its precursor 1-aminocyclopropane-1-carboxylic acid, putrescine, jasmonate, salicylic acid	plant growth regulation
Miscellaneous	unidentified acyl homoserine lactone mimics, saponin, scopoletin, reactive oxygen species, nucleotides, calystegine, trigonelline, xanthone, strigolactones	quorum quenching, plant growth regulation, plant defence, microbial attachment, microbial growth stimulation, initiate arbuscular mycorrhizal interactions

This stimulatory rhizosphere effect has been recognized for many years and was described for the first time by Hiltner (1904), who defined the rhizosphere as the zone of soil in which microbes are influenced by the root system. In turn, rhizosphere organisms also have a large impact on plants, because many microbes isolated from the rhizosphere are described to have root growth-stimulating or growth-inhibiting properties. The composition of the microbial population in the rhizosphere depends on the composition of the root exudates as well as on the plant species, root type, plant age, soil type, and history of the soil. It is known that the rhizosphere is dominated by gram-negative rods such as *Pseudomonas spp.* The presence and survival of beneficial rhizobacterial strains is, in contrast to the limited studies of rhizoremediation, described in more detail for processes such as biocontrol of soilborne plant diseases, biofertilization (e.g. by nitrogen fixing bacteria), and phytostimulation. The success of these beneficial processes is based on the rhizosphere competence of the microbes, which is reflected by the ability of the microbes to survive in the rhizosphere, compete for the exudate nutrients, sustain in sufficient numbers, and efficiently colonize the growing root system.

To study the patterns of microbial plant root colonization, different tools such as microscopy, microscopy can be combined with the use of marked strains, or strains equipped with reporter genes for DGGE/RFLP analysis. During in-vitro studies, it has been shown that, upon inoculation of seed or seedlings, cells of *Pseudomonas* mainly appear as microcolonies along the junctions of epidermal plant cells where nutrients are exuded. In addition, most bacterial cells were detected on the upper root parts and bacterial numbers decreased in the direction of the root tip.

7.3 INTRACELLULAR ACCUMULATION

The plants utilize number of metal specific strategies to combat high external metal concentrations which are mainly classified into two categories, Avoidance (restriction to metal uptake) and Tolerance. Transport of the metal across the cell membrane yields intracellular accumulation, which is dependent on the cell's metabolism. This means that this kind of biosorption may take place only with viable cells. It is often associated with an active defense system of the microorganism, which reacts in the presence of toxic metal.

A) Avoidance:

This mechanism limits the uptake of heavy metals and prohibits their entry in plant tissues through root cells. The avoidance strategy is the first line of defense which mainly works at extracellular level through various mechanisms like immobilization by mycorrhizal association, complexation by root exudates, and modification of rhizosphere pH, exudation of metal-binding organic acids or formation of redox barrier.

i. **Immobilization by Mycorrhizal Associations:** The presence of mycorrhizal associations between fungi and roots of host plants in metal contaminated soils indicate an important relationship in plant tolerance and accumulation. However, the exact role was not established. Ectomycorrhizas and Arbuscular Mycorrhizas are the two most common mycorrhizal associations in plants growing on heavy metal contaminated soils. Mycorrhizas adopt absorption, adsorption or chelation mechanisms to restrict the entry of heavy metals in to host plant. They provide an effective exclusion barrier to metal uptake.

ii. **Root Exudates and Siderophore-metal binding:** Roots releases variety of exudates (like amino or organic acids, water, inorganic ions, sugars etc.), excretions (like bicarbonates, protons, carbon dioxide, etc.) and secretions (like mucilage, siderophores, allelopathic compounds, etc.) which collectively termed as root exudates and helped the plant to survive in metal contaminated areas. The root exudates form stable heavy metalligand complexes in the vicinity of the root thus making them unavailable and lessening the toxicity. Some root exudates increase the pH of rhizosphere which precipitates metals and limits their bioavailabilty.

Siderophores are extracellular, low molecular weight metal chelating agents. These are either catechol or hydroxamate derivatives and contain certain reactive groups like dicarbxylic acids, polyhydroxy acids and phenolic compounds. In addition to iron, siderophores often bind to metal ions like aluminium, gallium, chromium, nickel etc. Specific extracellular metal-binding compounds can also be produced by microorganisms in response to low levels of metals, in order to facilitate the uptake of essential metals. The most studied system is the production of siderophores in response to low environmental iron concentrations. Siderophores are low-molecular-mass Fe(III) coordination compounds (500 ± 1000 Da.) produced by many micro-organisms and act by complexing and solubilizing insoluble Fe(III) in a form which can be transported into the cell using specific transport mechanisms. Although siderophores are iron(III)-binding compounds, they are also able to bind other metals such as magnesium, manganese, chromium(III), gallium(III) and radionuclides such as plutonium(IV). The efficiencies of siderophores can be increased by chemical modification like substituting Cl^-, NO_2^- on benzene ring. Modified siderophores can be used for removal of metals like cadmium, mercury, copper and even radionuclides like strontium or cesium from mixed effluent.

Siderophores are the largest class of known compounds that can bind and transport, or shuttle, Fe. They are highly specific Fe(III) ligands. Iron is essential for most forms of life. It is required for the catalytic activity of proteins mediating electron transfer and redox reactions, such as those involved in respiration, photosynthesis, DNA synthesis, and defense against reactive oxygen species. However, it is often unavailable because it is present as insoluble ferric hydroxide complexes in aerobiosis and at neutral pH. In its ferrous form, iron is more soluble and catalyzes the Fenton reaction in the presence of hydrogen peroxide, which leads to the formation of hydroxyl radicals, resulting in protein denaturation, DNA breaks, and lipid peroxidation. Therefore, iron acquisition, utilization, and storage are subject to different levels of homeostatic regulation.

These low-molecular-mass coordination molecules are excreted by a wide variety of fungi and bacteria to aid Fe assimilation. Although the mechanism could be used to acquire other metals, Fe is the only known essential element for which these specific organic shuttles operate. This is probably because Fe is needed in larger amounts by cells than other poorly soluble metals, and given the low solubility-product constant of ferric hydroxide, the concentration of free Fe^{3+} is too low to support microbial growth at pH values where most life exists. Organisms have most likely evolved mechanisms to ensure that Fe demand is met through the production of species-specific siderophores, or by attachment to a solid Fe mineral, e.g. Fe oxides, to shorten the pathway between the Fe substrate and cellular site of uptake. Siderophores can complex other metals apart from iron, in particular actinides. Because of such metal-binding abilities, there are potential applications for siderophores in medicine, reprocessing of nuclear fuel, bioremediation of metal-contaminated sites and treatment of industrial wastes. The production of siderophores by microorganisms is beneficial to plants, because it can inhibit the growth of plant pathogens.

In the last 10 years, much has been learned about the mechanism of iron acquisition and transport in Gram-negative bacteria, especially *Escherichia coli*. In these bacteria, the ferric-siderophore complex must cross the outer membrane (OM) and the cytoplasmic membrane (CM) before delivering iron within the cytoplasm. The ferric complexes are too large for passive diffusion or nonspecific transport across these membranes; ferric siderophore uptake is both receptor and energy dependent. Translocation of iron through the bacterial outer membrane as the ferric-siderophore requires the formation of an energy transducing complex with the proteins TonB, ExbB, and ExbD, which couple the electrochemical gradient across the cytoplasmic membrane to a highly specific receptor and so promote transport of the iron complex across the outer membrane. Once in the periplasmic space (PP), the ferric-siderophore binds to its cognate periplasmic binding protein and is then actively transported across the cytoplasmic membrane by an ATP-transporter system. Once in the cytoplasm of the cell, the Fe^{3+} – siderophore complex is usually reduced to Fe^{2+} to release the iron, especially in the case of "weaker" siderophore ligands such as hydroxamates and carboxylates. Siderophore decomposition or other biological mechanisms can also release iron, especially in the case of catecholates such as ferric-enterobactin, whose reduction potential is too low for reducing agents such as flavin adenine dinucleotide, hence enzymatic degradation is needed to release the iron (Fig. 7.7).

Although there is sufficient iron in most soils for plant growth, plant iron deficiency is a problem in calcareous soil, due to the low solubility of iron(III) hydroxide. Calcareous soil accounts for 30% of the world's farmland. Under such conditions graminaceous plants (grasses, cereals and rice) secrete phytosiderophores into the soil, a typical example being deoxymugineic acid.

Deoxymugineic acid

Nicotianamine

Phytosiderophores have a different structure to those of fungal and bacterial siderophores having two α-aminocarboxylate binding centres, together with a single α-hydroxycarboxylate unit. This latter bidentate function provides phytosiderophores with a high selectivity for iron(III). When grown in an iron deficient soil, roots of graminaceous plants secrete siderophores into the rhizosphere. On scavenging iron(III) the iron-phytosiderophore complex is transported across the cytoplasmic membrane using a proton symport mechanism. The iron(III) complex is then reduced to iron(II) and the iron is transferred to nicotianamine, which although very similar to the phytosiderophores is selective for iron(II) and is not secreted by the roots. Nicotianamine translocates iron in phloem of all plants.

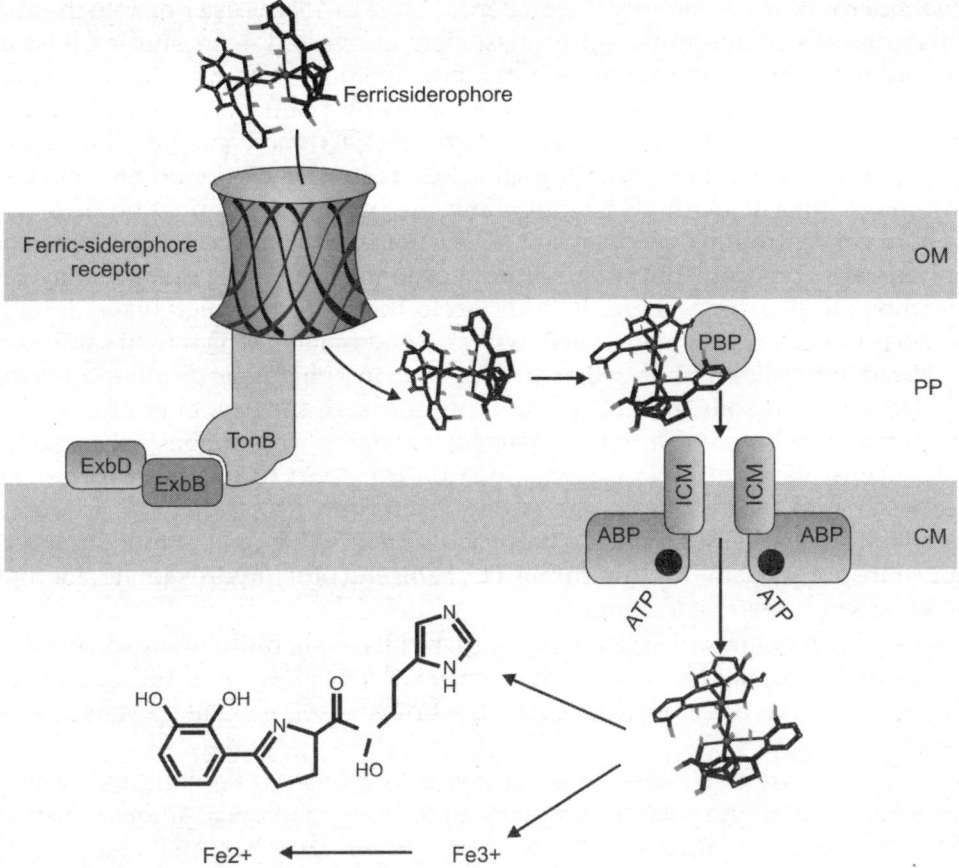

Fig. 7.7: Transformation of iron through siderophore

Over 80 different siderophores of bacteria and fungi have been isolated. Among these, plants use the hydroxamate siderophores, ferrichrome (FC), ferrioxamine B (FOB), and rhodotorulic acid (RA); the catechol siderophore, agrobactin; and the mixed ligand, catechol-hydroxamate-hydroxy-acid siderophores produced by Pseudomonas. However, not all siderophores may be used by plants, and individual plant species and varieties have different abilities to utilize specific siderophore types. Such differences suggest there are at least two mechanisms for siderophore-mediated iron transport. To acquire iron from siderophores, microorganisms employ receptor protein complexes that bind and transport the intact iron chelate into the cell or remove iron from siderophores at the cell surface. Though less understood, plants might also have siderophore-iron transport systems or, under reduced soil conditions, might acquire inorganic iron buffered by dissociation of siderophores in the plant rhizosphere. Prior research on plant iron nutrition has examined synthetic chelates that donate iron to plant roots by extracellular dissociation. Dissociation occurs when plants alter the redox and pH conditions of the rhizosphere to increase solubility of inorganic iron. Also, iron efficient dicotyledonous plants may remove ferric iron from synthetic chelates with a cell surface reductase that releases inorganic ferrous iron into the root apoplast. In contrast, plants appear to utilize microbial siderophores differently since siderophores do not release iron into the apoplast of dicotyledonous plants nor readily dissociate above pH 4. In studies postulating siderophore transport, Smarrelli and Castignetti (1986) have shown that inorganic iron may be released from hydroxamate siderophores by plant cytosolic NADH: nitrate reductase. Alternatively, Romheld and Marschner (1983) have suggested a model that involves a plasmalemma chelate binding site with a reductase-carrier protein for inorganic iron. Although current methods do not allow the precise determination of individual siderophore concentrations, bioassays of soil-water extracts indicate that hydroxamate siderophores are a predominant siderophore-type in soils and those they occur in elevated concentrations in the rhizosphere. In addition to the milieu of microbial siderophores, plants also produce organic acids, such as citrate and malate, which may solubilize iron at low pH and 'phytosiderophores' that may function in iron uptake by monocotyledonous grasses. However, in soils, microbial siderophores have much higher affinity for iron than plant-produced organic acids or phytosiderophores. This suggests that plants, like microorganisms, have multiple systems for iron transport to use the various chelates that sequester iron. For example, the enteric bacterium *Escherichia coli* produces and transports the catechol-siderophore, enterobactin, but also has at least four other separate receptor/transport systems for iron citrate, FC, FOB, and other hydroxamate siderophores produced by soil bacteria and fungi.

The total Fe content in soil is relatively high, but its availability to organisms is low in aerated soils because the prevalent form (Fe^{3+}) is poorly soluble. Plants and microorganisms have developed mechanisms to increase Fe uptake (Marschner 1995) (Fig. 7.8).

In plants, there are two different strategies in response to Fe deficiency. Strategy I plants (dicots and non-graminaceous monocots) release organic acid anions which chelate Fe. Iron solubility is also increased by decreasing the rhizosphere pH, and Fe uptake is enhanced by an increased reducing capacity of the roots ($Fe^{3+} \rightarrow Fe^{2+}$). Strategy II plants (Poaceae) release phytosiderophores that chelate Fe^{3+}. Iron is taken up in the chelated form as Fe-phytosiderophore. Phytosiderophores are released only for a few hours per day at the root tip. Under Fe deficiency stress, microorganisms release organic acid anions or siderophores that chelate Fe^{3+}. After movement of the ferrated chelate to the cell surface, Fe^{3+} is reduced either outside or within the cell.

The abundance of microbial siderophores in many soils, along with their outstanding Fe binding capacity, has raised the question of whether these siderophores may be used by plants as an Fe source. A number of researchers investigated the possibility of plant utilization of various microbial siderophores including ferrichrome A, agrobactin, pseudobactin and ferrioxamine B. A variety of plant species were shown to acquire Fe from ferrioxamine B, such as tomato, oat, sunflower and sorghum, and peanut and cotton. The mechanism by which plants utilize microbial siderophores has not yet been elaborated to a satisfying extent and it is a topic of controversy. Although some investigators proposed a specific Fe uptake mechanism, others did not support this hypothesis. Fe uptake from ferrioxamine B by maize (*Zea mays L.*) and cotton (*Gossypium spp.*) plants was investigated using the synthetic fluorescent ferrioxamine B analog NBD-desferrioxamine B (NBD-DFO), an analog of the natural siderphore ferrioxamine B as probe. NBD-DFO has been designed to be nonfluorescent when saturated with Fe^{3+}, but to gain fluorescence upon Fe^{3+} release. NBD-DFO, therefore, allows one to monitor the location and the dynamics of cellular Fe

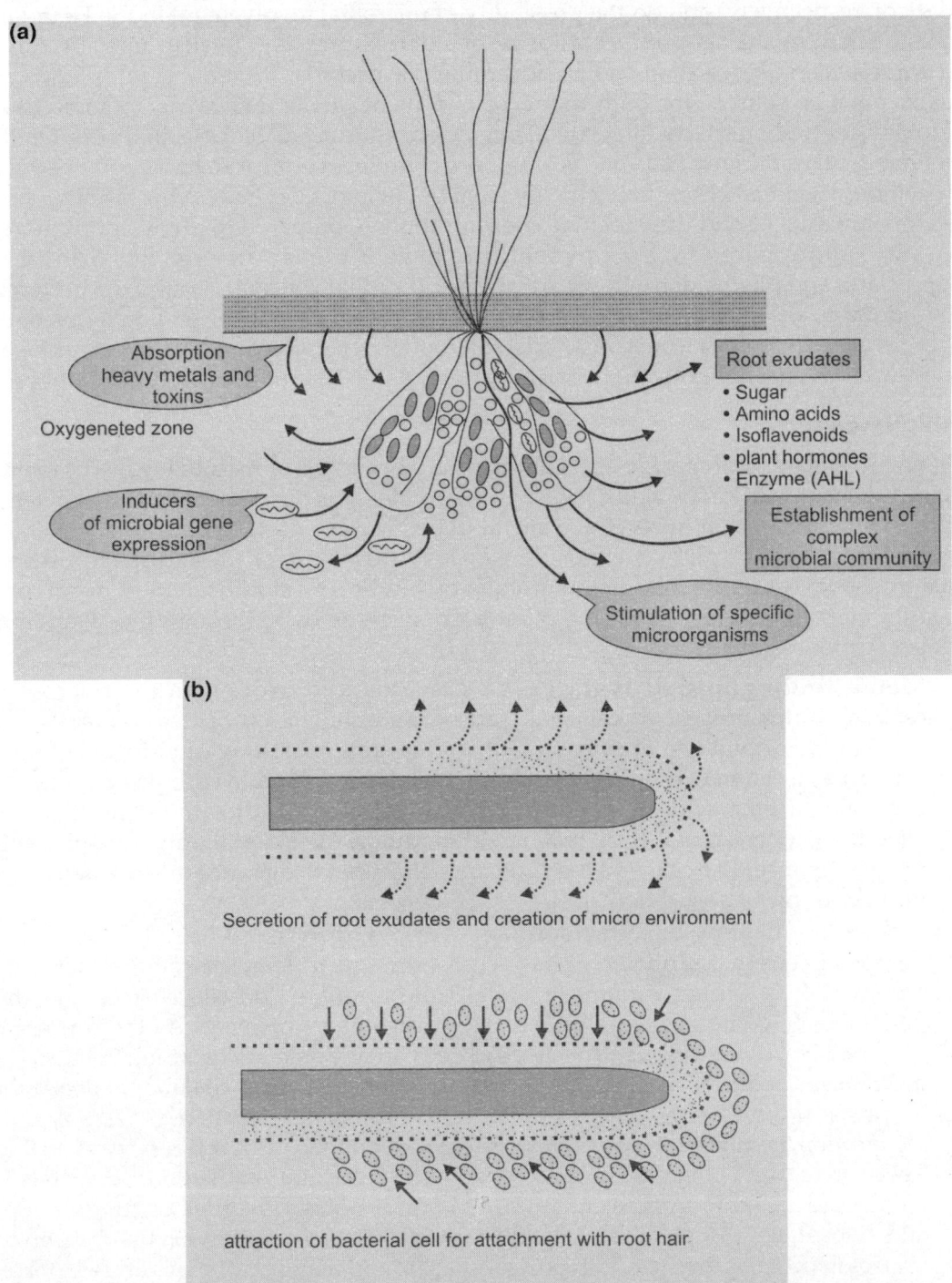

Fig. 7.8: (a) Creation of micro-environment in wetland plants (b) attraction of bacterial cell for attachment with root hair

uptake. This research focused on the uptake mechanism in maize and cotton (monocot and dicot, respectively) and on the possibility of microbial involvement in the Fe uptake process. Microorganisms produce a range of siderophores, e.g. ferrichromes by fungi, and enterobactin, pyoverdine and ferrioxamines by bacteria. Bacterial siderophores are usually poor Fe sources for both monocot and dicot plants. However, in some cases microbial siderophores have alleviated Fe deficiency-induced chlorosis in dicots. On the other hand, plant-derived Fe-phytosiderophore complexes appear to be a good Fe source for bacteria. The interactions between different Fe chelators depend on the affinity of the chelators towards Fe and their relative concentrations. Compared to phytosiderophores, bacterial siderophores such as pyoverdine have a higher affinity towards Fe. If siderophores and phytosiderophores are present at similar concentrations, Fe is preferentially bound to the siderophores, which may even remove Fe from the Fe-phytosiderophore complex. In contrast to many bacterial siderophores, rhizoferrin from the fungus *Rhizopus arrhizus* has only a slightly higher affinity towards Fe compared to phytosiderophores.

B) Tolerance:

Tolerance mechanism is capable of accumulating, storing and immobilizing heavy metals by binding them with amino acids, proteins or peptides. It is plant's second line of defense which mainly focuses on intracellular metal detoxification. Metal tolerance is generally achieved by two strategies on exposure to heavy metals plant firstly tries to prevent metal transport across the plasma membrane by binding or modification of metal ions. Secondly metal ions which enter the plant body were detoxify through inactivation or converted into less toxic form.

i. **Metal Binding to Cell Wall:** Pectic sites, histidyl groups and extracellular carbohydrates present in cell wall caused immobilization of heavy metals and prevent their uptake in to cytosol. Cell wall pectins consisting of polygalacturonic acid act as a cation exchanger. The heavy metals are bound to carboxylic groups of polygalacturonic acids restricting the plant uptake. Studies indicated that the chemical properties of the cell wall might modulate metal uptake and consequently metal tolerance. The cell wall has little impact on metal tolerance due to inadequate number of metal absorption sites.

Cell surface binding or rather sorption of heavy metals and of radionuclides by microbial cells is essentially a passive process and it does not require metabolic energy. Often a huge amount of metal is accumulated on cell surface and this phenomenon could easily be utilised for wastewater treatment. As the process is very passive, living cells as well as dead cells are capable of absorbing metals. The mechanism of biosorption is complex, mainly ion exchange, chelation, adsorption by physical forces, entrapment in inter and intrafibrilliar capillaries. This type of biosorption, i.e. non-metabolism dependent is relatively quick and can be reversible. There are several chemical groups that would draw and requisition the metals in biomass: acetamido groups of chitin, structural polysaccharides of fungi, amino and phosphate groups in nucleic acids, amido, amino, sulphhydryl and carboxyl groups in proteins, hydroxyls in polysaccharide and mainly carboxyls and sulphates in polysaccharides of marine algae. Different functional groups like amino, hydroxo, thio, phospho etc. groups on the surface of phytoplankton can interact with metal ions. Physical adsorption takes place with the help of van der Waals' forces uranium, cadmium, zinc, copper and cobalt biosorption by dead biomasses of algae, fungi and yeasts takes place through electrostatic interactions between the metal ions in solutions and cell walls of microbial cells. Electrostatic interactions have been demonstrated to be responsible for copper biosorption by bacterium *Zoogloea ramigera* and alga *Chiarella vulgaris*, for chromium biosorption by fungi *Ganoderma lucidum* and *Aspergillus niger*. Microorganisms may also generate organic acids (e.g. citric, oxalic, gluonic, fumaric, lactic and malic acids), which may chelate toxic metals result in the formation of metallo-organic molecules. These organic acids facilitate in the solubilisation of metal compounds and their leaching from their surfaces. Unicellular yeast *Saccharomyses cerevisae* adsorb uranium on cell wall as needle like fibrils in a 0.2 μm thick layer. Some chemical compounds of yeast cells can also act as ion exchangers with rapid reversible binding of cations. The bacteria make brilliant biosorbents because of their high surface-to-volume ratios and a high content of potentially active chemosorption sites such as on teichoic acid in their cell walls. The cell walls of gram-positive bacteria are efficient metal chelators. The carboxylic groups of glutamic acid and teichoic and teichuronic acids of peptidoglycan are metal binding sites. Filamentous bacteria *Thiothrix* sp. can adsorb metals in a very high rate on cell surface. In India, several good works on this aspect have been done on bacteria. In all the cases the cell-surface biosorbtion can be optimised by pH change, temperature, biomass concentration, pre-treatment and presence of other heavy metals. pH seems to be the most significant parameter in the biosorptive process: it affects the solution chemistry of the metals, the action of the functional groups in the biomass and the competition of metallic ions. The presence of Fe^{2+} and Zn^{2+} was found to manipulate uranium uptake by *Rhizopus arrhizus* and cobalt uptake by different microorganisms seemed to be entirely inhibited by the presence of uranium, lead, mercury and copper. Metal affinity to the biomass can be manipulated by pretreating the biomass with alkalies, acids, detergents and heat, which may amplify the amount of the metal absorbed. A variety of extra cellular polysaccharidic polymer or metallothionein like proteins are able to bind to metal ions. The purpose is mainly to reduce metal toxicity. Metals are reported to bind with these biopolymers by following methods-firstly, metals bind to the electronegative functional groups present in biopolymer. These groups include pyruvate, hydroxyl, phosphates groups; secondly, some functional groups like thiol group may form co-ordination complexes with metal ions. Expression of metallothioneins or metallopeptides to increase the affinity and biosorptive capabilities of bacterial cells for heavy metals is a promising technology for the development of bacterium-based biosorbents. Both naturally occurring metal-binding peptides such as metallothioneins (MTs) and synthetic peptides such as synthetic phytochelatins (ECs) have been expressed on the surface of bacterial cells for better uptake and biosorption of mercury. Various bacteria develop resistance to heavy metals by inducing the expression of an array of resistance proteins. Some microbially produced macromolecules can bind considerable amounts of potentially toxic metals. These include humic and fulvic acids arising from lignocelluloses degradation and extracellular polymeric substances (EPS), a mixture of polysac-charides, mucopolysaccharides and proteins produced by bacteria, algae and fungi. Some microorganisms are known to produce extracellular, specific metal-binding compounds. Examples include *Sarcina urea*, which has been shown to produce metal binding proteins, a *Pseudomonas sp.* which has been shown to produce cadmium-binding proteins in response to cadmium and *Vibrio alginolyticus*, which produces copper-binding proteins in response to high levels of copper. Many bacteria produce polysaccharidic exopolymers which are either loosely associated with cell wall or present in a gelatinous matrix surrounding the cell. These acidic polymers are polar, hydrophilic and bind metals very effectively. Exopolymer of *Zooglea* is one such instance, used for metal removal by activated sludge method. From fresh water system different bacteria were reported to immobilise metal ions by exopolymers in association with clay. Cyanobacteria often produce polysacchiridic biofloculants when challenged with metals. Cysteine rich polypeptides, i.e. metallothioneins bind essential and nonessential metals which are either stored as intracellular granules or transported out of the cell. An extracellular rhamnolipid biosurfactant has been shown to complex metals like cadmium, lead and zinc. *Desulfococcus multivorans* was shown to have the excellent extracellular copper-binding capacity. It binds copper reversibly and zinc irreversibly.

ii. **Active Efflux Pumping at Plasma:** To produce tolerance, membrane showed active efflux of the metal which lowers the intracellular concentration. This mechanism is well documented in bacteria and animal cells. P1B-ATPases which belongs to P-type ATPase superfamily and ATP-binding cassette (ABC) transporters are mainly worked as heavy metal efflux pumps in plants. The P-type ATPases, also known as E1–E2 ATPases, are a large group of evolutionarily related ion and lipid pumps that are found in bacteria, archaea, and eukaryotes. They are α-helical bundle primary transporters referred to as P-type ATPases because they catalyze auto (or self) phosphorylation of a key conserved aspartate residue within the pump. In addition, they all appear to interconvert between at least two different conformations, denoted by E1 and E2. While, ATP-binding cassette transporters (ABC transporters) are members of a protein superfamily that is one of the largest and most ancient families with representatives in all extant phyla from prokaryotes to humans. ABC transporters are transmembrane proteins that utilize the energy of adenosine triphosphate (ATP) hydrolysis to carry out certain biological processes including translocation of various substrates across membranes and non-transport-related processes such as translation of RNA and DNA repair. They transport a wide variety of substrates across extra and across extra and intracellular membranes, including metabolic products, lipids and sterols, and drugs. Proteins are classified as ABC transporters based on the sequence and organization of their ATP-binding cassette (ABC) domain(s). ABC transporters are involved in tumor resistance, cystic fibrosis and a range of other inherited human diseases along with both bacterial (prokaryotic) and eukaryotic (including human) development of resistance to multiple drugs. The two subfamilies of ABC transporters namely Multidrug Resistance associated Protein (MRP) and Pleiotropic Drug Resistance (PDR) are particularly active in the sequestration of chelated heavy metals. Only a few evident indications of plasma membrane efflux transporters are reported in plants.

iii. **Organic Acids:** Organic acids within cells act to detoxify metals by complexing and making them unavailable to the plant. It acts as metabolic intermediates in the formation of ATP from carbohydrates in nitrogen metabolism and in ionic balance.

Hence, metabolic abnormalities in any of these processes would be reflected by changes in the concentrations of organic acids. Therefore, an increase in organic acids with increasing supply of metals could imply a detoxification mechanism or conversely, disruption of metabolism resulting in the production of organic acids as a stress response to excess metal.

iv. **Inactivation of Toxic Metals:** At cytoplasmic level phytochelatins and/or metallothioneins play major role to provide metal tolerance. The metalphyto-chelatins/metallothioneins complex is actively transported from the cytosol across the tonoplast into the vacuole where it is store without any toxicity.

   a. **Phytochelatins (PCs):** Phytochelatins are derived from glutathione (GSH) and possess the general structure (g-Glu-Cys) n-Gly (where n = 2 to 11). On activation in presence of heavy metals PC synthase enzyme produced PCs by trans-peptidation of g-glutamyl-cysteinyl dipeptides from GSH. The cystein part of PCs carry out metal co-ordination while thiol group and gluatmic acid provides water solubility to PCs. Among the various heavy metals Cd is strongest inducer of PCs. Not all the metals which trigger the PCs synthesis are able to form complexes with it. Generally PCs are associated with both homeostasis and trafficking of essential metals as well as detoxification of heavy metals. Once the PC-metal complex was formed, it transported to vacuoles by metal/H+ antiporters or ABC transporters isolating the toxic metals from metal sensitive enzymes. In vacuoles, PC metal complexes become more resistant to proteolytic degradation on incorporation of inorganic sulfide and sulfite ions. Under favorable conditions metals from PC-metalsulfide complexes are liberated causing PC degradation.

   b. **Metallothioneins (MTs):** Metallothioneins are cysteine rich proteins induced by various abiotic stresses having molecular weight between 5 and 20 kDa and mercaptide groups which causes metal ion binding with thiols. Based on structure and content of cysteine they are divided in to MT1 and MT2. Class 1 MTs contain cysteine motifs that align with mammalian MTs, whereas Class 2 MTs contain similar cysteine clusters but they do not align with Class1 MTs.

   c. **Other Alternatives:** Under heavy metal stress plants also activate oxidative stress defense mechanisms and the synthesis of stress-related proteins such as heat shock proteins, reactive oxygen species and hormones.

      i. **Stress related proteins:** With heavy metal exposure, most of the plants trigger the synthesis of sets of novel proteins. Most of the proteins endowed plasma membranes to act as barrier for metal inflow which leads to metal homeostasis and detoxification. The common stress related Heat Shock Proteins (HSP), act as molecular chaperones and help in normal protein folding and assembly. It may also function in the protection and repair of proteins under stressful conditions. The synthesis of HSP under heavy metal stress has been observed in different plants but its role is largely unknown.

      ii. **Antioxidant Defense and Reactive Oxygen Species:** The increased synthesis of reactive oxygen species (ROS) like superoxide radicals (O_2^-), hydroxyl radicals (OH.) and hydrogen peroxide (H_2O_2) is one of the initial responses to heavy metal stress. These species are continuously produced at low level during normal metabolic processes. ROS, particularly H_2O_2 plays an important role as intermediate signaling molecules to regulate defense system. ROS have a dual function: higher concentrations, damages the cells; but at moderate levels, they help to adapt stress by induction of an antioxidant response. A complex ROS scavenging mechanism at the molecular and cellular levels decreases oxidative damage with increased resistance to heavy metals.

iii. **Hormones:** Peleg and Blumwald (2011) suggest that the regulation of hormone synthesis in presence of heavy metals indicates that plant hormones play a crucial role in the adaptation to metal stress. The hormones such as salicylic acid, jasmonic acid, ethylene, gibberellic acid are implicated in plant defense signaling pathways. Jasmonic acid treatment increased the capacity for glutathione synthesis which plays central role in protecting plants from heavy metal stress. Salicylic acid activates defense related genes either by H_2O_2 mediated signal transduction pathway or by directly affecting mechanisms of metal detoxification by inhibiting. Salicylic acid inhibits two major H_2O_2 scavanging enzymes catalase and ascorbate peroxidase which causes cellular H_2O_2 concentration to rise and subsequently acts as second messenger. Heavy metal stress induces ethylene biosynthesis acts as endogenous signal triggering the plant defense response.

7.4 OXIDATION-REDUCTION REACTIONS

Ferrous and manganese ions can be deposited by oxidation reactions of bacteria such as *Leptothrix*, *Sphaerotilus* and *Thiobacillus*. These reactions often solubilise certain metals, apparently increasing their biotoxicity. But however, it is only then possible for the bacteria to remove that metal from the surrounding environment. This process can be well utilised in recovery of metals like copper, cadmium, gold, etc. Micro-organisms can actively reduce toxicity or can even decrease bioavaillbility by reduction of metals. Such metals include mercury, iron, chromium, arsenic, selenium, etc. Several bacteria like *Pseudomonas putida*, *Escherichia coli* are known to reduce hexavalent chromium to the non-toxic trivalent form. Yeasts and fungi are recently reported to reduce chromium VI too.

7.5. ALKYLATION REACTIONS

Alkylation is another effective detoxification mechanism often adapted by micro-organisms. Tin, lead, selenium, etc. can be alkylated by incorporation of methyl groups which actually increases toxicity. But methylation facilitates diffusion of that metal away from the surroundings of the microbes. Methylation of mercury leading to volatilisation can be a significant mechanism of removal of mercury from contaminated pond. However, it bears serious health implications as methylation, in all cases, leads to high toxicity.

7.6 PRACTICAL APPLICATIONS OF MICROBIAL BIOTECHNOLOGY

It is clear that the interactions between microbial cells and heavy metals or nuclides are extensive. Bioremediational processes for metal removal or recovery from liquid effluent are based on these interactions, specially biosorption and biotransformation. Though in most of the cases field based trials for verification of effectiveness of these processes under practical circumstances have not been done due to infrastructural and financial constraints. Still many reports are available revealing more direct application oriented systems. These systems may incorporate both living and nonliving micro-organisms. The use of adsorbents of biological origin has emerged in the last two decade as one of the most promising alternatives to conventional heavy metal remediation strategies. In many path-breaking studies, metal removal abilities of various species of bacteria, algae, fungi and yeasts were investigated. Biosorption consists of several mechanisms as discussed above, mainly ion exchange, chelation, adsorption, and diffusion through cell walls and membranes, which differ depending on the species used, the source and processing of the biomass and solution chemistry. The exact mechanism by which micro-organisms take up metals may vary, but it has been demonstrated that both living and non-living fungal biomass may be utilised in biosorptive processes, as they often exhibit distinct tolerance towards metals and other adverse conditions such as low pH. There are many environmental factors affecting the biosorption process, such as temperature, pH, agitation rate and metal concentration. Some of these factors (e.g. pH and metal concentration) have greater influence on metal removal by this process. In living systems, biocatalysts like bacteria, algae or fungi; specially selected, enriched or genetically engineered are applied in a controlled artificial ecosystem in scientifically designed meander channels or impounds. This process is needed to be done at the site of production of such effluent, i.e. *in situ*. Biological agents may be individual strains or a complex microbial community. Biomass can come from (i) industrial spent biomass; (ii) organisms easily available in large amounts in nature; and (iii) organisms of quick growth, especially and easily cultivated or propagated for biosorption purposes. Cost effectiveness is the main attraction of metal biosorption, and it should be kept in mind for biomass election. Most studies of biosorption for metal removal have involved the use of either laboratory-grown micro-organisms or biomass generated by the pharmaceutical and food processing industries or wastewater treatment units. If, for any reason, by-products of industrial fermentation processes would not be available, biosorbents could be produced by using relatively unsophisticated and low-cost culture propagation techniques. Nutrients from readily available and economical sources such as carbohydrate-rich industrial wastewaters, which often pose pollution/treatment problems, such as food, dairy and starch industries, might be conveniently used. On the contrary, the costs of biosorbents especially produced could be higher and affect negatively the overall economy of their application. Some chemical compounds of yeast cells can act as ion exchangers with rapid reversible binding of cations. Volesky et al. (1993) working on cadmium biosorption by *Saccharomyces cerevisiae* established that this yeast is a reasonably effective biosorbent for cadmium. Some mucoralean fungi have shown fascinating metal biosorbent properties, particularly high for uranium and thorium whereby different metal deposition patterns could be clearly distinguished. Although microalgae are not unique in their bioremoval capabilities, selected microalgae strains, purposefully cultivated and processed for specific bioremoval applications, have the potential to provide significant improvements. The marine flagellate alga *Pavlova viridis* grown in an artificial seawater medium can uptake metals in the order Ni > Pb > Co > Hg > Cu > Cd > Ag at equilibrium. There is much evidence that algae could accumulate heavy metals in their tissues when grown in polluted waters, including the species *Ulva rigida* (Fe, Mn), *Padina gymnospora* (Zn), *Gracilaria tenuistipitata* (Cd), *Undaria pinnatifida* (Pb), *Cladophora* sp. (Cd), and *Cladophora glomerata* (Zn). The most relevant work on true bacterial biosorption has been done by the Brierley, who took the metal biosorption concept all the way to the commercial stage. Bacterial cell walls are negatively charged under acidic pH conditions and the functional groups of cell wall display a high affinity for metal ions in solution. Churchill et al. used two Gram-negative strains *Escherichia coli* K-12 and *Pseudomonas aeruginosa* and a Gram-positive strain *Micrococcus luteus* to demonstrate biosorption of Cu^{2+}, Cr^{3+}, Co^{2+} and Ni^{2+} ions. Their sorption binding constants suggested that *E. coli* cells were the most competent at binding copper, chromium and nickel and *M. luteus* sorbed cobalt most efficiently. In nonliving systems, only the biosorption capabilities of different micro-organisms have been used in specially designed bioreactors. The application of nonliving biomass for removal of metals by biosorption has several advantages. No nutrients or proper maintenance are required. Toxicity often cause loss of accumulation property in living cells, but in case of nonliving systems there is no scope of toxicity and hence no loss in biosorption rate. So we can use the same biomass for treatment purpose time and again. Hence, many works have been done to perfect the bioreactor system and to optimize efficiency. The complete design of the system varies with the process application, sorption agent and capital expenditure. The main target is to achieve maximum surface contact between metal and biosorbent material. System may operate in batch, semi-continuous or continuous flow in stirred tank or column reactor. Natural biomass has several disadvantages for use. So natural biomass of the selected strains are often modified in such a way so as to give it more mechanical strength, more hydrophilicity and uniformity in size and form. These modifications enable the biosorbent to be reused time and again. Immobilisation of biosorbent in an insoluble natural or synthetic polymer, is an effective solution. The materials used are polyurethane foam, silica gel, calcium alginate, polyacrylamide, polysulfone, polyethylenimine, formaldehyde and divinylsulfone. Several immobilized biosorbents are available in commercial scale, e.g. AMT-BIOCLAIM$_{TM}$, AlgaSORBTM, BIO-FIX$_{TM}$, etc. AMT-BIOCLAIM$_{TM}$, i.e. encapsulated *Bacillus* can simultaneously remove several heavy metals like U, Pb, Cd, Ni, Hg, Cr, Zn and Cu from effluent regardless of the initial low or high concentration of metals. An *Aspergillus oryzae* biosorbent is able to remove 90% cadmium from wastewater in only five minutes time in column reactor. Algal biosorbent like that of *Chlorella vulgaris* is also in use. Heavy metal recovery from biosorbents is of major importance in the assessment of competitiveness of biosorption processes. Understanding of the mechanisms of metal biosorption now allows the process to be scaled up and used in field applications, with packed-bed sorption columns being perhaps the most efficient for this purpose. Regenerating the biosorbents increases the process economy by allowing their reuse in multiple sorption cycles. Recovery allows metal recycling, leading to energy savings and materials conservation. Finally, biosorbent regeneration for use in multiple adsorption-desorption cycles, contributes to process cost effectiveness. The effectiveness of metal recovery depends on selection of eluant and elution conditions, as various eluants presenting different desorption mechanisms may be used. Lowering pH (e.g. with mineral acids) causes metal desorption, resulting from competition between protons and metal ions for binding sites. Mineral acids such as hydrochloric acid, sulphuric acid, acetic acid and nitric acid are proficient desorption agents. The strong chelating agent EDTA is another eluant commonly used. Despite several difficulties, living systems are also in use of metal removal in trickling filters or activated sludge. Living cells of bacteria, algae and fungi are also immobilised to form a fixed biofilm. Regeneration of living biomass in this case is very difficult as internal uptake of metals occur. The removal efficiencies of metals in trickling filters and activated sludge are comparable. But efficiency varies with metal type, e.g. removal of Cu, Pb, Zn is far better than that of Ni or Co.

7.A Ex-situ heavy metals phytoremediation potential of wetland plant: a case study

There is evidence that wetland plants such as cattail (*Typha latifolia*), common reed (*Phragmites australis*), and motha (*Cyperus malaccensis*) can accumulate heavy metals in their tissues. Cattail and common reed have been successfully used for the phytoremediation of Pb/Zn mine tailings under waterlogged conditions. Various other wetland plants are screened for heavy metals accumulation from the natural wetland by several workers. Further, the use of wetland plants in constructed wetland ecosystem for remediation of wastewater has recently drawn global attention for the systematic study of the potential wetland plants for environment management. The metal accumulation ultimately affects the growth and physiology of the plants, which has been reported by several workers. However, the higher concentration of metal can also adversely affect cellular parts of the plants. The heavy metals, i.e. Cd, Cu, Cr, Mn, Fe, Pb, Ni, and Zn are common pollutants, discharged from various industries of India. The frequent potential growth of wetland plants, i.e. *P. cummunis*, *T. angustifolia*, and *C. esculentus* on industrial wastewater contaminated site has given a strong evidence for phytoremediation of heavy metals from polluted environment. Considering the fact that *P. cummunis*, *T. angustifolia* and *C. esculentus* are resistant to polluted environment and have the capability to accumulate heavy metals in their tissue from contaminated wastewater. In the study the comparative accumulation potential of *P. cummunis*, *T. angustifolia* and *C. esculentus* in the mixed heavy metals solution has been shown to explore the safe possible use of these plants for phytoremediation of heavy metals from polluted wastewater. Phytotoxicity of heavy metals on physiology and histology of these plants were also observed (Sangeeta and Chandra, 2011).

7.A.1 Experimental design for ex-situ heavy metals accumulation

In the case study experiments were conducted by constructing the wetland ditch with the experimental units (0.60 m × 0.60 m × 1.20 m; width × length × depth) in open field. The bottom of experimental unit was sealed with cemented material, 30 cm of small stone pieces column was filled at the bottom of each unit followed by 30 cm of gravel in middle and 30 cm with sand at the uppermost layer for the plants support. Ten different young rhizome buds of three different wetland plants (*P. cummunis*, *T. angustifolia*, and *C. esculentus*) were planted in each experimental unit and acclimatized with Hoagland solution (Hoagland and Arnon 1938) until the height become higher than 0.3 m. The stock solutions of Cd, Cr, Cu, Fe, Mn, Ni, Pb, and Zn were prepared from analytical grade pure reagents (Merck, Darmstadt, Germany) in Hoagland using their salts, viz, $CdCl_2$, $K_2Cr_2O_7$, $CuSO_4.5H_2O$, $FeCl_3$, $MnCl_2.4H_2O$, $NiSO_4.6H_2O$, $Pb(NO_3)_2$ and $ZnCl_2$ respectively. The concentration of different heavy metals in the prepared solution was (mg/L) Cd, 8.00; Cr, 2.43; Cu, 48.00; Fe, 296.32; Mn, 20.54; Ni, 16.00; Pb, 33.92, and Zn, 26.30. The metals and their concentration were chosen on the basis of their presence and tolerance limit of these wetland plants growing in the natural wetland system contaminated with agro-based industrial wastewater (Kumar and Chandra, 2004). After the plant acclimatization, all experimental plots containing Hoagland media was replaced by metal solution (25 L). These solutions were replaced at 4 days interval to maintain the given metal concentration and water level up to 56 days. The plants grown in the pot without metal solution served as control.

7.A.2 Comparative heavy metal accumulation and Distribution of Heavy Metals in Different Parts of Wetland Plants

For heavy metal Analysis the Wetland Plants, Wetland plants were gently uprooted and washed thoroughly to remove sand clinging to the roots using distilled water followed by a 10 mmol/L solution of $CaCl_2$. The plants were separated into roots, shoots and leaves, chopped into small pieces. Heavy metals of the plants were wet-digested using HNO_3: $HClO_4$ (Allen, 1974). Concentration of Cd, Cr, Cu, Fe, Mn, Ni, Pb, and Zn were measured using Inductively Coupled Plasma-Atomic Emission Spectrophotometer (ICP-AES) (IRIS Interepid II XDL: Thermo Electron, Waltham, Mass., USA). Total metal concentration in the wetland plants has been presented in Tables 7.3 and 7.4. These were calculated by adding the mean values of the individual tissues i.e. roots, shoots and leaves. The metals concentration was found higher in wetland plants grown in metal solution than control. *P. cummunis* accumulated most of the metals (Cr, Fe, Mn, Ni, Pb, and Zn) in its shoots except Cd and Cu (Tables 7.3 and 7.4). The accumulation pattern of the total metals (µg/g) in *P. cummunis* was in the order of Fe (2813) > Mn (814.40) > Zn (265.80) > Pb (92.80) > Cr (75.75) > Cu (61.77) > Ni (45.69) > Cd (4.69) (Tables 7.3 and 7.4). During this study, *T. angustifolia* and *C. esculentus* accumulated most of the metals in their roots (Tables 7.1 and 7.3). Where, Mn translocation was noted maximum in leaves followed by roots and shoots of *T. angustifolia* and *C. esculentus*. While, *T. angustifolia* shoot accumulated more Zn than root and leaves. The following metals accumulation pattern in *T. angustifolia* and *C. esculentus* have been observed for various following metals

    T. angustifolia : Fe > Mn > Zn > Cr > Pb > Cu > Ni > Cd

    C. esculentus : Fe > Mn > Zn > Pb > Ni > Cu > Cr > Cd.

The translocation factor (TF) indicated the internal metal transportation of the plant. In this study, metals accumulation ratio (TF) in shoots/roots of *P. cummunis* was > 1 for Cr, Fe, Mn, Ni, Pb and Zn. This has strongly indicated that the *P. cummunis* was shoot accumulator for all tested metals except for Cd and Cu (Fig. 7.9a), where the TF values were < 1 (0.48 and 0.56). However, the TF of all tested metals from shoots to leaves were < 1 for *P. cummunis* except Cu (Fig. 7.9b). This had given strong evidence for the faster movement of different metals, i.e. Mn > Fe > Cr > Pb > Cd > Zn > Ni. Apart from this, the TF of Cd, Cr, Cu, Fe, Mn, Ni and Pb for the translocation from roots to shoots were < 1 except Zn in *T. angustifolia*. This indicated that these metals were restricted to roots. The higher metal accumulation in *T. angustifolia* roots might be due to strong roots development and complexation of metals with sulphydril (– SH) which resulted into less translocation of metals to upper parts of the plant. But, the higher translocation (>1) from shoots to leaves was observed for Mn, Ni and Pb. In contrast, *T. angustifolia* showed slow translocation from shoots to leaves for Cd, Cr, Cu, Fe, and Zn. In addition, the TF from roots to shoots for all tested metals was < 1 in *C. esculentus*, which was between 0.44–0.95, this indicated significant amount of metals accumulation in roots.

But, the TF from shoots to leaves for Fe and Mn was > 1, which indicated higher translocation of these metals from shoots to leaves and rest metals showed TF < 1 in *C. esculentus* (Fig. 7.9b). Higher Mn accumulation in leaves of *T. angustifolia* and

Table 7.3: Metal concentration (µg/g dw) in different parts of wetland plant irrigated with heavy metal solution

Samples		Cd C	Cd T	Cr C	Cr T	Cu C	Cu T	Fe C	Fe T
P. cummunis	Roots	0.14 0.000	2.73 0.050***	ND	27.38 1.821[anp]	0.24 0.001	25.59 0.600***	21.40 0.641	400.00 12.130***
	Shoots	0.07 0.000	1.31 0.030***	ND	32.18 1.109[anp]	0.12 0.000	14.29 0.330***	25.37 0.504	1965 78.600***
	Leaves	0.01 0.000	0.65 0.015***	ND	13.19 0.460[anp]	0.19 0.000	21.89 0.870***	22.70 0.632	448.27 10.530***
	Total	0.22 0.000	4.69 0.095***	ND	5.75 3.390[anp]	0.55 0.001	61.77 0.015***	69.47 1.777	2813 101.260***
Accumulation pattern			R > S > L		S > R > L		R > L > S		S > L > R
T.angustifolia	Roots	ND	0.82 0.020[anp]	ND	59.13 1.180***	0.16 0.003	35.14 0.700***	33.34 0.898	3327 186.450***
	Shoots	0.12 0.004	0.24 0.004***	0.28 0.000	11.55 0.231***	0.14 0.001	16.89 0.330***	26.70 0.876	941.00 36.130***
	Leaves	0.10 0.003	0.21 0.004***	0.15 0.003	5.14 0.102***	0.08 0.001	0.21 0.124[ns]	22.90 0.546	461.75 8.500***
	Total	0.22 0.007	1.27 0.015***	0.43 0.003	75.82 1.513***	0.38 0.005	52.24 0.015***	82.94 2.320	4729 231.080***
Accumulation pattern			R > S > L		R > S > L		R > S > L		R > S > L
C. esculentus	Roots	0.14 0.001	1.98 0.030***	0.15 0.001	17.03 0.400*	0.13 0.001	29.74 0.400***	31.30 0.387	513.25 17.450***
	Shoots	0.03 0.000	1.36 0.053***	ND	16.18 0.320[anp]	0.14 0.004	17.00 0.580***	28.60 0.165	382.20 10.700***
	Leaves	0.02 0.000	0.51 0.010***	0.04 0.000	14.63 0.340***	0.14 0.001	4.72 0.110***	43.90 0.343	539.55 11.830***
	Total	0.19 0.001	3.85 0.093***	0.19 0.001	47.84 0.015***	0.41 0.006	51.46 0.015***	103.80 0.895	1434 39.980***
Accumulation pattern			R > S > L		R > S > L		R > S > L		L > R > S

All values are mean of five replicates ± SD. (C) =Plant grown in Hoagland solution without heavy metals as control. (T) = Plant grown in Hoagland mixed heavy metals solution, ND = Not Detectable. Statistical significance between the values of control to their respected metal treated plant parts for individual metal evaluated by ANOVA. Significance levels: *** = $p < 0.001$; ** = $p < 0.01$; * = $p < 0.05$; ns = $p > 0.05$; anp = ANOVA is not possible because control value are not detectable. R: Roots; S: Shoots; L: Leaves.

Table 7.4: Metal concentration (µg/g dw) in different parts of wetland plant irrigated with heavy metal solution

Samples		Mn		Ni		Pb		Zn	
		C	T	C	T	C	T	C	T
P. cummunis	Roots	0.72 0.003	259.19 7.770***	0.15 0.000	14.45 0.289***	0.26 0.000	36.15 0.840***	0.38 0.001	60.46 1.209***
	Shoots	0.95 0.002	492.13 11.580***	0.26 0.000	17.09 0.340***	0.41 0.001	38.65 0.480***	0.44 0.001	133.38 3.145***
	Leaves	0.42 0.001	63.08 1.890***	0.02 0.000	14.15 0.330***	0.10 0.000	18.00 0.460**	0.58 0.001	71.96 1.690***
	Total	2.09 0.006	814.40 21.24***	0.43 0.000	45.69 0.959***	0.77 0.001	92.80 1.780***	1.40 0.003	265.80 6.044***
Accumulation pattern			S > R > L		S > R > L		S > R > L		S > L > R
T. angustifolia	Roots	0.51 0.001	340.63 4.250***	0.12 0.000	21.15 0.422***	0.22 0.007	50.82 1.190***	0.23 0.001	150.00 5.310***
	Shoots	0.40 0.002	153.23 3.590***	0.12 0.000	3.96 0.020***	ND	8.85 0.200^{anp}	1.28 0.004	176.67 9.120***
	Leaves	0.70 0.002	438.88 10.160***	0.10 0.000	4.09 0.096^{ns}	0.13 0.008	12.08 0.360***	0.24 0.001	23.48 0.469***
	Total	1.61 0.005	932.74 15.00***	0.34 0.000	29.20 0.538***	0.35 0.015	71.75 1.750***	1.75 0.006	350.15 14.899***
Accumulation pattern			L>R>S		R>L>S		R>L>S		S > R > L
C. esculentus	Roots	1.30 0.001	130.84 3.070***	0.21 0.008	55.85 1.316***	0.49 0.005	40.55 1.210***	0.68 0.007	58.56 1.380***
	Shoots	0.95 0.003	57.38 1.140***	0.07 0.001	11.10 0.490***	0.29 0.001	35.51 1.065***	0.48 0.002	35.86 0.860***
	Leaves	2.33 0.001	147.96 3.480***	0.11 0.000	4.66 0.149***	0.16 0.001	6.89 0.160***	0.18 0.002	17.36 0.500***
	Total	4.58 0.005	336.14 7.69***	0.39 0.009	71.61 1.955***	0.94 0.007	82.95 2.435***	1.34 0.011	111.78 2.74***
Accumulation pattern			L > R > S		R > S > L		R > S > L		R > S > L

All values are mean of five replicates ± SD. (C) = Plant grown in Hoagland solution without heavy metals as control. (T) = Plant grown in Hoagland mixed heavy metals solution. Statistical significance between the values of control to their respected metal treated plant parts for individual metal evaluated by ANOVA. Significance levels: *** = $p < 0.001$; ** = $p < 0.01$; * = $p < 0.05$; ns = $p > 0.05$; anp = ANOVA is not possible because control value are not detectable. R: Roots; S: Shoots; L: Leaves.

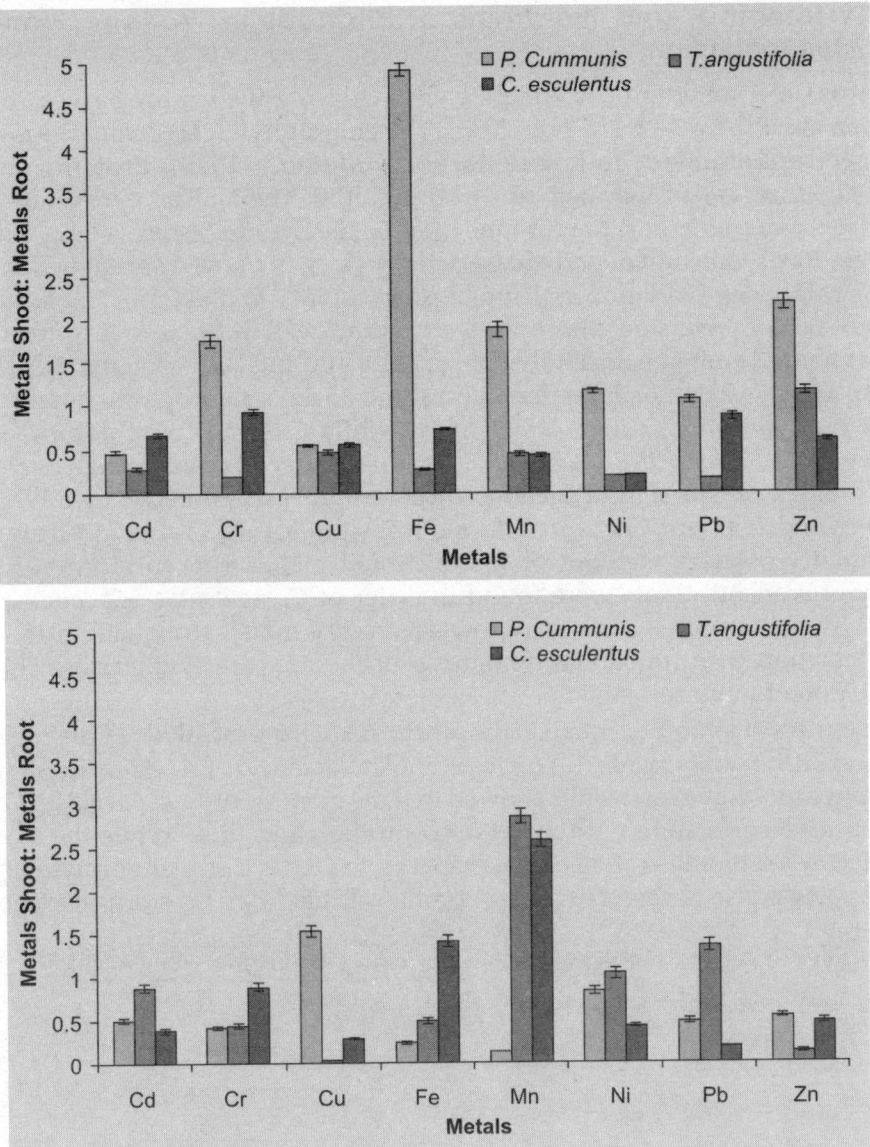

Fig. 7.9: (a and b) Showing the ratio of heavy metals accumulated in roots, shoots, and leaves of *P. cummunis*, *T. angustifolia*, and *C. esculentus*

C. esculentus was noted. This might be due to possible detoxification mechanism of wetland plants, Mn-malate complex is transported through the tonoplast membrane to the vacuole where Mn dissociates from malate and complexes with oxalate. Where, malate functions as a "transport vehicle" through the cytoplasm and oxalate as the "terminal acceptor" in the vacuole. The higher iron accumulation in roots of *P. cummunis* indicated either iron deposited on the iron oxyhydroxide plaque layer of roots or enhances accumulation of iron through iron oxyhydroxide plaque layer. The uptake of Mn and Cu in *Phragmites australis* is affected by pH and concentration of metal.

7.A.3 Physiological and Biochemical changes in Wetland Plant during Phytoremediation

Biomass was determined on dry weight basis by drying the freshly harvested plants in hot air oven at 80°C for 24 h (APHA, 2005). The chlorophyll (chl) content was estimated by the spectrophotometer following Arnon's Method (1949). Protein content was determined according to the method of Lowry et al. (1951). The peroxidase activities were measured in leaves at different time interval, i.e. 28 and 56 days along with control (Singh et al., 2006). One unit of peroxidase activity (U/g of tissue) represented the amount of enzyme catalyzing oxidation of 1 mmol guaiacol in 1 min at 25°C. Table 7.5 showed the effect of heavy metals on chlorophyll content of wetland plants at different time of metal treatment. Result suggested that chl a, chl b and total chl contents were increased initially up to 35 days of metal treatment. But, after 35 days chlorophyll content gradually decreases in *P. cummunis* and *T. angustifolia*, while *C. esculentus* chlorophyll content increased with time up to 28 days of metal treatment after 28 days gradual decrease were observed (Table 7.5). The significant effect on chlorophyll was noted in *C. esculentus* while *P. cummunis* and *T. angustifolia* were affected less significantly (Table 7.5). This indicated the potentiality of these wetland plants on tested metals concentration. Chlorophyll sensitivity of wetland plants were found in order of *C. esculentus* > *T. angustifolia* > *P. cummunis*. The chlorophyll sensitivity towards heavy metals in aquatic plant might be due to interaction with functional -SH group of the enzyme synthesizing chlorophyll. Metals are potent inhibitors of chlorophyll biosynthesis.

The major sites of inhibition are (i) in the formation of the proteolytic PC halide reductase complex and (ii) the synthesis of ALA (delta aminolevulinic acid), first characteristic precursor of the porphyrins. Moreover, results showed that the biomass of *T. angustifolia, P. cummunis* showed positive correlation with metal concentration and time. While the *C. esculentus* showed decreased biomass at 56 days compared to 28 days of plant growth (Table 7.5). Similar, trend was also observed in case of protein (Table 7.5). This study also showed that

Table 7.5: Effect of different metals on biomass and some physiological parameters of *P. cummunis*, *T. angustifolia* and *C. esculentus*

Wetland plants		Biomass (g dw)		Physiological Properties							
				Chlorophyll a ($\mu g\ g^{-1}\ fw$)		Chlorophyll b ($\mu g\ g^{-1}\ fw$)		Protein ($\mu g\ g^{-1}\ fw$)		Peroxidase ($U\ g^{-1}$ of tissue)	
		28 d	56 d	28 d	56 d	28 d	56 d	28 d	56 d	28 d	56 d
P. cummunis	Control	3.24	4.34***	215	200[ns]	114	97.00*	16.76	17.53[ns]	45.00	48.00[ns]
		(0.08)	(0.21)	(7.71)	(7.78)	(4.28)	(3.12)	(0.59)	(0.73)	(1.80)	(1.67)
	Treated	5.23	7.24***	468	546*	240	312*	21.34	28.98***	56.00	78.00***
		(0.19)	(0.35)	(18.16)	(18.84)	(6.13)	(11.70)	(0.14)	(1.26)	(1.77)	(3.53)
T. angustifolia	Control	2.14	3.87[ns]	229	214[ns]	146	110[ns]	11.98	13.56*	40.00	43.00[ns]
		(0.08)	(0.16)	(9.16)	(10.40)	(3.12)	(2.84)	(0.34)	(0.66)	(0.63)	(1.78)
	Treated	4.76	6.76***	531	460*	251	112*	18.78	19.45[ns]	65.00	87.00***
		(0.13)	(0.25)	(10.32)	(23.00)	(5.12)	(4.44)	(0.70)	(0.92)	(1.32)	(3.12)
C. esculentus	Control	1.12	1.87***	310	297[ns]	178	130***	9.34	10.12[ns]	38.00	41.00[ns]
		(0.06)	(0.05)	(9.53)	(14.79)	(3.17)	(3.12)	(0.37)	(0.35)	(1.74)	(1.44)
	Treated	2.21	1.67***	468	238***	245	96.00***	12.98	8.97***	73.00	98.00***
		(0.11)	(0.06)	(17.12)	(9.83)	(8.13)	(4.80)	(0.78)	(0.35)	(3.03)	(4.41)

All the data are mean value (N = 5) and values written in bracket are standard deviation. Statistical significance between the columns was valuated by ANOVA. Significance levels: *** = p < 0.001; * = p < 0.05; ns = p > 0.05. d: days the given metal concentration at observed time had not shown any visible morphological adverse effect. Significant increase of biomass and protein might be due to synthesis of stress protein. This revealed the defense mechanism of the plant in multi-metal solution.

7.A.4 Anatomical changes in Wetland plants during Phytoremediation

Shoots and roots of 5 mm length were excised from 2 cm above and 2 cm below the shoot-root intersection, respectively. Leaves of 5 mm length were excised from the middle portion of the third leaf (lower leaf) from the base of the plant. The roots, shoots and leaves samples were prepared for the transmission electron microscopy (TEM). All the samples were excised and quickly immersed in H_2S saturated water as pretreatment for 30 min at room temperature to precipitate Cd and Zn. For TEM Roots, shoots and leaves segment of approximately 3 mm length were collected. The samples were fixed in modified Karnovsky's fluid buffered with 0.1 M sodium phosphate buffer at pH 7.4. Fixation was done for 10–18 h at 4°C, after that the tissues were washed with fresh buffer and post fixed for 2 h in 1% osmium tetroxide in the same phosphate buffer. The tissues were dehydrated in graded acetone solution and embedded in CY 212 araldite. Ultrathin 60–80 mm section of the tissues was cut using ultra E (Reichert Jung). Ultra sections were stained with uranyl acetate and lead citrate for 10 minutes before examining the grid in transmission electron microscope (Phillips, M–10) operated at 60–80 KV transmission.

The TEM section showed apparent metals deposition near the cell wall, intercellular spaces and thinning of cell wall in *P. cummunis* and *T. angustifolia* (Fig. 7.10a to d). While, loss of cell shape and thickened cell wall was observed in *C. esculentus* roots due to deposition of metals as compared to control (Fig. 7.10e, f). This might be due to hyper

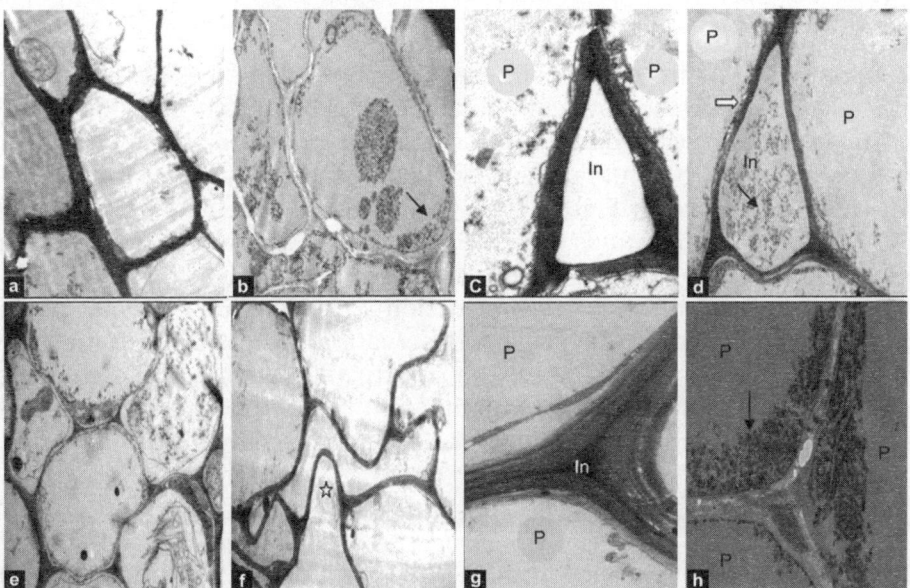

Fig 7.10(a to h): TEM micrographs of metal treated *P. cummunis* (b), *T. angustifolia* (d) and *C. esculentus* (f) roots along with control (a, c and e respectively) and treated shoots of *P. cummunis* (h) along with control (g) after 56 days of metal treatment. Metal granules (arrow), thinning of cell wall (hollow arrow), loss of cell shape (star), parenchyma (P) and intercellular spaces (In)

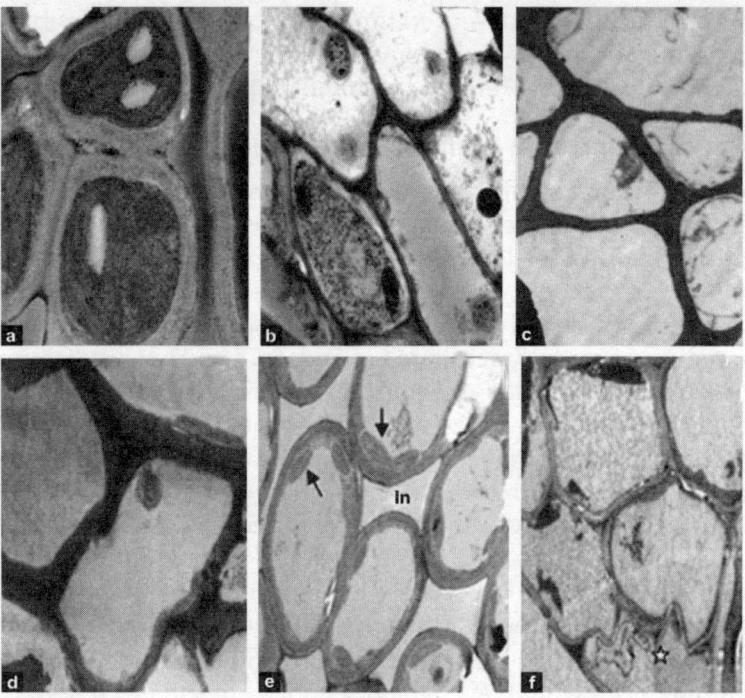

Fig. 7.11(a to f): TEM micrographs of metal treated *P. cummunis* (b), *T. angustifolia* (d) and *C. esculentus* (f) leaves along with control (a, c and e respectively) after 56 days of metal treatment. Loss of cell shape (star), decrease in intercellular spaces (In) and plastid (arrow)

accumulation of Cd, Cu in roots of all tested wetland plants besides these roots of *T. angustifolia* and *C. esculentus* also accumulated Fe, Ni, Pb, and Zn. Simultaneously, shoot cells of *P. cummunis* also showed deposition of metal granules near the cell wall under TEM observation (Fig. 7.10g, h). Further, TEM observations of leaves tissue of all tested wetland plants are shown in Figure 7.11a–f. Where no any histological damage was observed in leaves of *P. cummunis* except change in cell shape (Fig. 7.11a, b). Similarly, *T. angustifolia* leaves also showed no any apparent deformation (Fig. 7.11c, d). While, leaves tissue of *C. esculentus* showed reduction of intercellular spaces, plastids and irregular cell shape (Figure 7.11e, f). This observation indicated more metal tolerance behavior of *P. cummunis* and *T. angustifolia* than *C. esculentus*. The changes in the chloroplasts (a type of plastid) due to abiotic and biotic stresses are the results of an increase volume of the stroma and a disorganization of the thylakoid membranes.

7.B IN-SITU HEAVY METALS PHYTOREMEDIATION POTENTIAL OF WETLAND PLANT

7.B.1 Comparative heavy metal accumulation in various wetland plants growing at contaminated site

A cluster of tanneries and one major distillery are located in adjacent part of Unnao city, UP, India (26°32'0 "N, 80°30'0"E). Most of the tanneries and distillery discharge their partially treated and untreated effluents into a city drain, which outflows into a low lying sink to create a big waterlogged area as natural wetland and damages the surrounding natural flora and fauna. Two native common wetland plants, i.e. *T. angustifolia L.* and *C. esculentus* were only able to grow luxuriantly in the waterlogged natural wetland area of the industrial drain. Sampling site is shown in Fig. 7.12. The upstream flora in wetland area shows dominant growth of *T. angustifolia L.* while the downstream flora shows luxuriant growth of *C. esculentus*. This ultimately joins to the Ganga River, the most important river of the country. To determine the concentration of metals in these two potential hyperaccumulant plants, at least ten complete plants of each species were collected randomly from contaminated site and transported to the laboratory in polyethylene bags. Simultaneously, the contiguous soil was separately taken from the depth of 0–20 cm (root growing zone) in a plastic bag and plastic spatula, avoiding any contact with metals. Contaminated water samples were also collected in sterile Jerrycane from contaminated site to determine the heavy metals concentrations.

The physico-chemical characteristics of distillery and tannery wastewater prior to mixing at the wetland plant growing site are shown these values are far greater than the permissible limit of industrial effluent discharges (EPA 2002). The effluent showed high content of Fe, Cr, and Mn. The iron is released from distillery and chromium from tanneries discharges. Luxuriant growth of *T. angustifolia* and *C. esculentus* in highly polluted conditions shows strong evidence for its tolerance against high pollution load. The metals content in sediment was noted higher at inlet site of the wetland where *T. angustifolia*

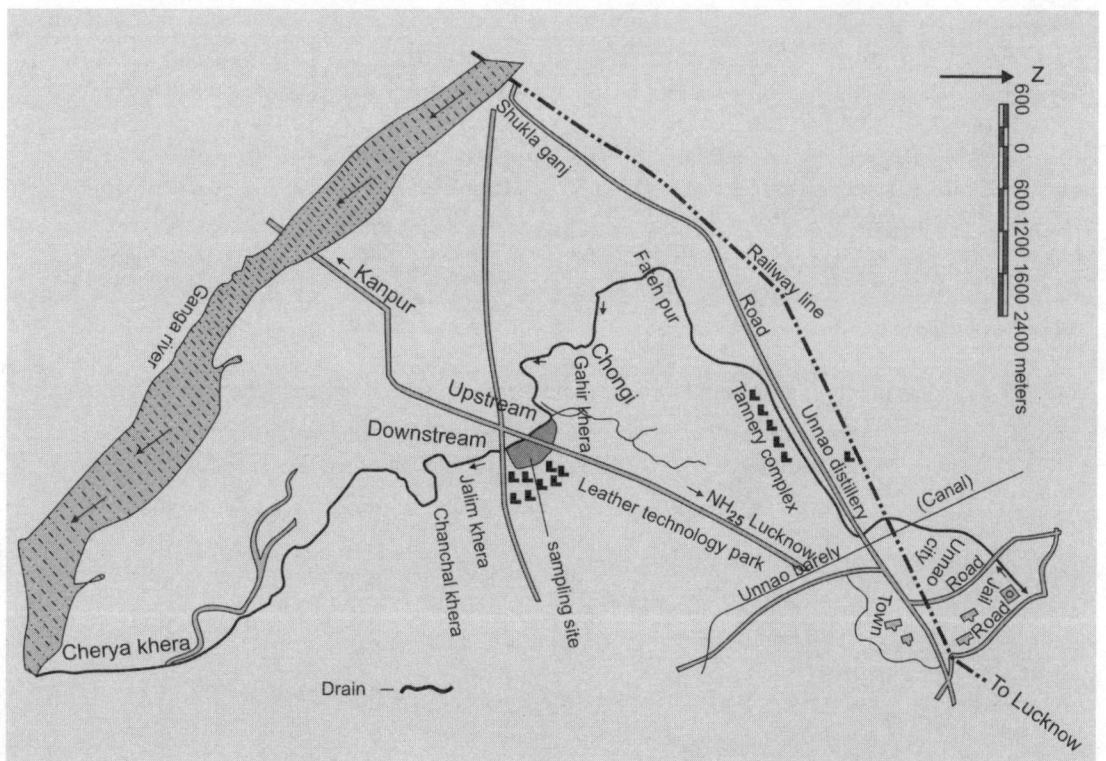

Fig. 7.12: Layout of Unnao industrial wastewater disposal drain

was growing as compared outlet site (downstream) where *C. esculentus* was growing. *T. angustifolia* was growing in the highest polluted area of the wetland, which indicated that *T. angustifolia* is more tolerant to the polluting metals. However, *C. esculentus* was growing at downstream of natural wetland system where metals content in sediment was comparatively low. The ratio between metal content of sediment:metal content of water at *T. angustifolia* and *C. esculentus* growing site for different metals were: Cd 217.00, 50.38; Cr 440.37, 71.39; Ni 50.36, 46.68; Pb 80.94, 15.19; Cu 24.58, 46.92; Fe 189.47, 23.37; Mn 98.82, 79.51; Zn 132.19, 65.72 (Table 7.6). This revealed that sediment contains more metals than wastewater. This might be due to adsorption of metals on the sediments because of continuous addition of distillery and tanneries wastewater to the site. The ability to accumulate metals from sediment to root was also measured as BCF. The BCF was noted '>1' in *T. angustifolia* and *C. esculentus* except Pb and Cr, respectively, as shown in Table 7.7. However, the translocation factor was observed '<1' except Zn and Cr. This revealed that most of metals were absorbed from sediment to root. Further, the translocation of metals from root to shoot was restricted due to protective mechanism of plant.

The accumulation of metals in the *T. angustifolia* and *C. esculentus* growing in tannery and distillery effluent polluted sites is shown in Fig. 7.13. Findings revealed that

Table 7.6: Metal contents in wastewater and sediments of natural wetland system

| Heavy Metals | *T. angustifolia growing site* | | | *C. esculentus growing site* | | |
	Sediment (mg/kg)	Water (mg/L)	Sediment/ Water	Sediment (mg/kg)	Water (mg/L)	Sediment/ Water
Cd	2.387 ± 0.008	0.011 ± 0.000	217.00	1.058 ± 0.040	0.021 ± 0.000	50.38
Cr	177.027 ± 3.30	0.402 ± 0.016	440.37	96.165 ± 2.560	1.347 ± 0.040	71.39
Ni	14.907 ± 0.130	0.296 ± 0.001	50.36	9.055 ± 0.530	0.194 ± 0.001	46.68
Pb	29.867 ± 0.586	0.369 ± 0.001	80.94	12.000 ± 0.360	0.790 ± 0.040	15.19
Cu	23.920 ± 0.500	0.192 ± 0.003	124.58	13.465 ± 0.350	0.287 ± 0.000	46.92
Fe	1872.00 ± 44.981	9.880 ± 0.130	189.47	1423.950 ± 20.990	60.940 ± 1.160	23.37
Mn	193.493 ± 3.060	1.958 ± 0.013	98.82	97.006 ± 4.520	1.220 ± 0.023	79.51
Zn	51.420 ± 0.972	0.389 ± 0.010	132.19	35.950 ± 0.850	0.547 ± 0.030	65.72

All values are mean (n = 3) ± S.D

Table 7.7: Accumulation and translocation of heavy metals in *T. angustifolia* and *C. esculentus*

| Metals | Bioconcentration Factor (BCF) | | Translocation Factor (TF) | |
	T. angustifolia	*C. esculentus*	*T. angustifolia*	*C. esculentus*
Cd	1.57	1.40	0.27	0.33
Cr	1.29	0.59	0.21	0.94
Ni	1.33	1.58	0.37	1.05
Pb	0.81	1.22	0.45	0.31
Cu	1.18	1.13	0.53	0.94
Fe	1.42	1.10	0.20	0.51
Mn	1.08	1.07	0.29	0.70
Zn	1.17	1.15	1.18	0.50

T. angustifolia and *C. esculentus* are root accumulators for Cd, Cu, Cr, Fe and Pb because these metals were found significantly higher in root than shoot or leaves. The order of metal accumulation in root of both plants were Fe > Cr > Pb > Cu > Cd. This might be due to high plant availability of the substrate metals as well as its limited mobility once taken up by the plants. Besides, higher metal accumulation in root of both wetland plants might be due to strong root development and complexation of metals with sulphydryl group (–SH) of soil constituent resulted into less translocation of metals to upper parts of plants. It is also possible that metal accumulation in root might be related with co-precipitation in the iron oxyhydrate plaque layers on the root surface of wetland plants. Figure 7.13 shows *T. angustifolia* is a shoot accumulator for Zn and *C. esculentus* have accumulated Ni maximum in shoot. *T. angustifolia* accumulated 70 mg/kg Zn while *C. esculentus* accumulated Ni 14 mg/kg in their shoots. The higher metal accumulation in *T. angustifolia* and *C. esculentus* indicated that internal metal detoxification tolerance mechanisms might exist in these wetland plants. The accumulation patterns of metals (sum of metals of all parts of wetland plants) in *T. angustifolia* and *C. esculentus* have been studied which are as follows:

T. angustifolia : Fe > Mn > Cr > Zn > Pb > Cu > Ni > Cd

C. esculentus was observed as Fe > Mn > Cr > Zn > Pb > Cu > Ni > Cd.

However, the leaves did not show any visual toxicity symptoms. This might be because of the tolerance mechanism of wetland plant for heavy metals.

7.B.2 Biochemical changes in Wetland Plants during in-situ Phytoremediation

The total chlorophyll, chlorophyll-a, and chlorophyll-b contents were observed higher in *T. angustifolia* as compared to *C. esculentus* (Fig. 7.14a). *T. angustifolia* grown in distillery and tannery effluent-contaminated site has shown significant increase (P < 0.05) in total chlorophyll (28.89%), chlorophyll-a (59.69%), and chlorophyll-b (76.92%) contents as compared to *C. esculentus*. Metal accumulation pattern for leaves of *T. angustifolia* and *C. esculentus* was observed as Fe > Mn > Cr > Zn > Pb > Cu > Ni > Cd. Metals might be essential for the synthesis of chlorophyll directly or indirectly. The metals are necessary for chlorophyll synthesis up to certain level. However, the higher induction of MDA in *C. esculentus* than *T. angustifolia* (Fig. 7.14b) revealed the lipid peroxidation of cell membrane in leaves and root. Simultaneously, this also indicates the sensitivity of plants. Further, induction of cysteine, ascorbic acid and protein in *T. angustifolia* (Fig. 7.14c) gives the plant strength against stress.

7.B.3 Anatomical changes in Wetland Plants during Phytoremediation

Anatomical observation through TEM in the root of *T. angustifolia* did not show any remarkable changes even after higher accumulation of various metals in the roots (Fig. 7.15a, b). This might be due to tolerance mechanism of plants with higher content of heavy metals for adaptation in adverse conditions. Naturally, all waterlogged soils have higher amounts of plant available Fe, Mn, and Cr due to redox potential. Thus, plants growing in the wetlands uptake more Fe, Mn, and Cr. However, the formation of multinucleolus in the shoot cells revealed the formation of extra protein for the plants protection (Fig. 7.15c, d). However, the histological view of root tissues of *C. esculentus* showed thin cell wall as compared to root tissues of *C. esculentus* collected from uncontaminated site, due to disturbance in lignifications (Fig. 7.16a, b).

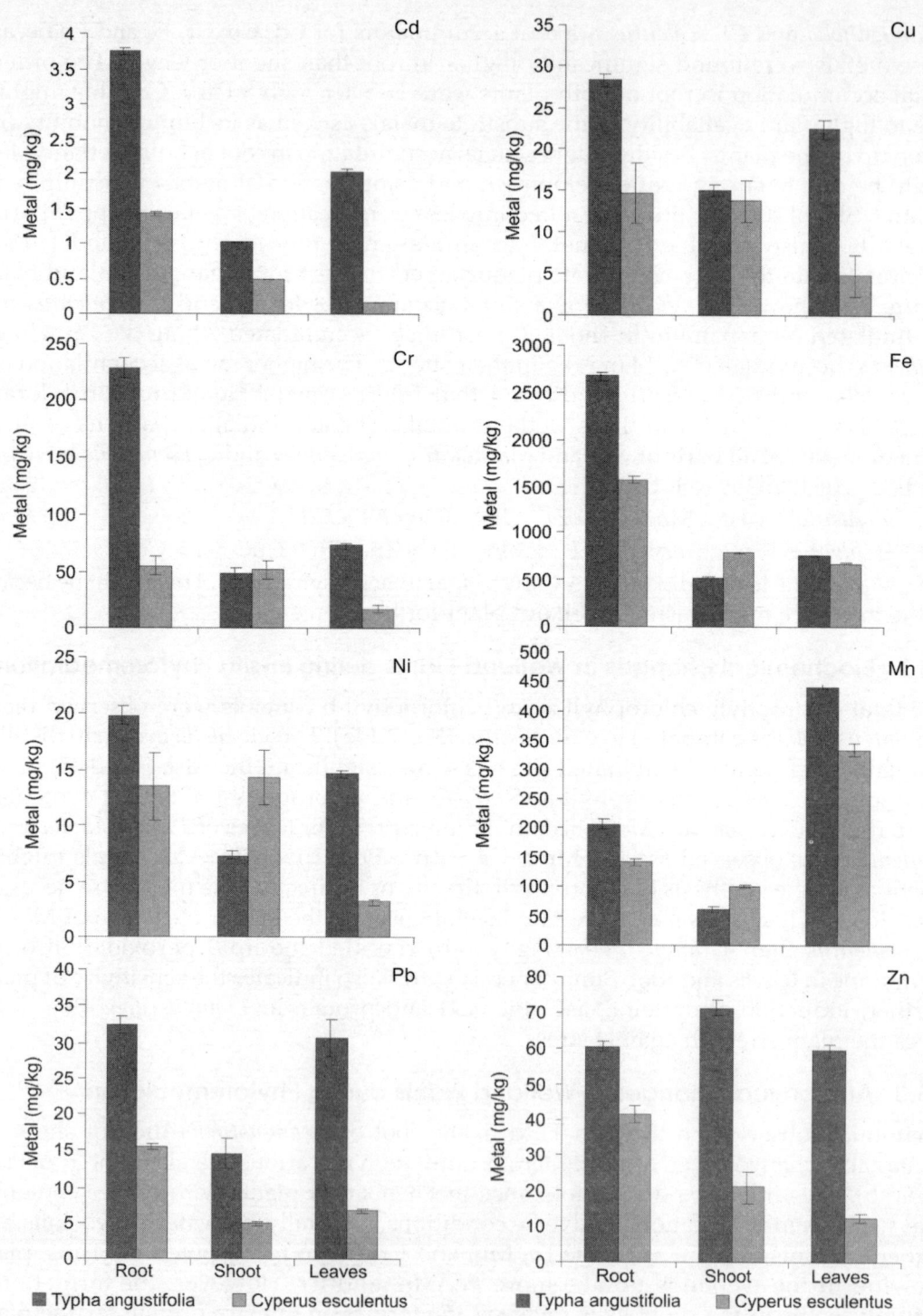

Fig. 7.13: Heavy metals (mg/kg) content in different parts of *T. angustifolia* and *C. esculentus*. Note: All the values are means of three replicates SD. Superscript indicate that they were significantly different at a probability level of 0.05 according to ANOVA test. a = p < 0.05, ns = p > 0.05

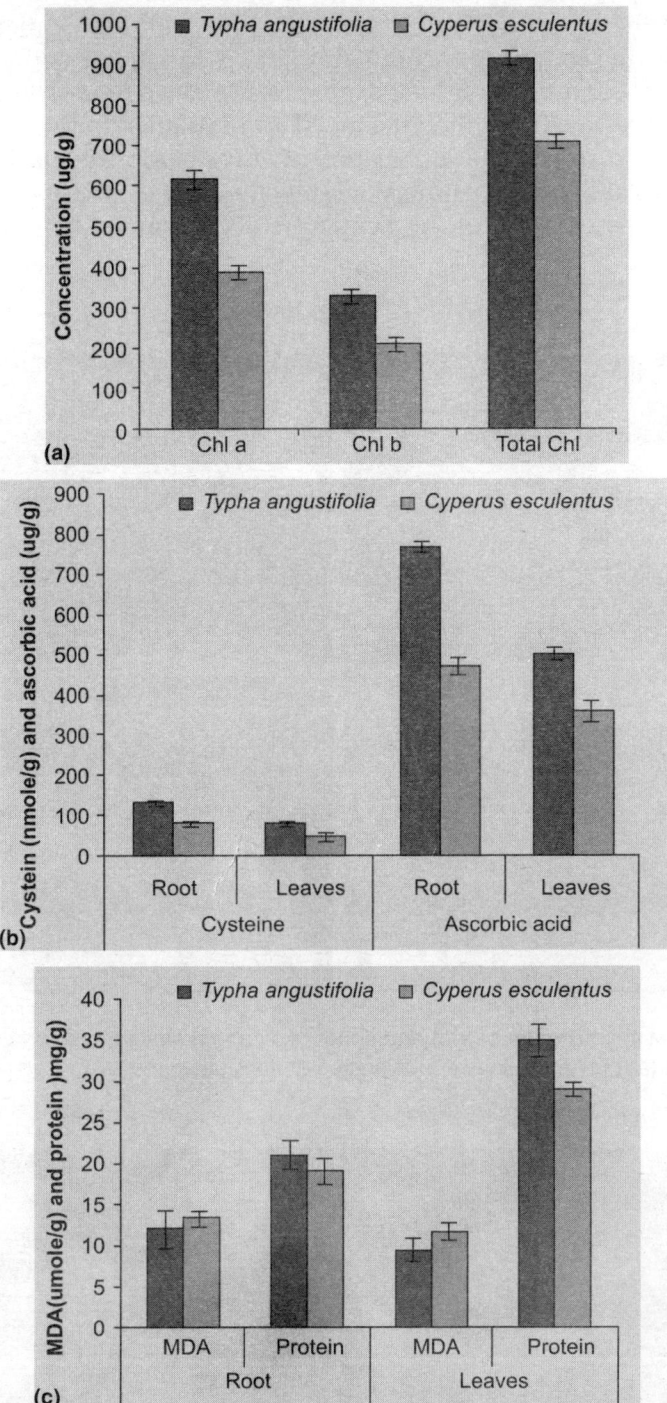

(a)

(b)

(c)

Fig. 7.14: The chlorophyll (Chl, A); malondialdehyde and protein (MDA, B); cysteine and ascorbic acid content (C) in *T. angustifolia* and *C. esculentus* grown in distillery and tannery effluent contaminated site. Note: All the values are mean of three replicates SD. Superscript indicate that they were significantly different at a probability level of 0.05 according to ANOVA test. a = $p < 0.05$, ns = $p > 0.05$

This supported the sensitivity of *C. esculentus* root with heavy metal accumulation. This might also be due to higher accumulation of Fe, Mn along with other metals, and pollutants in natural conditions for long exposure. Metals in the tissue can be trapped by negative charges of cell-wall, resulted the metal accumulation in the apoplast. The study concluded that *T. angustifolia* had higher potential for heavy metals accumulation than *C. esculentus* from distillery and tannery wastewater. Hence, these two plants could be used for the phytoremediation of the heavy metals contaminated swampy lands and wastewater.

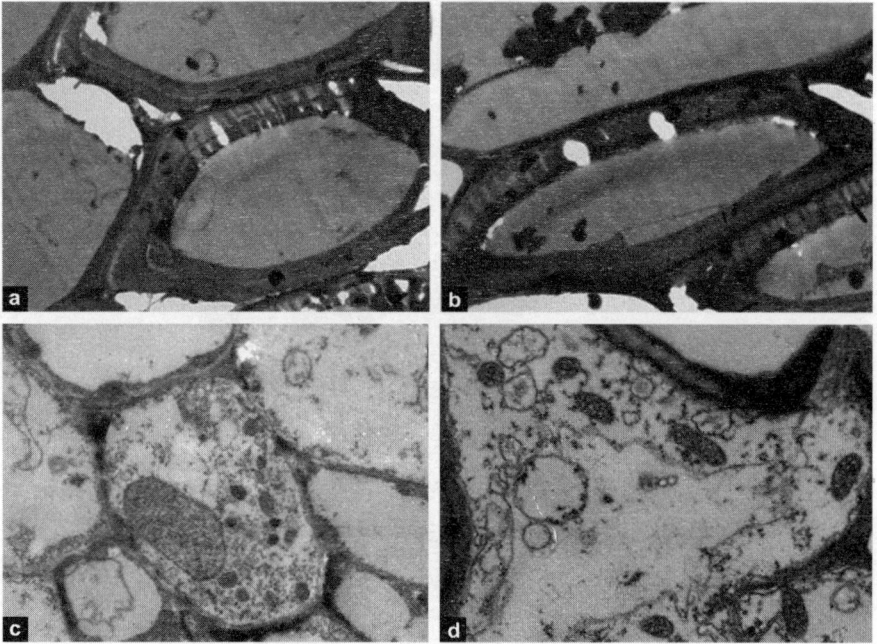

Fig. 15 (a to d): TEM micrographs of root and shoot of *T. angustifolia* grown on contaminated site (b, d) along with control (a, c). Arrow shows multinucleolus formation

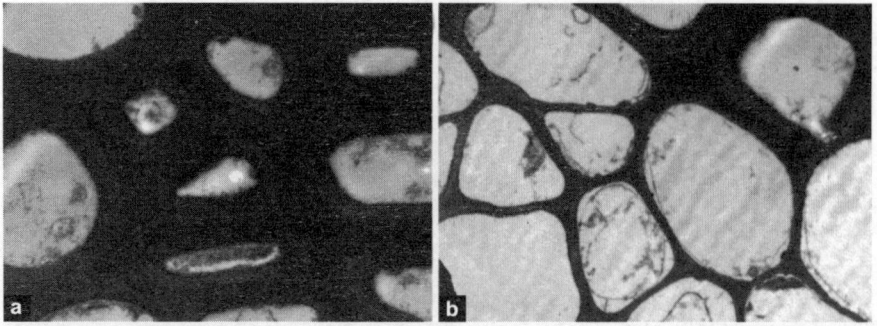

Fig. 7.16 (a and b): TEM micrographs of *C. esculentus* root grown on contaminated site (b) along with control (a). Arrow shows thinning of cell wall

7.C Advanced technique for qualitative and quantitative detection of heavy metals after phytoremediation

Particle-induced X-ray emission or proton-induced X-ray emission (PIXE): The technique was first proposed in 1970 by Sven Johansson of Lund University, Sweden, and developed over the next few years with his colleagues Roland Akselsson and Thomas B Johansson. PIXE is relatively high sensitive, multielemental and non-destructive analysis technique. They have shown that their system is capable to analyses the trace elements with a good resolution. Several analytical techniques have been used for trace element analysis, their sample preparation is generally complicated and takes a long time. The ion-beam techniques, especially PIXE is one of the most powerful techniques for material analysis, since its sample preparation techniques are generally simple and it requires a short measuring time. This powerful technique can easily analyze various elements with atomic number as low as 12 in the ppm range. Two stages procedure are followed in PIXE analysis. Firstly, elements in the target are identified from the energies of the characteristic peaks in the X-ray emission spectrum. Secondly, the quantity of a particular element in the target is determined from the intensity of its characteristic X-ray emission spectrum. The qualitative advantage of X-ray spectrometry was well established before PIXE (Particle Induced X-rays Emission). However, PIXE introduces both qualitative and quantitative advantages simultaneously in a single measurement. The heavy ions produce rather complicated X-ray spectra which contain the information of different elements of target matrices. X-rays can be produced by exciting the target atoms with an energetic incident ion beam of protons or alpha particles as shown in figure 7.17. The high-energy protons or alpha particles strike the target atoms and eject electrons from the innermost shell in atoms. As a result, a vacancy is created in the innermost shell. It is a common nature of an excited atom that it seeks to regain a stable energy state. Therefore, the created vacancy is filled by an electron coming from an outer shell, at that time an electromagnetic radiation in the form of characteristic X-rays is emitted. The de-excitation may also be possible by the emission of an electron, so-called Auger electron (Fig.7.18). Figure 7.18 shows the k-shell ionization, X-ray emission, and emission of Auger electron. The probability of the emission of an X-ray quantum (the fluorescence yield) is close to 1 for the heavy elements but only a few percent for the light ones. The X-ray spectrum is simply determined by the energy levels of the electrons in the atom. The energy level diagram of a medium-heavy element with the X-ray transitions is shown in figure 7.19. The transitions going to

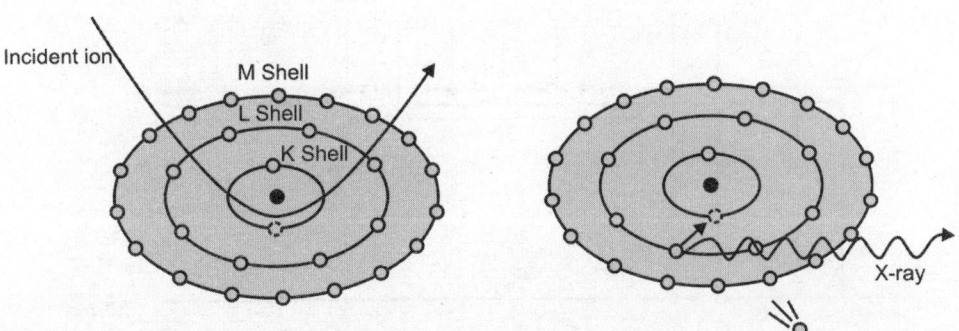

Fig. 7.17. Basic principle of PIXE. (a) Indicates ion interaction with inner shell electron. (b) Indicates emission of electron, fall of upper shell electron and radiation of X-ray

the K shell are indicated as K X-rays. When the vacancy is filled by an electron, comes from the L shell, the transition is denoted as K_α and when it comes from the M shell, Kβ. Similarly, the transitions to the L shell are indicated as L X-rays, and these have some components, especially for heavy elements. Generally, the light and medium-heavy elements are identified by their K X-rays and the heavy elements by L X-rays due to the effective detection of the K X-rays which can be obtained in the range $20 < Z < 50$ and of the L X-rays for $Z > 50$.

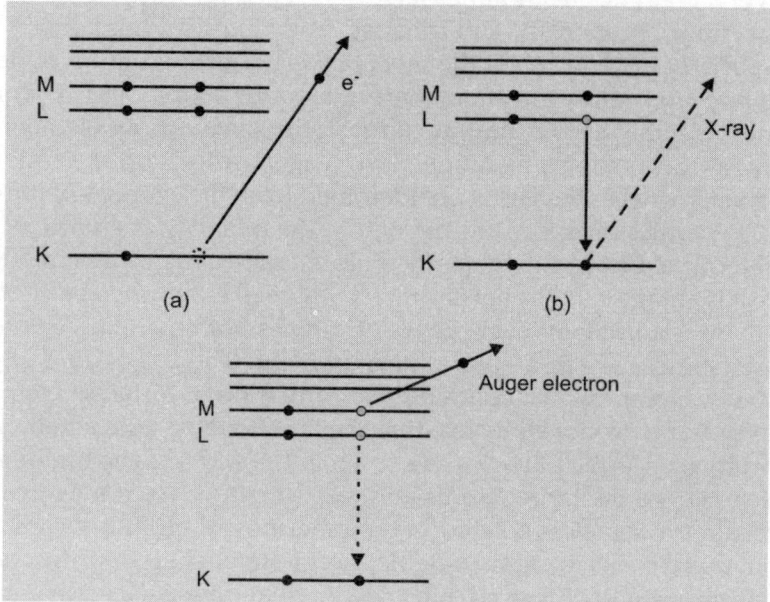

Fig.7.18: Ionization process; (a) k-shell ionization, (b) X-ray emission, and (c) Auger electron

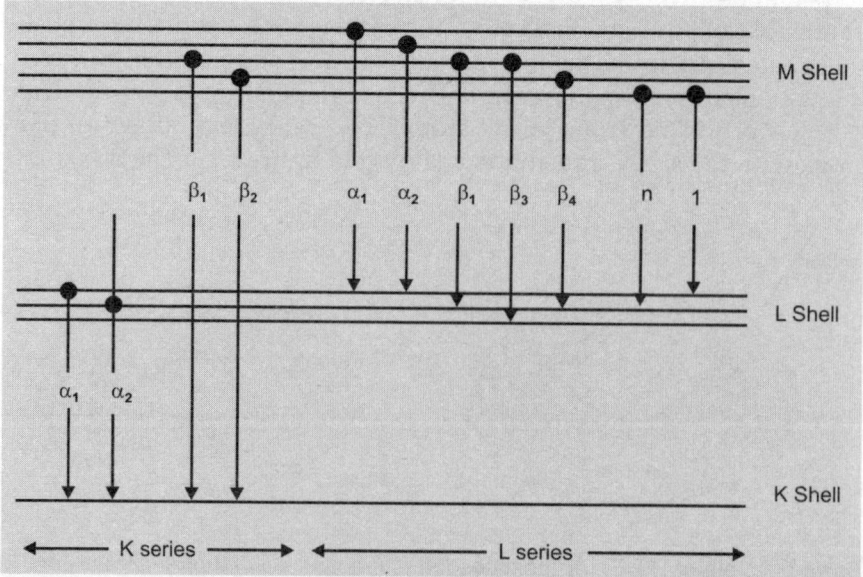

Fig.7.19: Energy levels and X-ray transitions in medium-heavy element

The energy of characteristic X-ray is equal to the difference between two shell electron binding energies those take part in the transitions express as

$$E_{X-ray} = E_1 - E_2$$

Where E_{X-ray} is the characteristic X-ray energy, E_1 is the vacant shell electron binding energy and E_2 is the donor shell electron binding energy.

For instance, Mn K_α X-ray energy

$$= \frac{\text{K- shell electron binding energy} = 6539.0 \text{ eV}}{\text{L (III) - shell electron binding energy} = 638.7 \text{ eV}}$$
$$\text{Difference} = 5900.3 \text{ eV}$$

Therefore, Mn K_α X-ray energy is 5900 eV, i.e. 5.9 keV

Using X-ray spectrum, energy level diagram and knowing X-ray energies, it can be possible to determine the elements those are in the specimen. Recent extensions of PIXE using tightly focused beams (down to 1 µm) give the additional capability of microscopic analysis. This technique, called microPIXE, can be used to determine the distribution of trace elements in a wide range of samples. A related technique, particle-induced gamma-ray emission (PIGE) can be used to detect some light elements. The line between the spatial distribution of elements and their quantification can be provided by the application of micro-proton X-ray emission (micro-PIXE), therefore improving our understanding of the detoxification mechanisms in different plant organs at the tissue level. Micro-PIXE is a highly sensitive technique for elemental mapping in 2D with particularly high sensitivity (1–10 µg/g) for elements with atomic numbers ranging from either 11 to 35 or 75 to 85 thus covering all biologically relevant elements it can also provide information on the localization of elements with 1 µm spatial resolution. Moreover, micro-PIXE is multielemental and the maps obtained are quantitative.

8

Phytoremediation Potential of Wetland Plants for Melanoidin Decolourization in Presence of Heavy Metals and Phenol

8.1 INTRODUCTION

To avoid the toxicity associated with hazardous chemicals that are present in the environment, researchers have developed strategies that employ plants to degrade, remove or stabilize a range of different compounds from polluted soils (i.e. phytoremediation). These environmental pollutants may include metals such as lead, zinc, cadmium, selenium, chromium, cobalt, copper, nickel and mercury; inorganic compounds such as arsenic, sodium, nitrate, ammonia and phosphate, radioactive compounds like uranium, cesium and strontium or organic compounds including chlorinated solvents like trichloroethylene (TCE), explosives such as trinitrotoluene (TNT) and 1, 3, 5-trinitro-1, 3, 5-hexahydrotriazine (RDX), petroleum hydrocarbons such as benzene, toluene and xylene (BTX), polycyclic aromatic hydrocarbons (PAHs), phenol and pesticides such as atrazine and bentazon. While, some organic compounds can be metabolized (remediated) by soil bacteria, in the absence of plants, this process is often slow and inefficient.

The biodegradation of recalcitrant organic compounds in the soil is often enhanced around the roots of plants. This is a direct consequence of the high level of nutrients (including sugars, amino acids and organic acids) that most plants release (exude) into the soil as root exudates, nutrients that typically support a bacterial concentration in the rhizosphere that is often 100 to 1000 fold greater than the bacterial concentration in the bulk soil. Some rhizosphere bacteria are directly involved in the degradation of the organic soil contaminants while others (plant growth-promoting bacteria) can positively affect plant growth and health, enhancing root development or increasing plant tolerance to various environmental stresses. As a direct consequence of their interaction with plant growth-promoting bacteria, plants grow larger and healthier, and are better able to phytoremediate a range of organic soil contaminants. Unfortunately, inorganic environmental pollutants cannot readily be degraded. They must be stabilized in the soil to make them less bioavailable and thereby reduce their spread in the environment; extracted, transported, accumulated and concentrated from the soil into plant roots and/or shoots (phytoextraction); removed from liquid effluents via the use of plant roots (rhizofiltration); or transformed into volatile forms (phytovolatilization). Following phytoextraction, plants may be harvested, dried and converted to ash to recover the concentrated metal. A serious impediment to more effective phytoextraction of metals is the tight binding of metals to soil particles so that often only a small fraction of the metal that is present in the soil can be mobilized and taken up by plant roots. As a result of the testing of numerous plants, several that are naturally able to accumulate large amounts of metal per unit of plant biomass have been identified and are being studied for possible use in the phytoremediation of metallic contaminants. These plants are called hyperaccumulators and are often found growing in soils with elevated metal concentrations. A practical limitation of using hyperaccumulators is that many of the plants that are most effective at removing metals from the soil, such as *Thlaspi caerulescens* (Alpine pennycress) and *Alyssum bertolonii*, are small, containing only a low level of biomass, and they are slow growing, thus reducing their potential for metal phytoextraction from soil (on a large scale) in the field. Moreover, the growth of metal-resistant metal-accumulating plants that are capable of hyperaccumulating metals can be severely inhibited when the concentration of available metal in the contaminated soil is very high. This results in a decrease in plant biomass and efficiency of phytoremediation. To be effective for the remediation of metal polluted soils, plants must be tolerant to one or more metals, highly competitive, fast growing, and produce a high aboveground biomass. Because of their high biomass and extensive root system, some species of trees (e.g. poplar) and wetland plants have been considered to be attractive for phytoremediation; however, metal accumulation by trees is generally low. Finally, a convergence of phytoremediation and bacterial bioremediation strategies has led to a more successful approach to remediation of contaminants, particularly organic compounds.

Given the above mentioned considerations, it is currently possible to develop phytoremediation strategies to clean up a large number of the sites contaminated with organic compounds (this process may require several field seasons depending on the particular plant, soil, bacteria, contaminants and climate involved). On the other hand, phytoremediation is not yet a practical approach for the removal of inorganic compounds from contaminated soil environments.

For removal of different hazardous compounds from contaminated soil and water, plant potentials have been exploited that resulted in several technological subsets (Fig. 8.1).

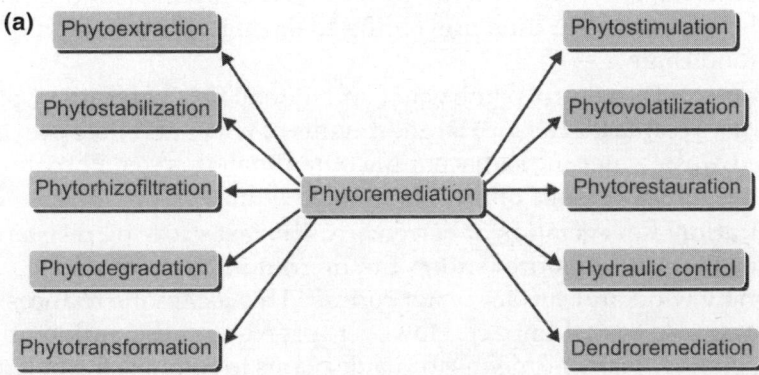

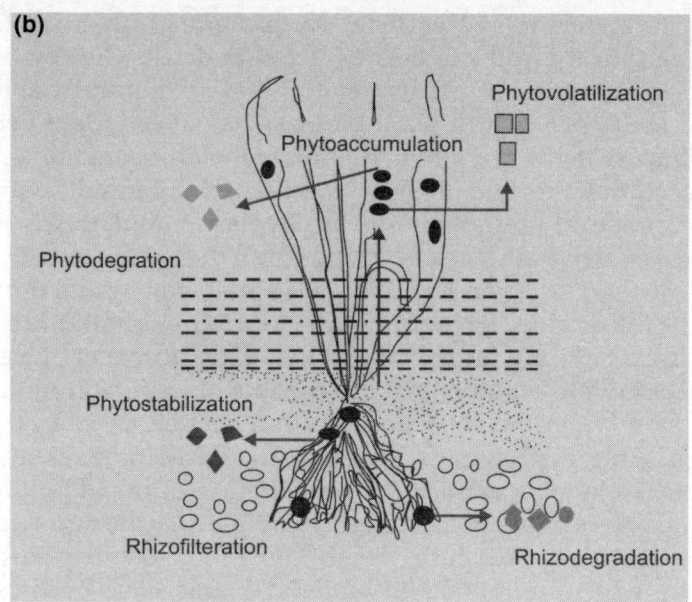

Fig. 8.1: Several process involved in phytoremediation

Phytoextraction: The use of pollutant accumulating plants to remove pollutants like metal organics from soil by concentrating them in harvestable plant parts,

Phytotransformation: The degradation of complex organic to simple molecules or the incorporation of these molecules into plants tissues,

Phytostimulation: Plant-assisted bioremediation or the stimulation of microbial and fungal degradation by release of exudates/enzymes into the root zone (rhizosphere),

Phytovolatilization: The use of plants to volatilize pollutants or metabolites,

Phytodegradation: Enzymatic breakdown of organic pollutants such as trichloroethylene (TCE) and herbicides, both internally and externally and through secreted plant enzymes,

Phytorhizofiltration: The use of plant roots to ab/adsorb pollutants, mainly metals, but also organic pollutants, from water and aqueous waste streams,

Phytostabilization: The use of plants to reduce the mobility and bioavailability of pollutants in the environment, thus preventing their migration to groundwater or their entry into the food chain.

Hydraulic control: The control of the water and the soil field capacity by plant canopies. This technique uses plants that absorb large amounts of water and thus prevent the spread of contaminated wastewater into adjacent uncontaminated areas. Phreatophytes can be used for cleaning saturated soils and contaminated aquifers.

Phytorestauration: Revegetation of barren areas by fast-growing resistant species that efficiently cover the soil, thus preventing the migration of contaminated soil particles and soil erosion by wind and surface water run-off. This technique reduces the spread of contaminants and also visual impact. However, previous soil conditioning is required (e.g. liming or berengeriteamendments) to enable plants to colonize the polluted substrate.

Dendroremediation (Pump and tree): The use of trees to evaporate water and thus to extract pollutants from the soil,

I. **Phytoextraction:** Phytoextraction represents the largest economic opportunities for phytoremediation. It is also called as phyto accumulation. It is considered as the best approach for removing and isolating the contamination from soil while keeping its structure and fertility intact. In metal polluted soil hyper accumulating plants are seeded/transplanted and are cultivated under established agriculture methodologies. Metals present in soil are absorbed by the plants and then translocated to the above ground shoots for accumulations (green colour in Fig. 8.1). When maximum plant growth and metal accumulation are achieved, plants from above ground levels are harvested that results permanent removal of metals from the site. The removed heavy metals can be recycled from the contaminated plant biomass. Phytoextraction is fit for the rehabilitation of large areas low to moderate levels of contaminated land at shallow depths. Possible plants for this technology must be tolerant of the focused metal or metals and be efficient in translocation to the harvestable portion of the plant. In addition, the plant should possess the ability to grow in difficult edaphic conditions like salinity, soil pH, soil structure and water content, to produce dense root system, element selectivity, ease of care and establishment and resistance to disease and insect problems. On the other hand, limitations in the selection of hyper accumulators are shallow root system, slow growth, small biomass production and final disposal. Plant growing naturally in mineralized soil are able to concentrate huge amount of essentials and nonessential metals in their foliage, are basically the cause of inspiration and development of phytoextraction. In hyperaccumulator species the extent of heavy metals like zinc, nickel and copper accumulation often reaches to 1–5%. It is also thought that cause of the evaluation of this uniform phenomena could be the prevention against the herbivores and disease. Various factors are responsible in success of phytoextraction as an environmental cleanup technology. These include level of soil pollution, bioavailability and plant ability to intercept, absorb and concentrate metals in harvestable parts. However, phytoextraction depends on the interaction among soil, metal and plant. This complex interaction in nature is controlled by the climatic conditions and genetic makeup for site specific phytoremediation. Several approaches have been used but the two basic strategies of phytoextraction, which have finally developed are; (i) Chelate assisted phytoextraction or induced phytoextraction, in which artificial chelates are added to increase the mobility and uptake of metal contaminant. (ii) Continuous phytoextraction in this the removal of metal depends on the natural ability of the plant to remediate; only the number of plant growth repetitions are controlled. Discovery of hyperaccumulator species has further boosted this technology. In order to make this technology feasible, the plants must, extract large concentrations of heavy metals into their roots, translocate the heavy metals to surface biomass, and produce a large quantity of plant biomass. The removed heavy metal can be recycled from the contaminated plant biomass. Factors such as growth rate, element selectivity, resistance to disease, method of harvesting, are also important.

The term 'chelate' denotes a complex between metal and a chelating agent and not the chelating agent itself. A shorter word for chelating agent is 'chelant' or 'chelator'.

It is therefore, suggested for using the term 'chelant enhanced phytoextraction'. Other terms such as 'chelant-induced' and 'chelant-assisted' phytoextraction can be used as synonyms to chelant-enhanced phytoextraction. 'Chelate' should be used whenever a metal chelating agent complex is meant, e.g. when talking about a specific complex in soil physical, chemical and biological properties. A large number but only a fraction of metals are readily available/bioavailability for transporting to the roots. To resolve this problem, chemically enhanced phytoextraction has been developed. This approach utilizes high biomass crops that are induced to absorb large quantity of metals whereas metals mobility is enhanced by the treatment of different chemicals. Research into the interaction of plants with chelating agents started in the 1950s with a view to alleviating deficiencies in the essential nutrient metals Fe, Mn, Cu and Zn. Initial results also showed that chelants such as EDTA enhanced plant uptake of Pb and Hg. Enhanced uptake was not only observed in nutrient solution and pot experiments but also in the field. Mainly, there are three factors that control the transportation of heavy metals from soil to plants. These include the total amount of bioavailable metals/elements (quantity factor), the activity and the ratio of elements present in soil as in ionic form (intensity factor) and the rate of elements transferred from solid to liquid phases to plant roots (reaction kinetics).

II. **Phytostabilization:** It is mostly used for the remediation of soil, sediment and sludges and depends on roots ability to limit contaminant mobility and bioavailability in the soil. Phytostabilisation can occur through the sorption, precipitation, complexaction, or metal valence reduction. The plants primary purpose is to decrease the amount of water percolating through the soil matrix, which may result in the formation of hazardous leachate and prevent soil erosion and distribution of the toxic metal to other areas. A dense root system stabilizes the soil and prevents erosion. It is very effective when rapid immobilisation is needed to preserve ground and surface water and disposal of biomass is not required. However the major disadvantage is that, the contaminant remains in soil as it is, and therefore requires regular monitoring. It is also termed as Phytorestoration. In this remedial technique, plant stabilizes wastes and prevents exposure pathway through wind and water erosion, enables hydraulic control that restricts the vertical migration of pollutants into ground water, and immobilizes the pollutants physically and chemically by root sorption and chemical fixation with different soil amendments. Selected plant for this technique should be poor translocators for metal contamination towards aerial parts likely to be consumed by humans or animals, easy to establish, quick to grow, having well developed canopies and root systems, and tolerant to metal pollution and other climatic and site stresses that could limit plant growth.

III. **Phytovolatilization:** Phytovolatilization involves the use of plants to take up contaminants from the soil, transforming them into volatile form and transpiring them into the atmosphere. Phytovolatilization occurs as growing trees and other plants take up water and the organic and inorganic contaminants. Toxic metals such as Se, As and Hg may exist as gaseous species in environment. Recently it is discovered that plants that absorb elemental form of metals from soil, could convert them biologically into gaseous species inside the plant, i.e. biomethylated to form volatile molecules and finally release them to the atmosphere. Some of these contaminants can pass through the plants to the leaves and volatilise into the atmosphere at comparatively low concentrations. Phytovolatilization has been primarily used for the removal of mercury; the mercuric ion is transformed into less toxic elemental mercury. The disadvantage of this technique is mercury released into the atmosphere is likely to be recycled by precipitation and then redeposit back into ecosystem. Gary Banuelos of USDS's Agricultural Research Service have found that some plants grow in high Selenium media produce volatile selenium in the form of dimethylselenide and dimethyldiselenide. Phytovolatilization has been successful in tritium (3_H), a radioactive isotope of hydrogen, it is decayed to stable helium with a half-life of about 12 years.

The mentioned process is controversial of all techniques due to its dubious nature that whether release of these volatilized elements in atmosphere is safe. The disadvantage is the volatilized element could be recycled by precipitation and then redeposit back into ecosystem. According to Brooks et al. (1998), the release of volatile Se compounds from higher plants was first reported by Lewis et al. (1999) whereas Terry et al. (1992) reported that members of the Brassicaceae are capable of releasing up to 40 g Se/ha/dayas various gaseous compounds. Volatile Se compounds, such as dimethylselenide, are 1/600 to 1/500 as toxic as inorganic forms of Se found in the soil. After genetic modification of *Arabidopsis thaliana L.* and *Nicotiana tobacum L.* with bacterial organomercurial lyase (Mer B) and mercuric reductase (Mer A) genes plants have developed abilities to absorb elemental Hg and methyl mercury from the soil and release volatile Hg from leaves to atmosphere. This technology does not require much management after plant seeding. In addition, it has advantage of minimum site disturbance, low erosion rate and there is no need for disposal of hazardous plant material.

IV. **Phytofiltration/rhizofilteration:** Phytofilteration is applicable for the treatment of surface water and groundwater, industrial and residential effluents, downwashes from power lines, storm waters, acid mine drainage, agricultural runoffs, diluted sludges, and radionuclide-contaminated solutions or we can say to all kinds of contaminated water can be treated with rhizofiltration. In phytofiltration plant roots (rhizofilteration) or seedlings (blastofilteration) are grown in aerated water from where they participate and concentrate toxic metals from contaminated effluents. The techniques involve growing plants hydroponically and transplanting into metal polluted water from where plants absorb and concentrate the metals in their roots and shoots. Ideal characteristics in plants for rhizofiltration are fast growing roots with capability for removing toxic metals from solution over extended period of time. After saturation with the metal contamination which forms precipitation over root surface, whole plants or roots are harvested for disposal. This precipitation is caused by the root exudates and changes in rhizospheres pH. Blastofilteration represents the second generation of plant based water treatment technology. According to data blaslofilteration is more efficient than rhizofiltertion for some metals. Due to the increase in surface to volume ratio after germination, seedlings tend to ab/adsorb large quantities of toxic metal ions. The mechanisms of phytofiltration are not necessarily similar for different metals.

Precipitation and exchangeable sorption are involved in the case of Pb. Biological processes are more important for Cd and Pb which are responsible for the slower components of metal removal, its deposition translocation to the shoots from the solution and intracellular uptake to the vacuole. The mechanism of rhizofiltration lies in physical and biochemical impacts of plant roots in wastewater treatment. Efficiency of mechanism of rhizofiltration lays in the efficiency of roots to synthesis certain chemicals which cause heavy metals to rise in plant body. Root exudates and changes in rhizosphere pH may cause metals to precipitate onto root surfaces. The root environment or root exudates may produce biogeochemical conditions that result in precipitation of contaminants onto the roots or into the water body. As they become saturated with the metal contaminants, roots or whole plants are harvested for disposal. Exudates such as simple phenolics and other organic acids can be released from living cells or from the entire cell contents during root decay. These exudates can change metals speciation (i.e. form of the metal), and the uptake of metal ions and simultaneous release of protons, which acidifies the medium and promotes metal transport and bioavailability. Basically certain genes in the plant body make them efficient metal accumulator. Glutathione and organic acids metabolism plays a key role in metal tolerance in plants. Other environmental conditions such as light, temperature, pH also play an important role in the uptake of heavy metals. The process involves raising plants hydroponically and transplanting them into metal-polluted waters where plants absorb and concentrate the metals in their roots and shoots. The root system provides an enormous surface area that absorbs and accumulates the water and nutrients essential for growth along with other non-essential contaminants. To acclimatize the plants, once a large root system has been developed, contaminated water is collected from a waste site and brought to the plants where it is substituted for their water source. The plants are then planted in the contaminated area where the roots take up the water and the contaminants along with it. As the roots become saturated with contaminants, they are harvested.

Initially most of aquatic plants were considered for rhizofiltration but now, Various terrestrial plant species have been found to effectively remove toxic metals such as Cu^{2+}, Cd^{2+}, Cr^{6+}, Ni^{2+}, Pb^{2+}, and Zn^{2+} from aqueous solutions. It was also found that low level radioactive contaminants can successfully be removed from liquid streams. A system to achieve this can consist of a "feeder layer" of soil suspended above a contaminated stream through which plants grow, extending the bulk of their roots into the water. The feeder layer allows the plants to receive fertilizer without contaminating the stream, while simultaneously removing heavy metals from the water. Trees have also been applied to remediation. Trees are the lowest cost plant type. They can grow on land of marginal quality and have long life-spans. This results in little or no maintenance costs. The most commonly used are willows and poplars, which can grow 6 – 8' per year and have a high flood tolerance. For deep contamination, hybrid poplars with roots extending 30 feet deep have been used. Their roots penetrate microscopic scale pores in the soil matrix and can cycle 100 L of water per day per tree. These trees act almost like a pump and treat remediation system. Some other terrestrial plants include sunflower species and brassica species as efficient rhizofiltrators. So we can say that Rhizofiltration is of great importance. It is a cost effective technology, with low labour cost, low or no operational or maintenance cost. It is good alternative to traditional practices of contaminated waste water treatment. It produces no secondary waste. So in nutshell we can say it is a permanent solution to get rid of contaminant from the environment. Advantage of this method is that it may be conducted in situ, with plants being grown directly in the contaminated water body. This allows for a relatively inexpensive procedure with low capital costs. Operation costs are also low but depend on the type of contaminant. This treatment method is also aesthetically pleasing and results in a decrease of water infiltration and leaching of contaminants. After harvesting, the crop may be converted to biofuel briquette, a substitute for fossil fuel. While, this treatment method has its limitation like any contaminant that is below the rooting depth will not be extracted. The plants used may not be able to grow in highly contaminated areas. Most importantly, it can take years to reach regulatory levels. This results in long-term maintenance. Also, most contaminated sites are polluted with many different kinds of contaminants. There can be a combination of metals and organics, in which treatment through rhizofiltration will not sufficient. Plants grown on polluted water and soils become a potential threat to human and animal health, and therefore, careful attention must be paid to the harvesting process and only non-fodder crop should be chosen for the rhizofiltration remediation method.

(a) **Genetic engineering in rhizofiltration:** Agricultural methods such as the application of fertilizers, chelators and pH adjustors can be utilize to further improve the potential for phytoremediation. By using biotechnologies we can enhance the metal uptake potential of the plants and can utilize them as better metal accumulators. Similarly, genetic modification can be used to over express the enzymes involved in the existing plant metabolic pathways or to introduce new pathways into plants. Glutathione and organic acids metabolism plays a key role in metal tolerance in plants. In plants, it is the major low molecular mass thiol compound. Glutathione occurs in plants mainly as reduced GSH. Its synthesis is mediated by the enzymes glutamylcysteine synthetase and glutathione synthetase. Glutathione metabolism is also connected with cysteine and sulphur metabolism in plants. Cysteine concentration limits gluthatione biosynthesis. Low-molecular thiol peptides phytochelatins (PCs) often called class III metalothioneins are synthetized in plants from glutathione induced by heavy metal ions These peptides are synthetized from glutathione by means of α-glutamylcysteine transferase enzyme which is also called phytochelatin synthase (PCS) catalyzing transfer reaction of (α-Glu-Cys) group from a glutathione donor molecule to glutathione, an acceptor molecule. PCS is a cytosolic, constitutive enzyme and is activated by metal ions viz., Cd^{2+}, Pb^{2+}, Ag^{1+}, Bi^{3+}, Zn^{2+}, Cu^{2+}, Hg^{2+}, and Au^{2+}. PCs thus, synthesized chelate heavy metals and form complexes and these complexes are transported through cytosol in an ATP-dependent manner through tonoplast into vacuole. Thus the toxic metals are swept away from cytosol.

(b) **Rhizofiltration limitations:** For effective application and better output there are some specific limitations for rhizofilteration the pH of the influent solution may have to be continually adjusted to obtain optimum metals uptake. The chemical speciation and interaction of all species in the influent have to be understood for proper application. A well-engineered system is required to control influent concentration and flow rate. Plants (especially terrestrial plants) may have to be grown in a greenhouse or nursery and then placed in the rhizofiltration system. Periodic harvesting and plant disposal are also required. Metal immobilization and uptake results from laboratory and greenhouse studies might not be achievable in the field.

(c) **Future research in rhizofilteration:** Modification or over expression of the enzymes that are involved in the synthesis of GSH and PCs might be a good approach to enhance heavy metal tolerance and accumulation in plants. In the process of attempting to improve rhizofiltration, it was discovered that young plant seedlings grown in aerated water (aquacultured) are often more effective than roots in removing heavy metals from water. The technology of using plant seedlings to remove toxic metals from water was termed blastofiltration (blasto is 'seedling' in Greek). Blastofiltration may represent the second generation of plant-based water treatment technology. It takes advantage of the dramatic increase in surface to volume ratio that occurs after germination and the fact that some germinating seedlings also ab/adsorb large quantities of toxic metal ions. This property makes seedlings uniquely suitable for water remediation. Seedling cultures used for blastofiltration can be produced in light or in darkness, and seeds, water and air are the only components required. Heavy metal hyperaccumulators have received increased attention in recent years, due to the potential of using these plants for phytoremediation of metal contaminated sites. However, there are some limitations for this technology to become efficient and cost effective on a commercial scale, as most of the metal hyperaccumulating plants identified have small biomass, and are not very adaptable to harsh environment. These limitations need to be overcome by achieving a good understanding of the mechanisms of metal hyperaccumulation in plants. In the past years, most researches focusing on the physiological mechanisms of hyperaccumulation have made great progress; however, the understanding of a range of molecular/cellular mechanisms will undoubtedly change our concept of metal acquisition and homeostasis in higher plants. With the completion of the Arabidopsis genome project, eventually followed by genome sequences for other plants, the full range of genes that are potentially involved in heavy metal homeostasis and accumulation will be identified. The problem of low biomass phytoremediators can be overcome by increasing plant yield and metal uptake by engineering common plants with hyperaccumulating genes. If non-native transgenic plants are used for phytoremediation, proper control of their dissemination has to be adopted to avoid the introduction of new weed species. Some key technical hurdles that must be overcome for an industry to develop and grow are:

- Identifying more species that have remediative abilities.
- Optimizing phytoremediation processes, such as appropriate plant selection and agronomic practices. Understanding more about how plants uptake, translocate,and metabolize contaminants.
- Identifying genes responsible for uptake and/or degradation for transfer to appropriate high-biomass plants.
- Decreasing the length of time needed for phytoremediation to work.
- Devising appropriate methods for contaminated biomass disposal, particularly

for heavy metals and radionuclides that do not degrade to harmless substances, and protecting wildlife from feeding on plants used for remediation.

- In addition to technical barriers, government regulations will also determine the overall success of phytoremediation.

So one can conclude that rhizofiltration is one of the important technologies, and when such technologies are merged with existing technologies they can be proved as efficient technologies. These are natural boon of nature where natural efficiency of plants can be utilized to treat contaminated sites. Sometimes certain modifications can be done in order to make these plants resistant to toxicants. So they can be greener technology and can help in reducing pollution and hence can help us to step towards a sustainable development.

V. **Siderophore-metal binding:** Siderophores are extracellular, low molecular weight metal chelating agents. These are often produced by bacteria in iron-limiting conditions to facilitate iron uptake. These are either catechol or hydroxamate derivatives and contain certain reactive groups like dicarboxylic acids, polyhydroxy acids and phenolic compounds. In addition to iron, siderophores often bind to metal ions like aluminium, gallium, chromium, nickel etc.. Specific extracellular metal-binding compounds can also be produced by microorganisms in response to low levels of metals, in order to facilitate the uptake of essential metals. The most studied system is the production of siderophores in response to low environmental iron concentrations. Siderophores are low-molecular-mass Fe(III) coordination compounds (500 ± 1000 Da) produced by many micro-organisms and act by complexing and solubilizing insoluble Fe(III) in a form which can be transported into the cell using specific transport mechanisms. Although siderophores are iron(III)-binding compounds, they are also able to bind other metals such as magnesium, manganese, chromium(III), gallium(III) and radionuclides such as plutonium(IV). The efficiencies of siderophores can be increased by chemical modification like substituting Cl^-, NO_2^- on benzene ring. Modified siderophores can be used for removal of metals like cadmium, mercury, copper and even radionuclides like strontium or cesium from mixed effluent.

8.2 PHYTOREMEDIATION POTENTIAL OF WETLAND PLANTS FOR MELANOIDIN DECOLOURISATION IN PRESENCE OF HEAVY METALS AND PHENOL IN EX-SITU CONDITION

Ex-situ removal requires removal of contaminated soil for treatment on or off site, and returning the treated soil to the resorted site. The conventional ex-situ methods applied for remediating the polluted soils relies on excavation, detoxification and/or destruction of contaminant physically or chemically, as a result the contaminant undergo stabilisation, solidification, immobilisation, incineration or destruction. Hence, an experimental study was conducted to evaluate the phytoremediation potential of *T. anguistifolia* for melanoidin decolourisation in presence of heavy metals and phenol. Melanoidin, heavy metals and phenol are common environmental pollutants and they are major source of aquatic pollution. These may lead to propose an effective phytoremediation technique for removal of heavy metals from post methanated distillery effluent containing phenol and melanoidin in a constructed wetland treatment system for environmental safety.

All the metal salts used in this study were purchased from Sisco Research Laboratories Pvt. Ltd. (SRL), India. Phenol was obtained from Sigma, USA. Synthetic melanoidin was prepared by the method of Kumar and Chandra (2006) and diluted with distilled water for preparation of different colour's unit melanoidin. The selections of heavy metals concentration for phytoremediation study were done on the basis of growth potential of *T. angustifolia* in aqueous solution at variable concentration of phenol and melanoidin. The concentration of metals Cd, 8.00; Cu, 48.00; Cr, 2.43; Zn, 26.30; Mn, 20.54; Ni, 16.00; Fe, 296.32, and Pb, 33.92 mg/L were observed optimum for the growth of *T. angustifolia*. Hence, this concentration was selected for further study. The study was conducted in thirteen plastic pots, installed in open natural environment. Pots 1–13 were designated as ST1, ST2, ST3, ST4, ST5, ST6, ST7, ST8, ST9, ST10, ST11, ST12, and ST13. Each pot having 45 L capacity (42 cm diameter and 55 cm depth) was filled from bottom to top with coarse gravel (particle size 4 cm), pea gravel (particle size 1 cm) and fine sand (particle size 0.1 cm). Each layering was done up to 10 cm. The ten healthy plantlets (rhizome) of *T. angustifolia*, collected from uncontaminated site with small amounts of attached native soil were planted in each pot with an average area 4.2 cm²/plant. Further, the plantlets were acclimatized in pots with Hoagland's solution (10 L) for 15 days (Chandra et al., 2008b). The water levels in pots were maintained 10 cm above the sand surface throughout the study. Treatment applications have been began in 15 days acclimatized plants. To evaluate the effect of phenol and melanoidin concentration on heavy metals phytoremediation, two sets of experiments were conducted as detail shown in Table 8.1.

Table 8.1: Experimental details of metal treatment on *T. angustifolia* in presence of various concentrations of phenol and melanoidin

Pots No.	Heavy metals (mg/L)	Phenol (mg/L)	Melanoidin (Co–Pt)
1st set of experiment (variable concentrations of phenol)			
ST1	Metal solution as described in section 2.1.	—	—
ST2	–do–	100	—
ST3	–do–	100	2500
ST4	–do–	200	2500
ST5	–do–	400	2500
ST6	–do–	600	2500
ST7	–do–	800	2500
2nd set of experiment (variable concentrations of melanoidin)			
ST8	–do–	—	2500
ST9	–do–	100	3000
ST10	–do–	100	4000
ST11	–do–	100	5500
ST12	–do–	100	7000
ST13	–do–	100	8500

8.3 PHYSICO-CHEMICAL CHANGES IN MELANOIDIN SOLUTION CONTAINING HEAVY METALS AND PHENOL

Physico-chemical changes of metal solution after addition of increasing concentration of phenol and melanoidin are shown in table 8.2 and 8.3. In 1st set of experiment, constant concentration of metals and melanoidin along with variable (increasing order) concentrations of phenol showed slight alkaline nature (pH 8.1 – 8.6) of ST1 – ST7 solution (Table 8.2). The increased phenol concentrations in metal solution along with melanoidin increased the pH, BOD, COD and colour of solution while nitrogen remained almost constant in first set of experiment (Table 8.2). In first set of experiment it was also observed that there was gradual decrease in all the physico-chemical parameters with pace of time and solution was found growth promoting (Table 8.2). This might be due to phytoremediation potential of *T. angustifolia* and interaction of metals with phenol and melanoidin.

In 2nd set of experiment there was an increase of melanoidin with fixed concentration of metals and phenol, developed acidic conditions (pH 3.0–6.5) with high colour, COD, BOD, and nitrate at 0 days (Table 8.3). The *T. angustifolia* of pots ST8–ST13 at 20 days showed no adverse effect (Fig. 8.2a). While, the plants growth was adversely affected by melanoidin of higher colour range 5500–8500 Co-Pt (ST11, ST12, and ST13) in 2nd set of

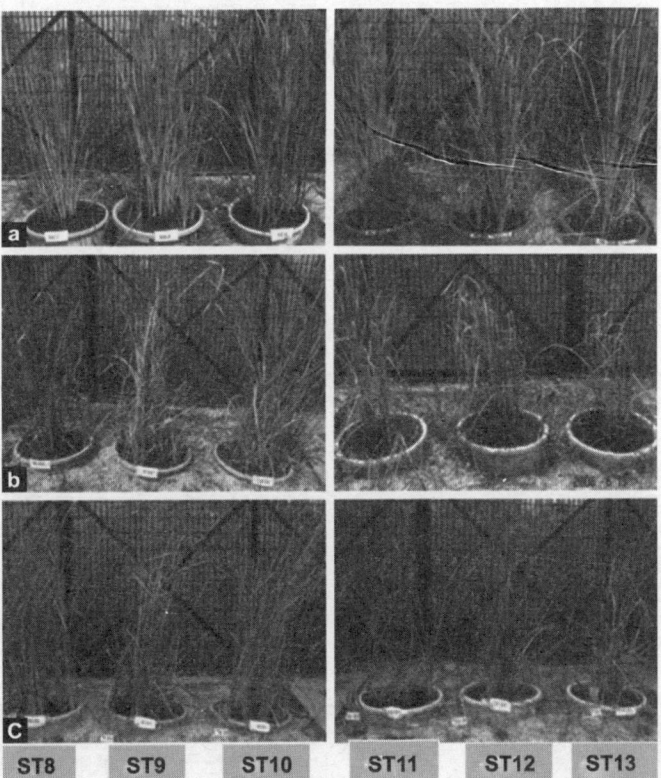

Fig. 8.2: Morphological effect of metals, phenol at variable concentration of melanoidin (Set II, ST8 – ST13 from left to right) on *T. angustifolia* during metal accumulation at 20 (a), 40 (b), and 60 days (c) incubation

Table 8.2: Physico-chemical analysis of water samples of set I

Time	Parameters	ST 1	ST 2	ST 3	ST 4	ST 5	ST 6	ST7
		Treatment pots						
0 day	pH	8.00 ± 0.11	8.10 ± 0.11	8.20 ± 0.10	8.30 ± 0.10	8.44 ± 0.10	8.53 ± 0.10	8.61 ± 0.12
	Colour	500 ± 10.12	900 ± 20.34	2400 ± 27.77	2450 ± 34.56	2450 ± 40.56	2500 ± 50.50	2550 ± 50.37
	COD	2500 ± 100	3200 ± 100	3500 ± 20.08	3600 ± 50.30	3800 ± 60.10	4000 ± 130	4200 ± 133
	BOD	1200 ± 34.45	1400 ± 20.00	1600 ± 80.56	1700 ± 400	1900 ± 100	2000 ± 50.40	2000 ± 20.45
	Phenol	ND	100 ± 5.23	100 ± 2.24	200 ± 4.54	400 ± 12.05	600 ± 20.02	800 ± 40.04
	Nitrogen	82.76 ± 3.64	96.00 ± 2.43	92.52 ± 1.57	94.67 ± 3.45	79.00 ± 2.87	74.00 ± 3.43	65.56 ± 1.87
	Nitrate	9.89 ± 0.34	10.42 ± 0.31	10.86 ± 0.42	11.36 ± 0.32	11.59 ± 0.40	11.97 ± 0.45	12.48 ± 0.36
20 days	pH	$7.77 \pm 0.08^{*}$	$7.84 \pm 0.05^{*}$	$7.85 \pm 0.06^{*}$	$7.87 \pm 0.05^{*}$	$7.83 \pm 0.04^{*}$	$8.00 \pm 0.21^{*}$	$8.00 \pm 0.30^{*}$
	Colour	$400 \pm 20.00^{*}$	$670 \pm 20.43^{*}$	$1180 \pm 20.00^{*}$	$1550 \pm 50.34^{*}$	$1800 \pm 60.12^{*}$	$2000 \pm 50.56^{*}$	$2180 \pm 22.45^{*}$
	COD	$1600 \pm 60.47^{*}$	$2200 \pm 40.78^{*}$	$2300 \pm 60.58^{*}$	$2600 \pm 50.00^{*}$	$2800 \pm 70.00^{*}$	$2900 \pm 30.48^{*}$	$3000 \pm 50.68^{*}$
	BOD	$800 \pm 50.66^{*}$	$1000 \pm 20.44^{*}$	1200 ± 20.67^{ns}	1300 ± 40.57^{ns}	1500 ± 30.74^{ns}	$1500 \pm 20.48^{*}$	$1600 \pm 20.46^{*}$
	Phenol	ND	80.55 ± 3.05^{ns}	82.00 ± 2.04^{ns}	138 ± 2.07^{ns}	164 ± 4.17	$493 \pm 7.20^{*}$	$689 \pm 13.31^{*}$
	Nitrogen	$67.00 \pm 0.76^{*}$	$75.00 \pm 0.84^{*}$	$72.00 \pm 0.92^{*}$	$84.00 \pm 2.38^{*}$	65.00 ± 2.91^{ns}	66.00 ± 2.52^{ns}	60.00 ± 1.76^{ns}
	Nitrate	$3.53 \pm 0.06^{*}$	$4.76 \pm 0.19^{*}$	$5.47 \pm 0.15^{*}$	$5.69 \pm 0.24^{*}$	$6.15 \pm 0.08^{*}$	6.17 ± 0.20^{ns}	$6.54 \pm 0.04^{*}$
40 days	pH	$7.30 \pm 0.10^{*}$	$7.40 \pm 0.00^{*}$	$7.50 \pm 0.13^{*}$	$7.60 \pm 0.00^{*}$	7.70 ± 0.10^{ns}	$7.80 \pm 0.00^{*}$	$7.90 \pm 0.00^{*}$
	Colour	$300 \pm 10.00^{*}$	$550 \pm 10.00^{*}$	$600 \pm 20.00^{*}$	$650 \pm 20.00^{*}$	$850 \pm 12.00^{*}$	$1800 \pm 40.00^{*}$	$2050 \pm 40.00^{*}$
	COD	$1000 \pm 50.00^{*}$	$1100 \pm 40.00^{*}$	$1200 \pm 50.00^{*}$	$1500 \pm 20.00^{*}$	$1600 \pm 45.00^{*}$	$1800 \pm 80.00^{*}$	$2000 \pm 80.00^{*}$
	BOD	$500 \pm 30.00^{*}$	$600 \pm 20.00^{*}$	$600 \pm 40.12^{*}$	$700 \pm 50.00^{*}$	$800 \pm 30.00^{*}$	$1000 \pm 70.00^{*}$	$1100 \pm 20.00^{*}$
	Phenol	ND	52.00 ± 15.00^{ns}	55.00 ± 20.23^{ns}	97.00 ± 20.00^{ns}	102 ± 5.09^{ns}	477 ± 30.00^{ns}	$596 \pm 10.00^{*}$
	Nitrogen	$54.00 \pm 6.00^{*}$	63.00 ± 2.00^{ns}	61.00 ± 3.87^{ns}	$70.00 \pm 3.00^{*}$	56.00 ± 1.50^{ns}	63.00 ± 4.00^{ns}	55.00 ± 2.85^{ns}
	Nitrate	3.06 ± 1.20^{ns}	4.15 ± 1.30^{ns}	$1.54 \pm 1.20^{*}$	$2.03 \pm 1.10^{*}$	2.32 ± 1.40^{ns}	$5.65 \pm 1.80^{*}$	$5.95 \pm 1.30^{*}$
60 days	pH	$7.10 \pm 0.00^{*}$	$7.20 \pm 0.00^{*}$	$7.20 \pm 0.00^{*}$	$7.30 \pm 0.00^{*}$	$7.3 \pm 0.00^{*}$	$7.30 \pm 0.00^{*}$	$7.30 \pm 0.00^{*}$
	Colour	$265 \pm 2.11^{*}$	$432 \pm 0.00^{*}$	$528 \pm 15.00^{*}$	$440 \pm 5.00^{*}$	$220 \pm 10.00^{*}$	$1675 \pm 10.00^{*}$	$1938 \pm 10.00^{*}$
	COD	$800 \pm 20.33^{*}$	$450 \pm 10.00^{*}$	$478 \pm 30.14^{*}$	$561 \pm 20.00^{*}$	$672 \pm 20.00^{*}$	$900 \pm 10.00^{*}$	1659 ± 10.00^{ns}
	BOD	$200 \pm 13.30^{*}$	$210 \pm 10.00^{*}$	220 ± 20.24^{ns}	266 ± 8.00^{ns}	$310 \pm 10.00^{*}$	$338 \pm 10.00^{*}$	360 ± 20.00^{3}
	Phenol	ND	30.00 ± 1.60^{ns}	42.00 ± 18.00^{ns}	62.00 ± 1.82^{ns}	68.00 ± 14.00^{ns}	$400 \pm 18.00^{*}$	579 ± 16.00^{ns}
	Nitrogen	$42.00 \pm 1.12^{*}$	51.00 ± 1.44^{ns}	53.00 ± 1.85^{ns}	$61.00 \pm 2.00^{*}$	55.00 ± 8.00^{ns}	59.00 ± 3.89^{ns}	55.00 ± 1.50^{ns}
	Nitrate	$1.28 \pm 1.12^{*}$	1.68 ± 0.34^{ns}	4.26 ± 1.33^{ns}	4.36 ± 1.63^{ns}	4.48 ± 0.80^{ns}	$2.88 \pm 1.50^{*}$	$4.14 \pm 1.70^{*}$

All values are mean (n = 3) ± SD in mg l^{-1} except pH and colour is in Co–Pt; ST1, heavy metals; ST2, heavy metals and phenol; ST 3–7, heavy metals, melanoidin (2500 Co–Pt) and increasing concentration of phenol (100, 200, 400, 600, and 800 mg l^{-1}). Statistical significance was evaluated within columns by mean of ANOVA. Significance level *p < 0.05; nsp > 0.05. ND = Not detectable.

Table 8.2: Physico-chemical analysis of water samples of set II

Time	Parameters	Treatment pots					
		ST 8	ST 9	ST 10	ST 11	ST 12	ST 13
0 day	pH	7.21 ± 0.10	7.30 ± 0.10	6.60 ± 0.10	6.50 ± 0.10	6.50 ± 0.20	6.50 ± 0.20
	Colour	2500 ± 50.65	2500 ± 26.74	4000 ± 50.56	5500 ± 67.34	7000 ± 80.23	8500 ± 100
	COD	5000 ± 150	6000 ± 200	6600 ± 100	7800 ± 200	8000 ± 150	9200 ± 100
	BOD	2400 ± 76.45	2800 ± 40.20	3200 ± 63.50	3600 ± 40.50	3800 ± 75.40	4200 ± 120
	Phenol	ND	100 ± 2.60	100 ± 3.70	100 ± 4.10	100 ± 5.00	100 ± 5.00
	Nitrogen	85.34 ± 1.25	68.45 ± 2.44	59.46 ± 2.53	71.17 ± 1.45	65.23 ± 2.41	88.38 ± 1.14
	Nitrate	4.88 ± 0.35	5.45 ± 0.62	5.26 ± 0.84	6.66 ± 0.64	6.35 ± 0.86	6.65 ± 0.95
20 days	pH	7.21 ± 0.21^{ns}	7.21 ± 0.14^{ns}	6.70 ± 0.16^{ns}	6.60 ± 0.23^{ns}	6.60 ± 0.28^{ns}	6.70 ± 0.25^{ns}
	Colour	$1400 \pm 50.00^{*}$	$1450 \pm 50.00^{*}$	$1800 \pm 83.00^{*}$	$3800 \pm 50.00^{*}$	$5800 \pm 200^{*}$	$7100 \pm 100^{*}$
	COD	$1800 \pm 50.00^{*}$	$2800 \pm 100^{*}$	$3000 \pm 100^{*}$	$3600 \pm 64.00^{*}$	$5200 \pm 100^{*}$	$6600 \pm 100^{*}$
	BOD	$1000 \pm 87.00^{*}$	$1400 \pm 20.00^{*}$	$1500 \pm 25.00^{*}$	$1800 \pm 30.40^{*}$	$2500 \pm 134^{*}$	$2600 \pm 156^{*}$
	Phenol	ND	82.37 ± 2.63^{ns}	86.27 ± 2.75^{ns}	87.18 ± 1.84^{ns}	$89.73 \pm 2.36^{*}$	92.26 ± 3.21^{ns}
	Nitrogen	$69.36 \pm 1.73^{*}$	$58.46 \pm 1.42^{*}$	55.48 ± 1.63^{ns}	63.36 ± 1.72^{ns}	$61.48 \pm 0.85^{*}$	80.75 ± 2.00^{ns}
	Nitrate	3.35 ± 0.35^{ns}	4.13 ± 0.21^{ns}	2.16 ± 0.75^{ns}	$1.24 \pm 0.26^{*}$	$2.79 \pm 0.32^{*}$	$3.24 \pm 0.15^{*}$
40 days	pH	7.00 ± 0.21^{ns}	$7.30 \pm 0.13^{*}$	7.00 ± 0.25^{ns}	6.70 ± 0.22^{ns}	$6.60 \pm 0.13^{*}$	6.60 ± 0.23^{ns}
	Colour	$1100 \pm 20.45^{*}$	980 ± 30.00^{ns}	1760 ± 40.46^{ns}	$3160 \pm 50.27^{*}$	$5200 \pm 60.25^{*}$	$6600 \pm 114^{*}$
	COD	$1500 \pm 50.21^{*}$	$2400 \pm 60.23^{*}$	$2800 \pm 30.35^{*}$	$3100 \pm 50.36^{*}$	$4600 \pm 40.53^{*}$	6500 ± 75.10^{ns}
	BOD	700 ± 10.10^{ns}	$1000 \pm 12.00^{*}$	$1000 \pm 38.36^{*}$	$1100 \pm 46.26^{*}$	$1200 \pm 50.34^{*}$	$1200 \pm 64.00^{*}$
	Phenol	ND	56.03 ± 1.35^{ns}	52.00 ± 1.84^{ns}	56.28 ± 2.01^{ns}	86.58 ± 2.42^{ns}	89.78 ± 1.51^{ns}
	Nitrogen	$55.00 \pm 3.45^{*}$	51.00 ± 2.14^{ns}	$46.35 \pm 2.34^{*}$	60.45 ± 1.66^{ns}	60.37 ± 1.95^{ns}	79.17 ± 1.33^{ns}
	Nitrate	3.49 ± 0.25^{ns}	3.39 ± 0.26^{ns}	3.44 ± 0.62^{ns}	2.65 ± 0.03^{ns}	1.25 ± 0.03^{ns}	2.57 ± 0.05^{ns}
60 days	pH	7.01 ± 0.20^{ns}	7.30 ± 0.10^{ns}	7.20 ± 0.11^{ns}	7.20 ± 0.13^{ns}	7.20 ± 0.32^{ns}	7.20 ± 0.31^{ns}
	Colour	$1000 \pm 20.00^{*}$	900 ± 15.00^{ns}	1680 ± 21.56^{ns}	$3080 \pm 53.45^{*}$	4970 ± 100^{ns}	6545 ± 132^{ns}
	COD	1380 ± 50.34^{ns}	2094 ± 60.02^{ns}	$2376 \pm 53.87^{*}$	3081 ± 38.65^{ns}	4528 ± 87.00^{ns}	6440 ± 145^{ns}
	BOD	$342 \pm 10.00^{*}$	425 ± 13.40^{ns}	$495 \pm 18.65^{*}$	581 ± 21.50^{ns}	$731 \pm 16.78^{*}$	834 ± 28.45^{ns}
	Phenol	ND	46.70 ± 4.56^{ns}	38.10 ± 2.18^{ns}	31.01 ± 1.85^{ns}	79.03 ± 1.84^{ns}	81.90 ± 2.12^{ns}
	Nitrogen	49.34 ± 1.84^{ns}	43.56 ± 2.13^{ns}	44.78 ± 1.43^{ns}	55.87 ± 1.65^{ns}	$56.45 \pm 1.34^{*}$	78.25 ± 2.14^{ns}
	Nitrate	2.40 ± 0.06^{ns}	1.48 ± 0.02^{ns}	1.75 ± 0.05^{ns}	0.68 ± 0.03^{ns}	1.13 ± 0.02^{ns}	1.52 ± 0.02^{ns}

All values are mean (n = 3) ± SD in mg/L except pH and colour is in Co-Pt; ST8, heavy metals and melanoidin (2500 Co-Pt); ST9–13, heavy metals, phenol (100 mg/L) and increasing concentration of melanoidin (3000, 4000, 5500, 7000, and 8500 Co-Pt). Statistical significance was evaluated within columns by mean of ANOVA. Significance level *$p < 0.05$; ns$p > 0.05$.

experiment by developing necrosis of leaves followed by chlorosis at 40 days of exposure (Fig. 8.2b). This might be due to the high metal accumulation at acidic pH. This might be also due to cumulative effect of phenol, melanoidin, and other factors, i.e. depletion of O_2 and rhizospheric bacterial community disturbance. However, after 40 days plants acclimatization, there was an emergence of new plantlet from rhizome of the same plants. This showed that at initial stage, the plants were shocked by the higher concentration of pollutants. Eventually, due to innate mechanism of tolerance in plants there was arrival of new leaves (Fig. 8.2c). This was also observed that the new plantlets further accumulated heavy metals as nutrient from the given solution. Consequently, there was reduction in all pollution parameter (e.g. colour, BOD, COD, and phenol) in pots ST1–ST13 at 20, 40, and 60 days of incubation compared to 0 day (Table 8.2 and 8.3).

The most parameters, i.e. colour, COD, BOD, phenol, and nitrogen reduced maximum being respectively 86.33, 86.33, 86.25, 82.90 and 46.88% in set I (ST1–ST 7), calculated from Table 8.2. Tables 8.2 and 8.3 showed rapid reduction in pollution parameters (colour, COD, BOD, and phenol) in 1st set of experiment as compared to 2nd set of experiment. These observations revealed that high contents of melanoidin along with constant concentration of phenol and metals were more inhibitor to *T. angustifolia* for heavy metals phytoremediation than aqueous solution containing higher phenol and constant melanoidin and metals.

8.4 HEAVY METALS ACCUMULATION IN DIFFERENT PARTS OF WETLAND PLANTS

After 20, 40 and 60 days of experiment startup a plant has been rooted out from each pots of both sets. The plant samples were carefully washed with tap water followed by 10% $CaCl_2$ solution. Finally, all the samples were ringed using deionized water and dried at 80°C for > 72 h. Further, all samples were ground to a <40 BSS mesh in a Wiley mill. Plant tissues for metals analysis were digested as per AOAC international methods (AOAC, 2002). The heavy metals in pots were also analyzed following the standard method for the examination of water and wastewater (APHA, 2005). Thereafter, Fe, Pb, Cu, Zn, Mn, and Ni were analyzed by using Inductively Coupled Plasma-Atomic Emission Spectrophotometer (ICP–AES) (IRIS Interepid II XDL: Thermo Electron, Waltham, Mass.,USA). The percent metal accumulation in *T. angustifolia* was measured using formula:

$$\% \text{ accumulation} = [(Fm - Im) / Im] \times 100$$

Where, Fm is final and Im is initial metal contents in plant.

Figure 8.3 showed the percent accumulation of different heavy metals in set I and set II. The percent accumulation of different metals, i.e. Cu = 22.32, Pb = 9.30, Ni = 11.10, Fe = 2.28, Mn = 5.06, Zn = 15.59% was noted in ST1. However, ST2 containing heavy metals and phenol showed higher accumulation of different metals than ST1 except Cu (e.g. Cu = 18.67, Pb = 11.58, Ni = 12.90, Fe = 9.29, Mn = 7.09, Zn = 20.00%. Furthermore, in ST3–ST7 showed decreasing pattern of metal accumulation by *T. angustifolia* (Fig. 8.3) after 20 days incubation. Decreased heavy metal accumulation by *T. angustifolia* might be due to increasing concentration of phenol, where melanoidin also contributed toxic effect on metabolic activity of plants. Further, the metal accumulation was increased in *T. angustifolia* at 40 and 60 days incubation compared to 20 days. This indicated the growth dependent metal accumulation in *T. angustifolia*. The heavy metals accumulation pattern from pot

ST1 was in order of Zn > Cu > Ni > Mn > Pb > Fe and in presence of heavy metal and phenol (ST2) the accumulation pattern was noted Pb > Zn > Cu > Ni > Mn > Fe after 60 days incubation. While, in ST3–ST7 the metal accumulation showed variable pattern of metals removal from solution.

Pot ST8 (solution containing heavy metals and melanoidin) accumulated Zn > Cu > Ni > Pb > Mn > Fe after 20 days incubation. Melanoidin along with heavy metals (ST8) more adversely affected the phytoremediation property of *T. angustifolia* as compared to phenol along with heavy metals (ST2) (Fig. 8.3). This might be due to negative

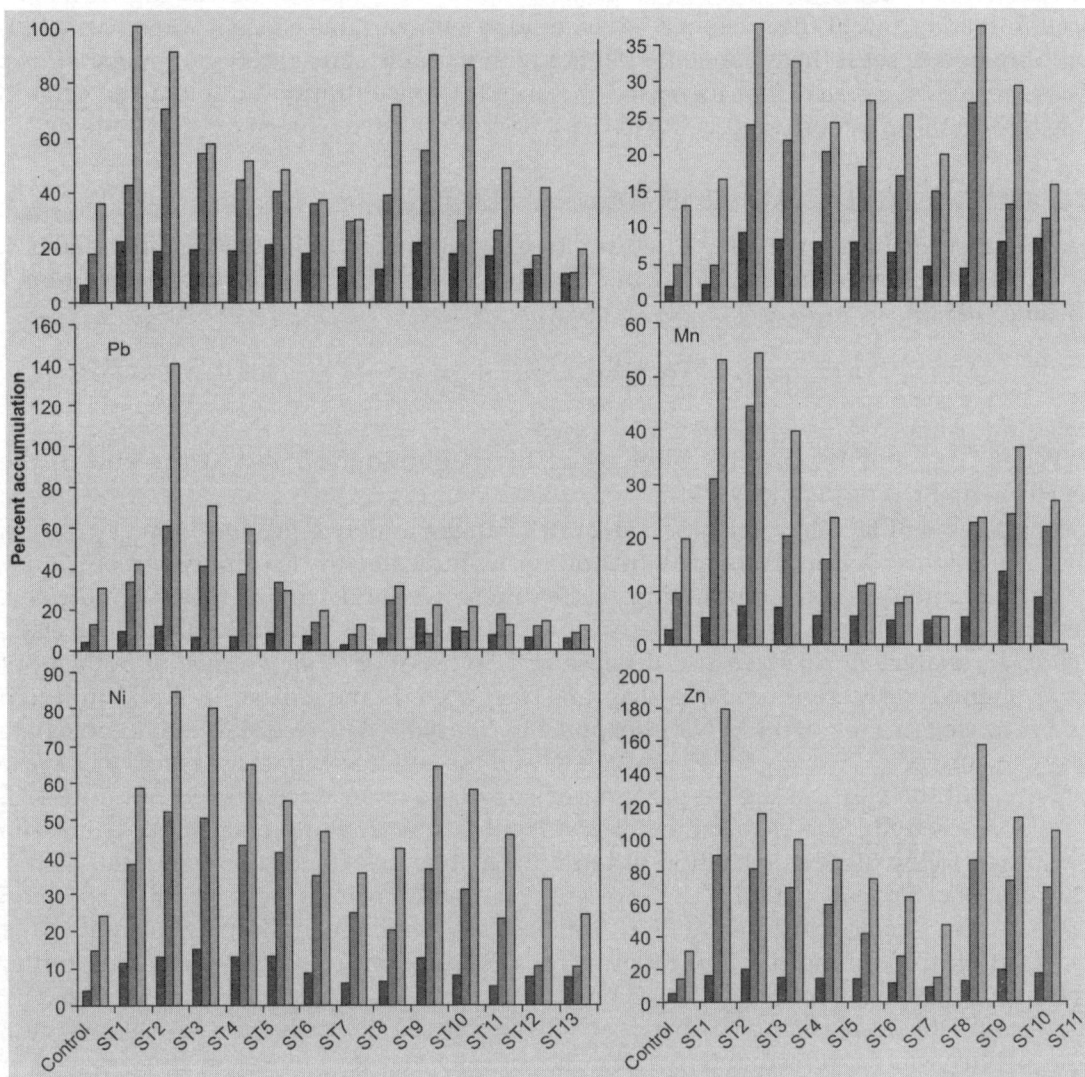

Fig. 8.3: Percent accumulation of different heavy metals in *T. angustifolia* at 20, 40, and 60 days of treatments. ST1, heavy metals; ST2, heavy metals and phenol; ST3-7, heavy metals, melanoidin (2500 Co-Pt) and increasing concentration of phenol (100, 200, 400, 600, and 800 mg/L); ST8, heavy metals and melanoidin (2500 Co-Pt); ST9-13, heavy metals, phenol (100 mg/L) and increasing concentration of melanoidin (3000, 4000, 5500, 7000, and 8500 Co-Pt)

charge of melanoidin which leads to melanoidin-metal complexes formation. Further, the increasing concentrations of melanoidin with constant heavy metals and phenol (ST9–ST13) inhibited the metal accumulation potential of plant after 60 days treatments. But, in all the experimental observation Cu and Zn accumulated maximum in *T. angustifolia* compared with other heavy metals (Fig. 8.3). This might be due to maximum bioavailability of these metals in solution. Further, the metal accumulation in *T. angustifolia* at 40 and 60 days were higher than 20 days accumulation but higher concentration of phenol and melanoidin inhibited metal accumulation (Fig. 8.3). Hence, these results supported time dependent heavy metals accumulation. Figure 8.3 showed higher metals accumulations in set I (heavy metal + melanoidin with variable concentration of phenol) as compared to set II (heavy metal + phenol with variable concentration of melanoidin). These results suggested that increased melanoidin concentration inhibited the growth due to osmotic effect on plant cell and depletion of dissolve oxygen.

8.5 BIOCHEMICAL CHANGES IN WETLAND PLANTS

The plants were harvested after 20, 40 and 60 days of treatment for biochemical analysis. The relative growth rate (RGR) of plant materials was determined using equation 1 (Beadle, 1982),

$$RGR = \frac{W_1 - W_0}{t_1 - t_0}$$

Where, W_0 and W_1 are dry biomass at the beginning (t_0) and at the end of the experimental (t_1) respectively.

The RGR will facilitate the assessment of changes in above-ground biomass during the experimental period. Plant growth monitoring indicated the feasibility and efficiency of each planted pots for employing in treatment wetlands under tested conditions. Ascorbic acid content was determined according to our previous work (Bharagava et al., 2008). To analyze peroxidase and catalase activity, root of *T. angustifolia* of both set of experiments, at different time 20, 40 and 60 days were homogenized in liquid nitrogen and extracted in 2 mL of 0.1 M Na-phosphate buffer pH 7. The resulting cell homogenate was centrifuged at 7000 × g for 15 min at 4°C and the cell free extract assayed for enzyme activity. All the steps in the preparation of enzyme extract were carried out at 0–4°C. Peroxidase activity was determined at 25°C with a spectrophotometer (GBC Cintra-40, Australia) following the formation of tetraguaiacol as described by Singh et al. (2006). One unit of peroxidase activity (U) represents the amount of enzyme catalyzing oxidation of 1 mmol of guaiacol in 1 min at 25°C. Catalase was estimated following the method of Singh et al. (2008). The reaction mixture contained 0.6 mL of 0.1 M phosphate buffer (pH 7.0), 0.3 mL of 70 mM H_2O_2 and 0.1 mL of root extract. The activity was estimated by monitoring the decrease in absorbance at 240 nm due to H_2O_2 reduction (e = 39.4 M/cm). The activity was expressed in terms of mmol of H_2O_2 reduced min/g at 25°C.

The RGR can be employed to confirm the health of plants during treatment period. The RGR values are shown in Fig. 8.4. RGR of control and ST1 grown *T. angustifolia* (heavy metal solution) was 0.034, 0.038 g/day respectively while it was 0.044 g/day for pot ST2 (treated with metals and 100 mg/1 phenol). This indicated *T. angustifolia* growth is not affected by 100 mg/L of phenol. Later in pot ST3 to ST5 the RGR values ranges

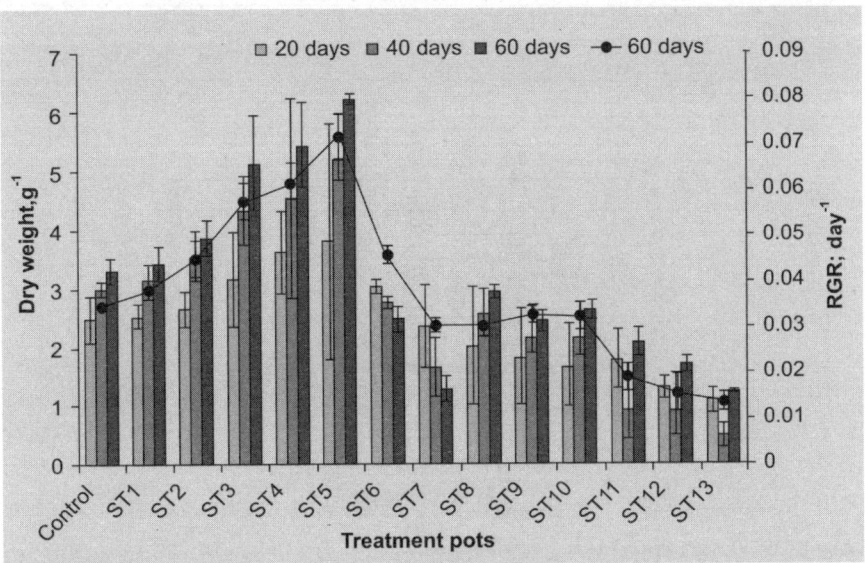

Fig. 8.4: Dry weight and Relative growth rate (RGR) of *T. angustifolia* at 20, 40 and 60 days of treatment

between 0.057 and 0.071 g/day. This revealed that the provided concentrations of phenol (400 mg/L) facilitated the growth of plants. However, RGR values in ST6, ST7 being respectively 0.039 and 0.030 g/day this revealed higher concentrations of phenol inhibited the plants growth. Further pot ST8 (treated with metals and melanoidin) RGR value was 0.030 g/day, which is comparatively lesser than above values. Similarly in pot ST11 to ST13, RGR values concomitantly decreases from 0.018 to 0.014 g/day. Study showed that heavy metals with higher concentration of phenol and melanoidin, adversely affected the growth of plant by reducing the total dry biomass and RGR values. Similar, observation for *T. angustifolia* has been reported earlier for tropical zones receiving municipal wastewater in constructed wetlands treatment system. The heavy metals accumulation and growth in presence of phenol and melanoidin indicated the capability of *T. angustifolia* for distillery effluent pollutants (heavy metals, phenol, and melanoidin) in a mixed solution. Since, toxic levels of heavy metal enhanced production of reactive oxygen species (ROS), which damage cell membranes, nucleic acids, and chloroplast pigments. Accumulation of ROS may be the consequence of disruption of the balance between their production and the antioxidative system activity, composed of enzymic antioxidants such as catalase, peroxidases and superoxide dismutases, and non-enzymic scavengers, e.g. glutathione, carotenoids and ascorbic acid. Under normal circumstances, concentration of oxygen radicals remains low because of the activity of these antioxidative enzymes. In stress condition, the free radical species may be increased, which will enhance the activities of these detoxifying enzymes. Hence, to evaluate the combined effects of phenol and melanoidin on enzymic, e.g. catalase, peroxidase and non-enzymic activity e.g. ascorbic acid in root of *T. angustifolia* were observed in this study. Ascorbic acid content in root of *T. angustifolia* increased as compared to control during 20, 40, and 60 days incubation except ST11-ST13 at 60 days incubation (Fig. 8.5). But, ascorbic acid content was drastically reduced in pot ST11-ST13 where pH was 6.5, BOD ranging between 3600 and

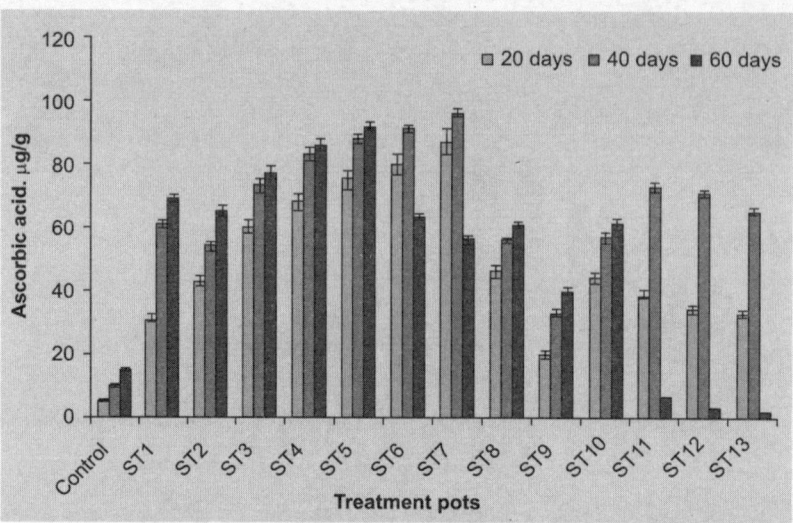

Fig. 8.5: Ascorbic acid contents in *T. angustifolia* root. ST1, heavy metals; ST2, heavy metals and phenol; ST3-7, heavy metals, melanoidin (2500 Co-Pt) and increasing concentration of phenol (100, 200, 400, 600 and 800 mg/ L); ST8, heavy metals and melanoidin (2500 Co-Pt); ST9-13, heavy metals, phenol (100 mg/L) and increasing concentration of melanoidin (3000, 4000, 5500, 7000 and 8500 Co-Pt)

4200 mg/L and COD ranging between 7800 and 9200 mg/L, during 60 days incubation in comparison to set I. Pots ST1-ST5 of set I and ST8-ST10 of set II up to 60 days of plant growth and incubation, ascorbic acid content was gradually increased. This might be due to plants cope up with the oxidative stress induced by free radicals generated under stress condition of heavy metals, phenol and melanoidin. However, in pot ST6–ST7 and ST11–ST13 ascorbic acid content was decreased at 60 days incubation as comparison to 40 days incubation. This indicated the decrease in stress protein after long term incubation (60 days) was probably due to the elevated concentrations of heavy metals and lipid peroxidation. This also indicated the crossing of threshold limit of plant for its tolerance.

The presence of melanoidin along with increased concentration of phenol (ST3–ST5) induced peroxidase activity in root of *T. angustifolia* as compared to control and ST1 after 20 and 40 days incubation (Fig. 8.6). This indicated stress environment around the root. But at 60 days of plant growth, peroxidase activity was decreased along with melanoidin, colour and phenol as shown in Fig. 8.6 and Table 8.2. It was also observed that 57.70–82.90% of phenol disappeared from solution (ST3–ST5) within 60 days of plant growth. However, the peroxidase activity in root of ST6-ST7 *T. angustifolia* decreased compared to ST3–ST5. Similarly, melanidin (2500 Co-Pt) in metal solution (ST8) also induced the peroxidase activity. Further, the presence of phenol along with increasing concentration of melanoidin (ST9–ST11) induced peroxidase up to 20–40 days of growth. But, in ST12 and ST13 peroxidase activity were decreased as compared to ST9–ST11 (Fig. 8.6). This indicated the failure of *T. angustifolia* detoxification mechanism at higher concentration of melanoidin. But, interestingly the catalase activity decreased in root where peroxidase was higher (Fig. 8.7). This might be due to generation of high H_2O_2 and reactivity oxygen species. The peroxidant activity at cellular level by phenoxyl radicals of dietary flavonoids has been previously reported.

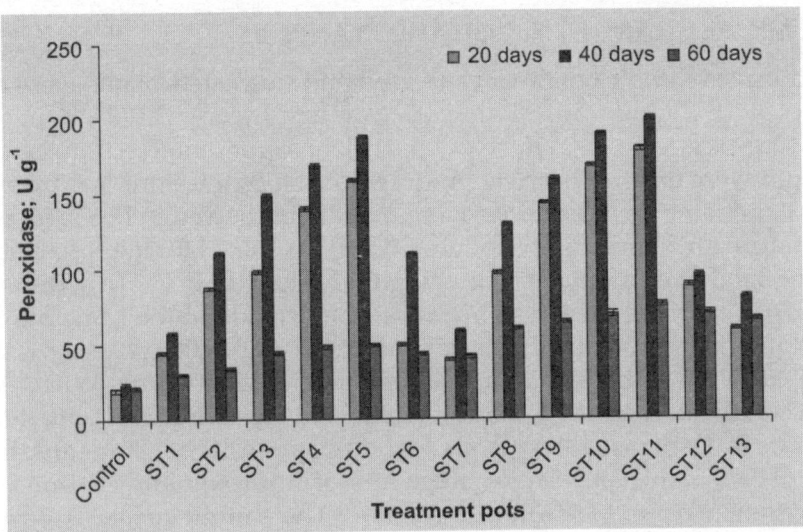

Fig. 8.6: Peroxidase contents in *T. angustifolia* root in presence of metal, melanoidin, and phenol. ST1, heavy metals; ST2, heavy metals and phenol; ST3-7, heavy metals, melanoidin (2500 Co-Pt) and increasing concentration of phenol (100, 200, 400, 600 and 800 mg/L); ST8, heavy metals and melanoidin (2500 Co-Pt); ST9-13, heavy metals, phenol (100 mg/L) and increasing concentration of melanoidin (3000, 4000, 5500, 7000, and 8500 Co-Pt)

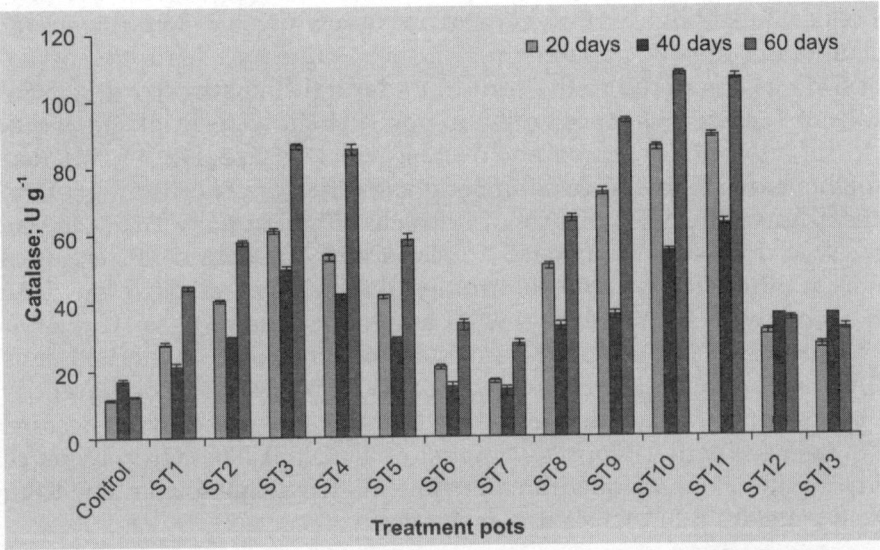

Fig. 8.7: Catalase contents in *T. angustifolia* root in presence of metal, melanoidin, and phenol. ST1, heavy metals; ST2, heavy metals and phenol; ST3-7, heavy metals, melanoidin (2500 Co-Pt) and increasing concentration of phenol (100, 200, 400, 600, and 800 mg/L); ST8, heavy metals and melanoidin (2500 Co-Pt); ST9-13, heavy metals, phenol (100 mg/L) and increasing concentration of melanoidin (3000, 4000, 5500, 7000, and 8500 Co-Pt)

8.6 ANATOMICAL CHANGES IN WETLAND PLANTS

The root and leaves samples of 60 days treated and untreated plants were analyzed by light microscopy (LM) and transmission electron microscopy (TEM). Root samples of 5 mm length were excised from 2 cm below the rhizome-root intersection. Leaves samples of 5 mm length were excised from the middle portion of the third leaf from the base of the plants. All the samples were excised and quickly immersed in H_2S saturated water at room temperature for 30 min to precipitate Cd and Zn. For Light microscopy (LM) Plants samples were fixed overnight at 4°C in 3% glutaraldehyde in 0.1 M sodium cacodylate buffer (pH 6.9), post-fixed for 2 h in 1% osmium tetroxide in the same buffer and then processed as described previously (Rascio et al., 1991). For light microscopy, thin sections (1 μm) were cut with an ultra microtome (Ultracut, Reichert-Jung, Wien, Austria) and stained with equal volumes of 1% toluidine blue and 1% sodium tetraborate, being then examined (Phase Contrast microscope; Nikon; Japan). For Transmission electron microscopy (TEM) Root, and leaves segments of approximately 3 mm length were collected for transmission electron microscopic. The samples were fixed in modified Karnovsky's fluid (David et al., 1973) buffered with 0.1 M sodium phosphate buffer at pH 7.4. Fixation was carried out for 10–18 h at 4°C. After fixation, tissues were washed by fresh buffer and post fixed for 2 h in 1% osmium tetroxide in same phosphate buffer. The tissues were dehydrated in graded acetone solution and embedded in CY 212 araldite. Ultrathin sections of tissue having 60–80 nm thickness were cut using ultra E (Reichert Jung). Ultra sections were stained with uranyl acetate and lead citrate for 10 min before examining the grid in a transmission electron microscope (Phillips, M–10) operated at 60–80 kv transmission.

The anatomical observation of different parts of *T. angustifolia* also showed physiological and biochemical linked deformities in the plant tissue (Fig. 8.8–8.9). *T. angustifolia* being as root accumulator showed apparent metallic deposition and disruption of parenchyma cell in pot ST11 as compared to the control under light microscopy (Fig. 8.8a–b). TEM micrographs of *T. angustifolia* root grown in pot ST11 showed shrinkage of cell, resulted in formation of intercellular spaces, and decrease nucleus size (Fig. 8.8c–d). Reduction of nucleus might be due to tolerance failure of plant at higher concentration of melanoidin. Fig. 8.9a has shown the TEM picture of untreated *T. angustifolia* palisade parenchyma while Fig. 8.9b, c, d has shown damaged trend in spongy tissue and de-shaping of palisade parenchyma at different time. The deformities of cell shape indicated the disturbance in lignifications of cell wall might be due to hyper-accumulation for Cu, Zn and Ni in *T. angustifolia* resulted into induced peroxidase activity in 20, 40 and 60 days of plant growth. In *Brassica juncea* the breakdown of spongy and palisade parenchyma cells followed by loss of cell shape due to reduced lignifications and hyper-accumulation of Zn and Cd has been earlier reported (Sridhar et al., 2005). The parenchyma cell of root which converted to elongate quadrangular this might be due to combine toxic effect of phenol, heavy metals, BOD, COD, and melanoidin.

The study concluded that *T. angustifolia* grown on metal solution containing variable concentration of phenol (200–800 mg/L), melanoidin (2500–8500 Co-Pt) are well adapted due to increase level of antioxidants, which minimized the damage cause by reactive oxygen species, high level of colour, COD, and BOD. However, toxicity emerges when the concentration of pollutants are exceeded the quenching capacity of natural protection system of *T. angustifolia*. *T. angustifolia* showed optimum tolerance for various heavy metal bioremediation in presence of phenol (200–400 mg/L) and melanoidin (3000–4000 Co-Pt). Hence, it can be concluded that, *T. angustifolia* is effective for heavy metals bioremediation from metal, melanoidin, and phenol containing industrial wastewater at optimized condition. Advantages and disadvantages of phytoremediation are shown in table 8.4.

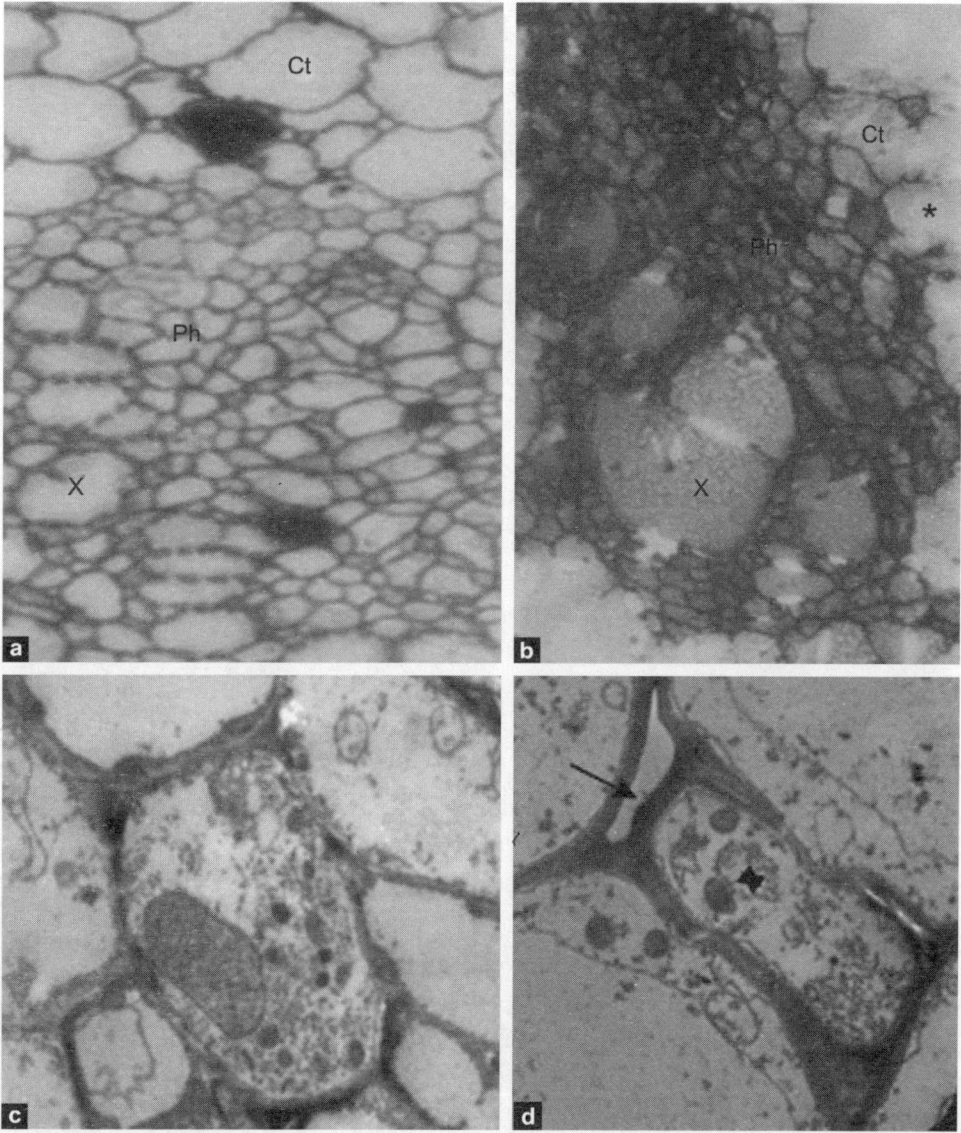

Fig. 8.8: Light micrograph of *T. angustifolia* root shows metal deposition (dark staining) and disruption of cortex cell (b vs a; *) and TEM micrograph shows intercellular space (→) and nucleus size reduction (→) (d) in ST11 as compared to control (c) during metal accumulation in 60 days. Cortex (Ct), phloem (Ph) and xylem(X)

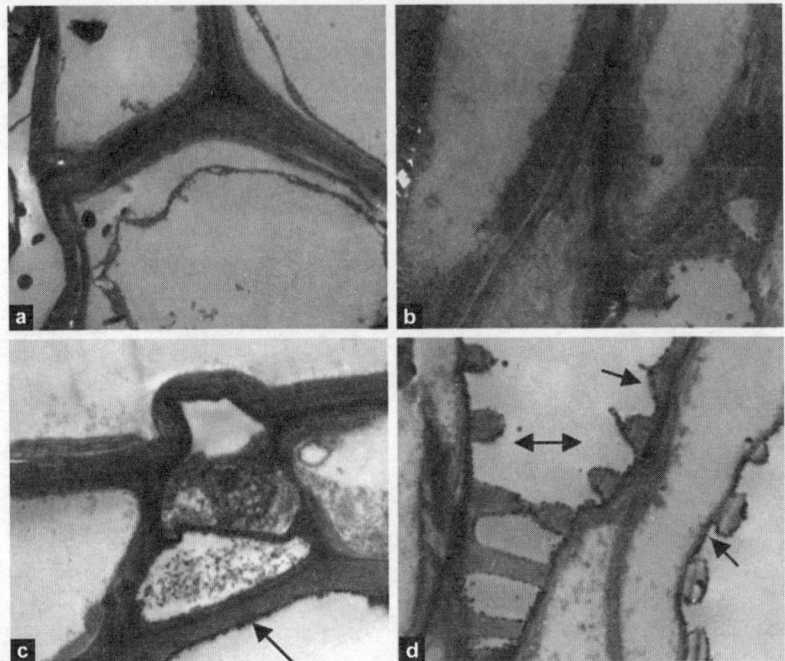

Fig. 8.9: TEM micrograph of *T. angustifolia* leaves shows gradual change and breakdown of cell (b, c, d; ↔) in presence of phenol (100 mg/L), melanoidin (5500 Co-Pt) as compared to control (a) during metal accumulation at different period [20 (b), 40 (c) and 60 (d) days]. Arrow showing metals granules deposition

Table 8.4: Advantages and disadvantages of phytoremediation

S.No.	Advantages	Disadvantages/Limitations
1.	Amendable to a variety of organic and inorganic compounds	Restricted to sites with shallow contamination within rooting zone of remediative plants.
2.	In *Situ/Ex Situ* Application possible with effluent/soil substrate respectively.	May take up to several years to remediate a contaminated site.
3.	In Situ applications decrease the amount of soil disturbance compared to conventional methods.	Restricted to sites with low contaminant concentrations.
4.	Reduces the amount of waste to be landfilled (up to 95%), can be further utilized as bio-ore of heavy metals.	Harvested plant biomass from phytoextraction may be classified as a hazardous waste hence disposal should be proper.
5.	In Situ applications decrease spread of contaminant via air and water.	Climatic conditions are a limiting factor.
6.	Does not require expensive equipment or highly specialized personnel.	Introduction of nonnative species may affect biodiversity.
7.	In large scale applications the potential energy stored can be utilized to generate thermal energy.	Consumption/utilization of contaminated plant biomass is a cause of concern.

Table 8.4: Advantages and disadvantages of phytoremediation (*Contd.*)

S.No.	Advantages	Disadvantages/Limitations
8.	Amendable to a broad range of organic and inorganic contaminants including many metals with limited alternative options.	Restricted to sites with shallow contamination within rooting zone of remediative plants; ground surface at the site may have to be modified to prevent flooding or erosion.
9.	In Situ/Ex Situ application possible with effluent/soil substrate respectively; soil can be left at site after contaminants are removed, rather than having to be disposed or isolated.	A long time is often required for remediation; may take up to several years to remediate a contaminated site.
10.	In Situ applications decrease the amount of soil disturbance compared to conventional methods; it can be performed with minimal environmental disturbance; topsoil is left in a usable condition and may be reclaimed for agricultural use; organic pollutants may be degraded to CO_2 and H_2O, removing environmental toxicity.	Restricted to sites with low contaminant concentrations; the treatment is generally limited to soils at a meter from the surface and groundwater within a few meters of the surface; soil amendments may be required.
11.	Reduces the amount of waste to be landfilled (up to 95%), can be further utilized as bio-ore of heavy metals.	Harvested plant biomass from phytoextraction may be classified as a hazardous waste hence disposal should be proper.
12.	In Situ applications decrease spread of contaminant via air and water; possibly less secondary air and/or water wastes are generated than with traditional methods.	Climatic conditions are a limiting factor; climatic or hydrologic conditions may restrict the rate of growth of plants that can be utilized.
13.	Does not require expensive equipment or highly specialized personnel; it is cost-effective for large volumes of water having low concentrations of contaminants; it is cost-effective for large areas having.	Introduction of non-native species may affect biodiversity.
14.	In large scale applications the potential energy stored can be utilized to generate thermal energy; plant uptake of contaminated groundwater can prevent off-site migration.	Consumption/utilization of contaminated plant biomass is a cause of concern; contaminants may still enter the food chain through animals/insects that eat plant material containing contaminants.

Future of Phytoremediation: One of the key aspects to the acceptance of phytoextraction pertains to the measurement of its performance, ultimate utilization of by-products and its overall economic viability. To date, commercial phytoextraction has been constrained by the expectation that site remediation should be achieved in a time comparable to other clean-up technologies. So far, most of the phytoremediation experiments have taken place in the lab scale, where plants grown in hydroponic setting are fed heavy metal diets. While these results are promising, scientists are ready to admit that solution culture is quite different from that of soil. In real soil, many metals are tied up in insoluble forms, and they are less available and that is the biggest problem, said Kochian. The future of phytoremediation is still in research and development phase, and there are many technical barriers which need to be addressed. Both agronomic management practices and plant genetic abilities need to be optimised to develop commercially useful practices. Many hyperaccumulator plants remain to be discovered, and there is a need to know more about their physiology. Optimisation of the process, proper understanding of plant heavy metal uptake and proper disposal of biomass produced is still needed. The use of wetlands to control pollution is considered to be technologically, economically and environmentally acceptable; the retention of heavy metals in wetlands accumulates problems for the future. A wetland limits the spread of heavy metals, which are stored in the wetland instead. The destruction or harvesting of wetland biomass will release the heavy metals into the environment with the risk of the metals entering the food chain. The long term control of heavy metal pollution control, therefore, lies in the use of other technologies at the extraction, smelting and usage stages.

9

Different Bioremediation Technique for the Decolourization and Detoxification of PMDE

Distilleries are one of the highly polluting industries with reference to the water pollution and the quantity of the wastewater generated. In general about 6–15 L of spentwash is generated per liter of alcohol produced depending on the type of the process thus creating enormous environmental problems. The characteristics of distillery wastewater vary considerably according to the fermentation feed stock, location and the fermentation process adopted. The wastewater is characterized by a high dissolved solid loading (of which 50% may be present as reducing sugars), high ash content, high temperature, low pH and high percentage of dissolved organic and inorganic matter. The biochemical oxygen demand (BOD) and chemical oxygen demand (COD), the index of its polluting character, typically range between 35000–50000 and 100000–150000 mg/L, respectively. Due to stringent requirements being imposed by regulations concerning the discharge of the effluents, there is a growing interest in the development of new technologies and procedures for the treatment of distillery wastewater. The discharge of coloured effluents, though less toxic is presented by the public on the assumption that colour is an indication of pollution. The colour of water, polluted with organic colourants, reduces when the cleavage of the –C = C– bonds, the –N = N– bonds and heterocyclic and aromatic rings occurs. The absorption of light by the associated molecules shifts from the visible to the ultraviolet or infrared region of the electromagnetic spectrum. Disposal of conventionally treated distillery spentwash is harmful as it contains mostly recalcitrant compounds, which are toxic to aquatic biota. Therefore, a comprehensive treatment strategy is required for decolourization of distillery spentwash to meet the discharge safety standards. Following techniques are reported for the treatment of distillery effluent.

9.1 CONVENTIONAL TREATMENT TECHNIQUES

Three popular methods are employed by distilleries to handle their wastewaters: (1) collection of wastewater in storage tanks, followed by irrigation, (2) wastewater treatment in ponds, primarily for the settling of solids, evaporation processes and application of resultant sludge on land, and (3) discharge of the wastewater to a local municipal treatment facility. These three methods have their associated problems and environmental risks. Treatment of distillery wastewaters at municipal facilities is very expensive and is often not a feasible, practical or viable option. Distillery wastewaters were thought to have some beneficial impacts on crop yields, as land application or irrigation is a common method of disposal. The wastewater is first screened, settled in ponds and then distributed over land containing trees, grass and crops using a sprinkler system or channels. Distillery wastewater disposal by irrigation has tremendous potential for polluting ground water and other fresh water bodies due to the presence of high concentrations of phenolic compounds, salinity, phosphates, nitrates and ammonia, which can lead to toxic effects and eutrophication. Hence, the irrigation of fields by distillery effluent can be done only if the concentration of nutrients is within set limits. Till the early 1970s, land disposal was practiced as one of the main treatment options, since it was found to enhance yield of certain crops. For example, in Brazil, vinasse generated from sugarcane juice fermentation is mainly used as a fertilizer due to its high nitrogen, phosphorus and organic content. Its use is further reported to increase sugarcane productivity; furthermore under controlled conditions, the effluent is capable of replacing application of inorganic fertilizers. However, for the high strength molasses-based spentwash, the odor, putrefaction and unpleasant landscape due to unsystematic disposal are concerns in land application. In addition, this option is subject to land availability in the vicinity of the distillery; also, it is essential that the disposal site be located in a low-medium rainfall area. More recent investigations have indicated that land disposal of distillery effluent can lead to groundwater contamination. Deep well disposal is another option but limited underground storage and specific geological location limits this alternative. Other disposal methods like evaporation of spentwash to produce animal feed and incineration of spentwash for potash recovery have also been practiced. The land application of distillery effluent has been advocated due to the presence of high amount of nitrogen and phosphorous. But its indiscriminate application may create adverse effect on crop and contamination of underground water. Kaushik et al. (2005) reported that a long-term application of PMDE caused significant increase in the total organic carbon (TOC), total Kjeldahl nitrogen (TKN), potassium and phosphorous but the harmful concentration of sodium ion (Na^+) accumulated in soil. The short-term application of 50% PMDE along with bioamendments proved most useful in improving the properties of sodic soil and also favored successful germination and improved seedling growth of pearl millet. Similar observation have been recorded by Chandra et al. (2004), where the impact of pre-methanated and post-methanated distillery effluent irrigation on soil microflora and growth of *Phaseolus aureus* was assessed in pot-grown experiment. It was concluded that lower concentration of raw distillery effluent (1–5%) and PMDE (1–10%) stimulated the growth of *P. aureus* and soil micoflora except soil bacteria (inhibited by all the concentrations of raw effluent). The sludge of PMDE also contains high quantity of organic and inorganic pollutants. The sludge amended soil above 10% W/V concentration causes adverse effect on legume crop. The sludge concentration above 50% causes toxicity to plant and even inhibits seed germination (Ramana et al., 2002b). Continuous irrigation with PMDE affects the underground water quality by increasing 40% TDS with salt. A study by Hati et al. (2007) on soil properties and crop yield on a vertisol in India shows increase in organic carbon and salinity. The study shows judicious application of distillery effluent as an amendment to the agricultural field can be considered as available option for safe disposal. These limits were established after some researchers reported that the high salt concentrations in distillery wastewaters resulted in severe inhibitory effects on plants during irrigation. Investigations showed that there were differing responses to varying concentrations of wine distillery wastewater in irrigation water with regard to the percentage of seeds sown that germinated and the speed of germination. At low concentrations of wine distillery wastewater, all crops that were tested showed no inhibition of seed germination, except for tomatoes. However, the percentage germination and the germination speed were inhibited by irrigation with water containing increased concentrations of distillery effluent. The inhibitory effects of distillery wastewater on plant growth can be attributed to the high percentage of organic compounds and salts, and thus its high electrical conductivity, which makes water uptake by seeds difficult and causes retardation of germination. It was found that at concentrations of distillery effluent > 25% (v/v), there was significant fungal growth on the seeds, which was inhibited seed germination. Conversely, an increase in the grain yield of maize, associated with larger cob sizes, higher numbers of seeds per cob and increased grain weight upon irrigation with wine distillery wastewater. It was found that the positive effect on maize crops was observed at low concentrations of distillery wastewater. At these low concentrations, grain yields equivalent to those achievable when using the recommended NPK + FYM (nitrogen, phosphate, potassium and farmyard manure) level of fertilization could be obtained. The concentration of distillery wastewater used to irrigate maize crops could not be increased to greater than 25%, as this would have resulted in problems of salinity. Instead it was recommended that a nonsaline fertiliser be used to supplement distillery wastewater for increased maize grain yields. A similar effect was observed for groundnut. It was concluded that soil and crop types are important when choosing to irrigate land with distillery wastewater, as its effect is both soil dependent and crop specific). The organic components of distillery wastewater leached through the sandy soil and reached the groundwater table, receiving at least partial treatment on the way. Groundwater recharge by high-rate infiltration is a common method of renewing water sources with wastewater in the arid regions of the USA, and water shortages in areas surrounding alcohol and distilleries in South Africa could be partially ameliorated by the reuse of treated wine distillery wastewater to replace potable water for irrigation purposes wherever possible, for example in vineyard irrigation. However, distillery wastewater disposal by irrigation could potentially cause a large-scale environmental problem to which little attention has been paid by this industry until recently. One historical alternative to broad surface irrigation disposal of stillage was deep well disposal. Even though deep well disposal is a cheaper method than land disposal, limited underground storage and very specific geological formations interfere with any wide scale stillage disposal. Again, ferti-irrigation and biocomposting with sugarcane press mud were also found to be popular methods for wastewater disposal. However, these methods are highly energy intensive and hence financially and environmentally expensive. These disadvantages emphasised the need for further research using novel solid/liquid separation methods. As a result, membrane-based separation techniques, such as reverse osmosis (RO) and nanofiltration (NF), were investigated and yielded excellent results when applied to wine distillery wastewater. The effectiveness of NF membrane processes in water and wastewater treatment is generally acknowledged and has now become the most reliable standard technique in combination with biological treatment.

9.2 PHYSICO-CHEMICAL TREATMENT

9.2.1 Adsorption

Among the physico-chemical treatment methods, adsorption on activated carbon (AC) is widely employed for removal of colour and specific organic pollutants. Activated carbon is a well known adsorbent due to its extended surface area, microporus structure, high adsorption capacity and high degree of surface reactivity. Activated carbon is a widely used adsorbent for the removal of organic pollutants from wastewater but the relatively high cost restricts its usage. Decolourization of synthetic melanoidin using commercially available activated carbon as well as activated carbon produced from sugarcane bagasse was investigated. The adsorptive capacity of the different activated carbons was found to be quite comparable. Chemically modified bagasse using 2-diethylaminoethyl chloride hydrochloride and 3-chloro-2 hydroxypropyltrimethylammonium chloride was capable of decolourizing diluted spentwash. 0.6 g of chemically modified bagasse in contact with 100 mL 1 : 4 (v/v) spentwash:water solution resulted in 50% decolourization after 4 h contact with intermittent swirling. Significant decolourization was observed in packed bed studies on anaerobically treated spentwash using commercial activated charcoal with a surface area of 1400 m^2/g. Almost complete decolourization (499%) was obtained with 70% of the eluted sample, which also displayed over 90% BOD and COD removal. In contrast, other workers have reported adsorption by activated carbon to be ineffective in the treatment of distillery effluent. Adsorption by commercially available powdered activated carbons resulted in only 18% colour removal; however, combined treatment using coagulation-flocculation with polyelectrolyte followed by adsorption resulted in almost complete decolourization. Low cost adsorbents such as pyorchar (activated carbon both in granular and powdered form, manufactured from paper mill sludge) and bagasse fly ash have also been studied for this application.

9.2.2 Coagulation and Flocculation

Coagulation is the destabilization of colloids by neutralizing the forces that keep them apart. Cationic coagulants provide positive electric charges to reduce the negative charge (zeta potential) of the colloids. As a result, the particles collide to form larger particles (flocs). Flocculation is the action of polymers to form bridges between the flocs, and bind the particles into large agglomerates or clumps. Bridging occurs when segments of the polymer chain adsorb on different particles and help particles aggregate. Generally coagulation seems to be an expensive step taking into account expenses of chemicals and sludge disposal. Thus, there is a need for development of low cost alternatives for post biomethanated effluent. Some researchers reported that coagulation with alum and iron salts was not effective for colour removal. They explored lime and ozone treatment with anaerobically digested effluent. The optimum dosage of lime was found to be 10 g/L resulting in 82.5% COD removal and 67.6% reduction in colour in a 30 min period. These findings are in disagreement with those of Migo et al. (1993) who used a commercial inorganic flocculent, a polymer of ferric hydroxysulfate with a chemical formula $[Fe_2(OH)n (SO_4)_{3-n/2}]m$ for the treatment of molasses wastewater. The treatment resulted in around 87% decolourization for biodigested effluents; however an excess of flocculent hindered the process due to increase in turbidity and TOC content. $FeCl_3$ and $AlCl_3$ were also tested for decolourization of distillery effluent and showed similar removal efficiencies. About 93% reduction in colour and 76% reduction in TOC were achieved when either $FeCl_3$ or $AlCl_3$ was used alone. The process was independent of chloride and sulfate ion concentration but was adversely affected by high fluoride concentration. However, in the presence of high flocculent concentration (40 g/L), addition of 30 g/L CaO enhanced the decolourization process resulting in 93% colour removal. This was attributed to the ability of calcium ions to destabilize the negatively charged melanoidins; further, formation of calcium fluoride (CaF_2) also precipitates the fluoride ions. Almost complete colour removal (98%) of biologically treated distillery effluent has been reported with conventional coagulants such as ferrous sulfate, ferric sulfate and alum under alkaline conditions. Coagulation studies on spentwash after anaerobic-aerobic treatment have also been conducted using bleaching powder followed by aluminum sulfate. The optimum dosage was 5 g/L bleaching powder followed by 3 g/L of aluminum sulfate that resulted in 96% removal in colour, accompanied by up to 97% reduction in BOD and COD.

9.2.3 Oxidation Process

Ozone is a powerful oxidant for wastewater treatment. Once dissolved in water, ozone reacts with a great number of organic compounds in two different ways: by direct oxidation as molecular ozone or by indirect reaction through formation of secondary oxidants like free radical species, in particular the hydroxyl radicals. Both ozone and hydroxyl radicals are strong oxidants and are capable of oxidizing a number of compounds. Ozone destroys hazardous organic contaminants and that have been applied for the treatment of, dyes, phenolics, pesticides, etc. Oxidation by ozone could achieve 80% decolourization for biologically treated spentwash with simultaneous 15–25% COD reduction. It also resulted in improved biodegradability of the effluent. However, ozone only transforms the chromophore groups but does not degrade the dark coloured polymeric compounds in the effluent. Similarly, oxidation of the effluent with chlorine resulted in 497% colour removal but the colour reappeared after a few days. Ozone in combination with UV radiation enhanced spentwash degradation in terms of COD; however, ozone with hydrogen peroxide showed only marginal reduction even on a very dilute effluent. A combination of wet air oxidation and adsorption has been successfully used to demonstrate the removal of sulfates from distillery wastewater. Studies were done in a counter current reactor containing 25 cm base of small crushed stones supporting a 20 cm column of bagasse ash as an adsorbent. The effluent was applied from the top of the reactor and air was supplied at the rate of 1.0 L/min. The treatment removed 57% COD, 72% BOD, 83% TOC and 94% sulfates. Wet air oxidation has been recommended as part of a combined process scheme for treating anaerobically digested spentwash. The post-anaerobic effluent was thermally pre-treated at 150°C under pressure in the absence of air. This was followed by soda-lime treatment, after which the effluent underwent a 2 h wet oxidation at 225°C. 95% colour removal was obtained in this scheme. Another option is photocatalytic oxidation that has been studied using solar radiation and TiO_2 as the photocatalyst. Use of TiO_2 was found to be very effective as the destructive oxidation process leads to complete mineralization of effluent to CO_2 and H_2O. Up to 97% degradation of organic contaminants was achieved in 90 min. Humic compounds and lignin derivatives constitute the major portion of this dark brown wastewater. The distillery wastewater was diluted with municipal wastewater in the ratio of 3 : 4, irradiated with electron beam and then coagulated with $Fe_2(SO_4)_3$. The optical absorption in UV region was decreased by 65–70% after this treatment. The cost was found to be less than the existing method wherein the effluent was transported about 20 km via pipeline to a facility for biological treatment followed by sedimentation. The Fenton's oxidation technology is based on the production of hydroxyl radicals OH, which have an extremely high oxidation potential. Fenton's reagent, which involves homogeneous reaction and is environmentally acceptable, is a mixture of hydrogen peroxide and iron salts (Fe^{2+} or Fe^{3+}) which produces hydroxyl radicals which ultimately leads to decolourization of the effluent.

9.2.4 Membrane treatment

Pre-treatment of spentwash with ceramic membranes prior to anaerobic digestion is reported to halve the COD from 36,000 to 18,000 mg/L. The total membrane area was 0.2 m^2 and the system was operated at a fluid velocity of 6.08 m/s and 0.5 bar transmembrane pressure. In addition to COD reduction, the pre-treatment also improved the efficiency of the anaerobic process possibly due to the removal of inhibiting substances. Kumaresan et al. (2003) employed emulsion liquid membrane (ELM) technique in a batch process for spentwash treatment. Water-oil-water type of emulsion was used to separate and concentrate the solutes resulting in 86% and 97% decrease in COD and BOD, respectively. Electrodialysis has been explored for desalting spentwash using cation and anion exchange membranes resulting in 50–60% reduction in potassium content. In another study, Vlyssides et al. (1997) reported the treatment of vinasse from beet molasses by electrodialysis using a stainless steel cathode, titanium alloy anode and 4% w/v NaCl as electrolytic agent. Up to 88% COD reduction at pH 9.5 was obtained; however, the COD removal percentage decreased at higher wastewater feeding rates. In addition, reverse osmosis (RO) has also been employed for distillery wastewater treatment. A unit in western India is currently processing effluent obtained after anaerobic digestion, followed by hold-up in a tank maintained under aerobic conditions, in a RO system. 290 m^3/d of RO treated effluent is mixed with 300 m^3/d of fresh water and used in the process for various operations like molasses dilution (290 m^3/d), make-up water for cooling tower (178 m^3/d), fermenter washing (45 m^3/d), etc. Yet another unit in southern India is employing disc and tube RO modules for direct treatment of the anaerobically digested spentwash. The permeate is discharged while the concentrate is used for biocomposting with sugarcane pressmud.

9.2.5 Evaporation/Combustion

Molasses spentwash containing 4% solids can be concentrated to a maximum of 40% solids in a quintuple- effect evaporation system with thermal vapor recompression. The condensate with a COD of 280 mg/L can be used in fermenters. The concentrated mother liquor is spray dried using hot air at 180°C to obtain a desiccated powder with a calorific value of around 3200 kcal/kg. The powder is typically mixed with 20% agricultural waste and burnt in a boiler. The use of recirculating fluidized bed (RCFB) incinerator is recommended to overcome the constraints due to stickiness of spentwash and its high sulfate content. Combustion is also an effective method of on-site vinasse disposal as it is accompanied by production of potassium-rich ash that can be used for land application.

9.2.6 Photodegradation

This phenomena is ascribed to active oxygen species generated by a photo initiated electron-transfer reaction photo-degradation is generally accelerated in the presence of transition metal ions. The photochemical processes are useful for industrial units like effluents from distilleries, pesticide industries, paper mills, pharmaceutical companies and chemical industries, but are in developing stage. The oxidation and reduction reactions are the basic mechanisms in photo catalytic treatment of water/air in their remediation and photo catalytic hydrogen production. A simplified mechanism for photo catalytic process on a semiconductor is presented in equation 1.

Photo catalyst (e.g. TiO_2) + hv = eCB + hVB   ...(1)

For photo catalytic water/air remediation as an environmental application, valence band (VB) holes are the important elements. These induce the oxidative decomposition of environmental pollutants in which the positive hole can oxidize pollutants directly, but mostly reacting with water constituent like hydroxide ion, (OH^-) to produce the hydroxyl radical ($\bullet OH$), which is the very powerful oxidant with the oxidation potential of 2.8 V. This $\bullet OH$ rapidly attacks pollutants at the surface of semi-conducting material and in solution as well and can mineralize them into CO_2, H_2O, etc. The photo catalysts have a potential to completely oxidize a variety of organic compounds, including many highly persistent organic pollutants. The reducing conduction band (CB) electrons are more important when photo catalytic reaction is applied for hydrogen production in water splitting.

9.3 BIOLOGICAL TREATMENT

9.3.1 Anaerobic process

Anaerobic digestion is the transformation of organic matter by a consortium of anaerobic micro-organisms. The main product is biogas. It is composed of a mixture of methane and carbon dioxide as main components, and di-hydrogen, carbon monoxide and di-hydrogen sulphur as minor components. This biological reaction is widespread in natural environment. It happens in anaerobic environment such as marches, rumen, landfills, grounds, etc. Anaerobic digestion of industrial wastewater is commonly used all over the world. It is used as a tool for pollution reduction but also to produce energy. Anaerobic processes usually operate at a temperature of 35°C. In order to maintain this temperature, the methane gas generated in the process is used to heat the reactor. The first step is the organic matter fermentation into volatile fatty acids, alcohol, acetic acid, di-hydrogen and carbon dioxide by fermentative and acidogenic bacteria. The second step is the transformation of volatile fatty acids and alcohol into acetic acid, hydrogen and carbon dioxide by acetogenic bacteria; the last step is methane production from acetic acid by acetoclastic methanogens, and from hydrogen and carbon dioxide by hydrogenophilic methanogens. The temperature could be psychrophilic (5 to 25°C), mesophilic (20 to 45°C) or thermophilic (50 to 70°C). The mesophilic temperature range is generally used for distillery wastewater treatment. The redox potential in the medium is very low (under - 300 mV) and the pH range is between 6.5 to 8. Growth of anaerobic bacteria is low and only a little amount of sludge is produced compare to activated sludge processes. The anaerobic digestion is particularly suitable for winery and distillery wastewater. Their COD/N/P ratio is unbalanced for aerobic treatment, which need phosphorus and nitrogen addition. The COD/N/P ratio for anaerobic digestion could be in the order of 800/5/1. A typical COD/BOD ratio of 1.8–1.9 indicates the suitability of effluent for biological treatment. The high organic content of molasses spentwash makes anaerobic treatment attractive in comparison to direct aerobic treatment. Therefore, biomethanation is the primary treatment step and is often followed by two-stage aerobic treatment before discharge into a water body or on land for irrigation. Aerobic treatment alone is not feasible due to the high energy consumption for aeration, cooling, etc. Moreover, 50% of the COD is converted to sludge after aerobic treatment. In contrast, anaerobic treatment converts over half of the effluent COD into biogas. Anaerobic treatment can be successfully operated at high organic loading rates; also, the biogas thus generated can be utilized for steam generation in the boilers thereby meeting the energy demands of the unit. Further, low nutrient requirements and stabilized sludge production are other benefits of this technique. The performance and treatment efficiency of anaerobic process can be influenced both by inoculum source and feed pre-treatment. In particular, thermal treatment of wastewaters can result in rapid degradation of organic matter leading to lower hydraulic residence time (HRT), higher loading rate and BOD reduction. Moreover the methane content and calorific value of biogas produced from thermophilic systems was higher.

On the contrary, in shake flask studies with an initial COD loading of 20,000 mg/L using biogas plant sludge, Dhar et al. (1998) observed a COD removal of just 6.5% with thermally pre-treated spentwash whereas untreated influent displayed 30.4% COD removal. Significant improvement was observed using inoculum from anaerobic lagoon with 27.2% COD reduction with thermally pre-treated wastewater and 51% reduction with untreated wastewater. Thermal pre-treatment changes the biodegradability of wastewater; thus, it acts as an entirely new feedstock for which the inoculum has to be acclimatized afresh. Further, the temperature of thermal pre-treatment is also important. After 150 d adaptation period, wastewater treated at lower temperature (170°C) showed 66% COD reduction. This was nearly twice the removal obtained with treatment at 230°C. Further, addition of micronutrients (iron, boron and molybdenum) eliminated the long adaptation periods. Anaerobic lagoons are the simplest option for the anaerobic treatment of distillery spentwash. However, large area requirement, odor problem and chances of ground water pollution restrict its usage. Though anaerobic lagoons are still employed in Indian distilleries, high rate anaerobic reactors are more popular. These reactors offer the advantage of separating the hydraulic retention time (HRT) from solids retention time (SRT) so that slow growing anaerobic microorganisms can remain in the reactor independent of wastewater flow.

9.3.1.1 Suspended bed reactor

The Upflow anaerobic sludge blanket (UASB) reactor is a methanogenic (methane-producing) digester that evolved from the anaerobic clarigester. A similar but variant technology to UASB is the expanded granular sludge bed (EGSB) digester. UASB uses an anaerobic process whilst forming a blanket of granular sludge which suspends in the tank. Wastewater flows upwards through the blanket and is processed (degraded) by the anaerobic microorganisms. The upward flow combined with the settling action of gravity suspends the blanket with the aid of flocculants. The blanket begins to reach maturity at around 3 months. Small sludge granules begin to form whose surface area is covered in aggregations of bacteria (Fig. 9.1).

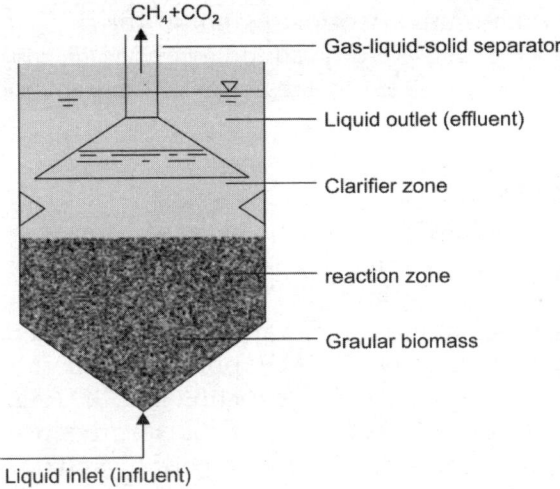

Fig. 9.1: Upflow anaerobic sludge blanket (UASB) reactor

In the absence of any support matrix, the flow conditions create a selective environment in which only those microorganisms, capable of attaching to each other, survive and proliferate. Eventually the aggregates form into dense compact biofilms referred to as "granules". Biogas with a high concentration of methane is produced as a by-product, and this may be captured and used as an energy source, to generate electricity for export and to cover its own running power. The technology needs constant monitoring when put into use to ensure that the sludge blanket is maintained, and not washed out (thereby losing the effect). The heat produced as a by-product of electricity generation can be reused to heat the digestion tanks. UASB reactor is the most popular high rate digester that has been utilized for anaerobic treatment of various types of industrial wastewaters. Treatment by a UASB reactor resulted in 75% COD removal in sugarcane molasses spentwash and 90% COD reduction in whisky pot ale. However, dilution is required before treatment due to the presence of some inhibitory substances such as sulfur compounds, potassium and calcium ions and free hydrogen ions left in solution after pH correction. Most of the practical UASB systems are operated under mesophilic conditions; however, thermophilic operation results in higher methanogenic activity. Mesophilically grown sludge utilized in thermophilic UASB as a seeding material leads to prompt start up and stable operation with 85% COD removal efficiency at a loading of 30 kg COD/m^3d. There are also reports on the cultivation of thermophilic granular sludge for the seeding of thermophilic UASB reactor. Wiegant et al. (1985) reported the cultivation of thermophilic sludge on sucrose for a period of 4 months. The system after adaptation was able to take high COD loadings (86.4 kg/m^3 d) and resulted in 60% COD removal efficiency. In another study in a thermophilic UASB reactor, Harada et al. (1996) reported 39–67% COD removal, with a corresponding BOD removal of over 80%. The results suggested that the wastewater contained high concentration of refractile compounds; this, in turn, affected the microbial population in the sludge granules. Generally, the predominant genera of methanogens in granular sludge are Methanobacterium, Methanobrevibacter, Methanothrix and Methanosarcina; however, the predominance of

Methanothrix in granular sludge is most essential for the establishment of a high performance UASB process. In this study, abundance of *Methanosarcina sp.* was observed whereas *Methanothrix sp.* was present to a lesser extent thereby indicating that the latter are more sensitive to refractile compounds.

9.3.1.2 Fixed bed reactor

This involves immobilization of microorganisms on some inert support to limit the loss of biomass and enhance the bacterial activity per unit of reactor volume. Moreover it provides higher COD removal at low HRT and better tolerance to toxic and organic shock loadings. In anaerobic contact filters, various packing materials, viz. polyurethane, clay brick, granular activated carbon (GAC), polyvinyl chloride (PVC) plastic media have been employed resulting in 67–98% reduction in COD (Fig. 9.2). GAC as support media is relatively expensive but because of its adsorptive properties, it contributes towards improved process stability. The interference by sulfate, unionized sulfite and total hydrogen sulfide in anaerobic filters is reported to be negligible. It was observed that the percentage sulfate removal increased with increasing HRT from 2 to 5 d. This may be due to the utilization of sulfate as a nutrient by microorganisms present in anaerobic contact filter and their conversion to sulfide by sulfate reducing bacteria (SRB) under anaerobic conditions. However at higher sulfate concentration (426 mg/L), the removal decreased, possibly due to low SRB population in comparison to methanogens. Also the removal of sulfide was explained by stripping of hydrogen sulfide from liquid to vapor phase by the carbon dioxide and methane generated during the anaerobic process.

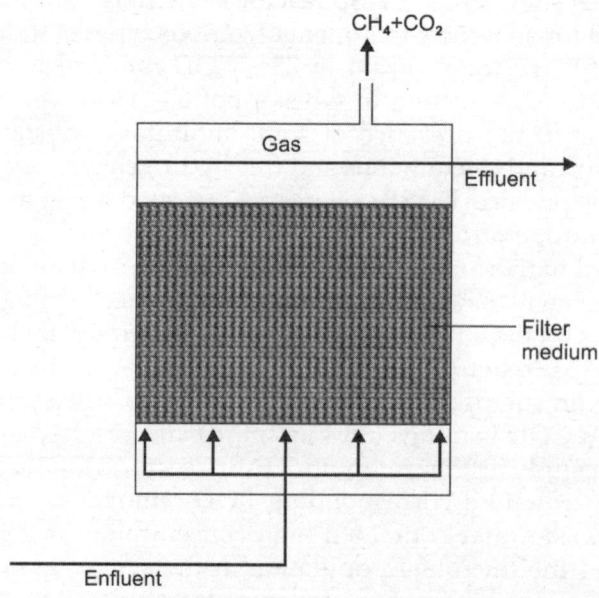

Fig. 9.2: Fixed bed reactor

The pilot system also consisted of a decanter, dephosphatation or magnesium ammonium phosphate (MAP) ($MgNH_4PO_4$) reactor, denitrification reactor, nitrification reactor and sedimentation tank for the reduction of nitrogen and phosphate. Down flow filter using plastic PVC as support material has been employed for the treatment of beet molasses wastewater. The system resulted in 55–85% reduction in COD. Also, though high sulfide concentration (4250 mg) was inhibitory to the system, it was not toxic at higher loadings (44 kg COD/m^3 d) probably due to high stripping of H_2S.

9.3.1.3 Fluidized bed reactor (FBR)

In this type of reactor, a fluid (gas or liquid) is passed through a granular solid material (sand, gravel or plastics for bacterial attachment and growth) at high enough velocities to suspend the solid and cause it to behave as though it were a fluid. This process, known as fluidization, imparts many important advantages to the FBR. As a result, the fluidized bed reactor is now used in many industrial applications. The reactors can be operated either in the upflow or downflow modes, with fluidization being realized by applying high fluid velocities, normally by effluent recycling. The solid substrate material in the fluidized bed reactor is typically supported by a porous plate, known as a distributor (Fig. 9.3).

The fluid is then forced through the distributor up through the solid material. At lower fluid velocities, the solids remain in place as the fluid passes through the voids in the material. This is known as a packed bed reactor. As the fluid velocity is increased, the reactor will reach a stage where the force of the fluid on the solids is enough to balance the weight of the solid material. This stage is known as incipient fluidization and occurs at this minimum fluidization velocity. Once this minimum velocity is surpassed, the contents of the reactor bed begin to expand and swirl around much like an agitated tank or boiling pot of water. The reactor is now a fluidized bed. Depending on the operating conditions and properties of solid phase various flow regimes can be observed in this reactor. Studies on distillery effluent treatment in a down flow fluidized bed system using ground per liter (an expanded volcanic rock) resulted in 75–95% reduction in carbon content. The utilization of ground perlite as a carrier material was advantageous in the down flow configuration because it requires low fluidization velocities which preclude the possibility of clogging.

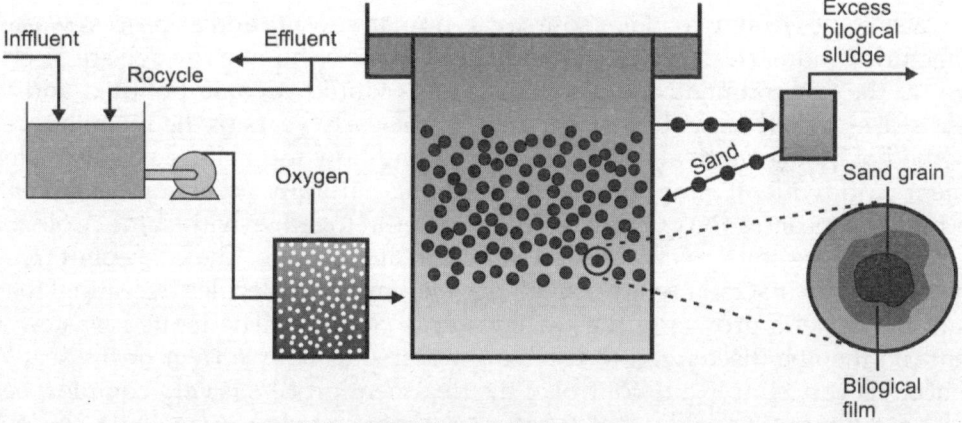

Fig. 9.3: Fluidized bed reactor

9.3.1.4 Two-stage processes and hybrid reactors

A two stage process with an anaerobic filter followed by a UASB reactor was investigated by Blonskaja et al. (2003). The acidogenic and methanogenic phases were clearly separated ensuring better conditions for the methanogens. COD reduction was 54% and 93% in the first and second stage, respectively. In another study on a two-phase thermophilic system, 65% COD reduction combined with a three-fold increase in biogas yield over a single phase system was observed. Boopathy and Tilche (1991) studied the anaerobic digestion of 3–12 times diluted beet molasses wastewater, without pH adjustment, in a hybrid anaerobic baffled reactor (HABR). Additional nitrogen and phosphorus were provided in the form of urea (0.007 g/g of COD) and diammonium hydrogen phosphate (0.0006 g/g of COD). Anaerobic baffled reactor (ABR) consists of a number of UASB reactors connected in series. UASB reactors connected in series. The reactor consisted of three chambers and a final settler. 77% COD removal at a loading rate of 20 kg COD/m3d was obtained. Several variations of the UASB reactor have been investigated for distillery wastewater treatment. In large scale operations, highly variable process wastewater flows makes it difficult to maintain suitable inlet UASB flow rate; further, prevention of the loss of low density granules is also important. To overcome these problems, Akunna and Clark (2000) used granular-bed anaerobic baffled reactor (GRABBR). The reactor consisted of 10 equal compartments, each of which was further divided into two with suitable baffles. Acidogenesis was found to be predominant in the compartments near the inlet and methanogenesis in those located near the outlet. 82–90% COD reduction was observed at a HRT of 4 d. Yet another modification is upflow blanket filter (UBF) in which the packing is limited to 5–10% height of the reactor. This configuration resulted in 70% COD removal in sugarcane molasses distillery spentwash. Vlissidis and Zouboulis (1993) have investigated the thermophilic anaerobic treatment of wastewater from the processing of beet molasses. The process consisted of two stages: anaerobic digestion in upflow sludge bed reactor followed by coagulation-flocculation with lime. The HRT was 11 d in the bioreactor and 2.5 h in the flocculator-precipitator tank. On an average, the overall treatment scheme resulted in 86% BOD and 71% COD removal. Biogas produced in the anaerobic reactor had a methane content of 76%. This configuration was reported to be efficient in treating undiluted wastewaters.

9.3.2 Biological Treatment by Activated Sludge

Wastewater comes from two major sources: as human sewage and as process waste from manufacturing industries. If untreated industrial waste or human sewage are discharged directly to the environment, the receiving waters would become polluted and water-borne diseases would be widely distributed. In the early years of the twentieth century the method of biological treatment was devised and now forms the basis of wastewater treatment worldwide. It simply involves confining naturally occurring bacteria at very much higher concentrations in tanks. These bacteria, together with some protozoa and other microbes, are collectively referred to as activated sludge. The concept of treatment is very simple. The bacteria remove small organic carbon molecules by 'eating' them. As a result, the bacteria grow and the wastewater is cleansed. The treated wastewater or effluent can then be discharged to receiving waters-normally a river or the sea. Whilst the concept is very simple, the control of the treatment process is very complex, because of the large number of variables that can affect it. These include changes in the composition of the bacterial flora of the treatment tanks, and changes in the effluent passing into the plant. The influent can show variations in flow rate, in chemical composition and pH, and temperature. Many municipal plants also have to contend with surge flows of rainwater following storms. Those plants receiving industrial wastewater have to cope with recalcitrant chemicals that the bacteria can degrade only very slowly, and with toxic chemicals that inhibit the functioning of the activated sludge bacteria. High concentrations of toxic chemicals can produce a toxic shock that kills the bacteria. When this happens the plant may pass untreated effluent direct to the environment, until the dead bacteria have been removed from the tanks and new bacterial 'seed' introduced. Globally, the composition of effluents discharged to receiving waters is regulated by the national environment agencies. In Europe the regulatory legislation is the Urban Waste Water Treatment Directive (1991) and the more recent Water Framework Directive (2000). In the USA, the Environmental Protection Agency (EPA) ensures compliance with the Clean Water Act (1977). The legislation is concerned with the prevention of pollution, and therefore sets concentration limits on dissolved organic carbon (as BOD or COD), nitrogen and phosphates- which cause eutrophication in receiving waters. It also attempts to limit the discharge of known toxic chemicals by setting allowable concentration limits in the effluent. Recently, in recognition that effluents contain unknown toxic chemicals, a more pragmatic approach to regulation is being introduced in Europe, using Direct Toxicity Assessment (DTA) tests. In the US these have been in use for many years and are known as Whole Effluent Toxicity (WET) tests. These tests are used to measure the toxic effects of effluents on representative organisms from the receiving waters. Any toxicity detected in the effluents will obviously have been present in the sewage entering the plant. Surprisingly, direct toxicity assessment of influents to wastewater treatment plants that could impact on the functioning of the bioprocesses is not yet included in legislation.

9.3.2.1 Degradable and non-degradable carbon

For control of the biological processes in a treatment plant, it is necessary to have some knowledge of the organic strength, or organic load, of the influent wastewater. Three different measures of this are available, and they each have their merits and weaknesses. The Total Organic Carbon (TOC) is analytically straightforward to measure. It involves oxidation by combustion at very high temperatures and measurement of the resultant CO_2. However, TOC values include those stable organic carbon compounds that cannot be broken down biologically. Organic carbon can also be measured by chemical oxidation. The sample is heated in strong sulphuric acid containing potassium dichromate, and the carbon oxidised is determined by the amount of dichromate used up in the reaction. The result is expressed in units of oxygen, rather than carbon, and the procedure is referred to as the COD. Again it is an analytically simple method. However, its weakness is that a number of recalcitrant organic carbon compounds that are not biologically oxidisable, are included in the value obtained. Conversely, some aromatic compounds, including benzene, toluene, and some pyridines, which can be broken down by bacteria, are only partly oxidised in the COD procedure. Overall however, COD will overestimate the carbon that can be removed by the activated sludge. The current method used to determine the biodegradable carbon, is the 5-day Biological Oxygen Demand (BOD_5). This is a measure of the oxygen uptake over a 5-day period by a small 'seed' of bacteria when confined, in the dark, in a bottle containing the wastewater. During this time the biodegradable organic carbon is taken up, and there is a corresponding decrease in the dissolved oxygen, as some of the carbon is used for the respiration of the bacteria. Respiration is a form of biological oxidation, and will be explained later. Rather unhelpfully, the biodegradable carbon, as in the COD test, is expressed in oxygen units. This is because the test was originally introduced to measure the oxygen depletion in receiving waters caused by the residual degradable carbon in the effluent. Its main value is in regulating the composition of effluents from the treatment water. For process management, where knowledge of the organic loading of the influent is required, BOD_5 is of limited value, because of the 5 days required to make the measurement. There are now moves afoot to replace the use of BOD_5 as a measure of influent strength, with a short-term test (BODST), which can be carried out over a timescale of 30 minutes to several hours.

The values obtained for BOD_5 are always lower than those for COD, for 2 reasons:
- Activated sludge bacteria cannot degrade some of the compounds oxidized chemically in the COD test.
- Some of the carbon removed during the BOD test is not oxidised, but ends up in new bacterial biomass. So the BOD is only measuring the biodegradable carbon that is actually oxidised by the bacteria.

The ratio of BOD_5/COD will depend on the composition of the wastewater. For distillery industries, the ratio is about 0.5–0.6. However, for effluent leaving the treatment plant, it is closer to 0.2. This is because the readily biodegradable organic carbon has been removed during treatment, leaving behind the compounds that are not readily broken down by the bacteria – 'hard' BOD. These will be readily measured by chemical oxidation, but will not be readily degraded and removed by the bacteria in the BOD bottle.

9.3.2.2 'Soft' and 'Hard' BOD

The time-course for the removal of the organic carbon varies with the ability of the activated sludge bacteria to ingest it. Small molecular weight compounds will start to be removed from the sewage immediately after it has entered the activated sludge tanks. Their removal may be completed in 1– 2 hours. This group of compounds is often referred to as the readily biodegradable or 'Soft' BOD. Other, higher molecular weight compounds will take several hours to be degraded and removed. Yet other compounds are more recalcitrant, and may still be present after several days. This less readily biodegradable BOD is often referred to as 'Hard' BOD. The mechanism of their degradation and removal by the bacteria will be dealt with later. The net result is that larger, complex organic carbon molecules may be not be degraded because the treatment time available (the hydraulic retention time) is not sufficiently long, and they will therefore pass out in the effluent. To summarise, the organic carbon in wastewater may be represented as Fig. 9.4.

9.3.2.3 Activated sludge bacteria

The activated sludge of the aeration basin of a wastewater treatment works is a complex ecosystem of competing organisms. The dominant organisms are the bacteria, of which there may be 300 species present. Bacteria are amongst the smallest and most abundant living organisms. Each comprises a single cell varying in size from about 0.5 – 2 μm. On the outside, the cell is bounded by a membrane that regulates the inflow of ions and molecules from the surrounding water. This, in turn is surrounded by a rigid cell wall, made of a sugar polymer. The interior of the cell contains the cytoplasm and the thousands of different chemicals whose reactions are regulated by enzymes. The bacterial cell does not have a nucleus (Fig. 9.5). Most bacteria are spherical, but some may be rod shaped or have a spiral form. Filamentous bacteria comprise long chains of small bacterial cells, sometimes surrounded by a tubular sheath, and can reach lengths of 100 μm. Small

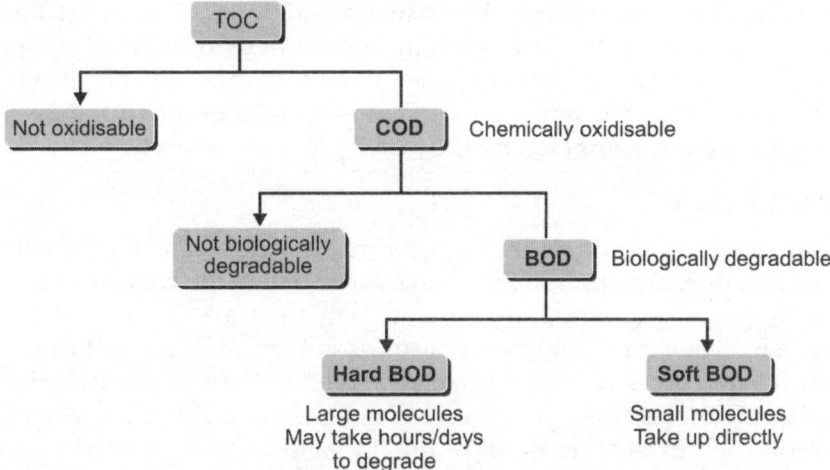

Fig. 9.4: The relationship between the organic carbon fractions in sewage

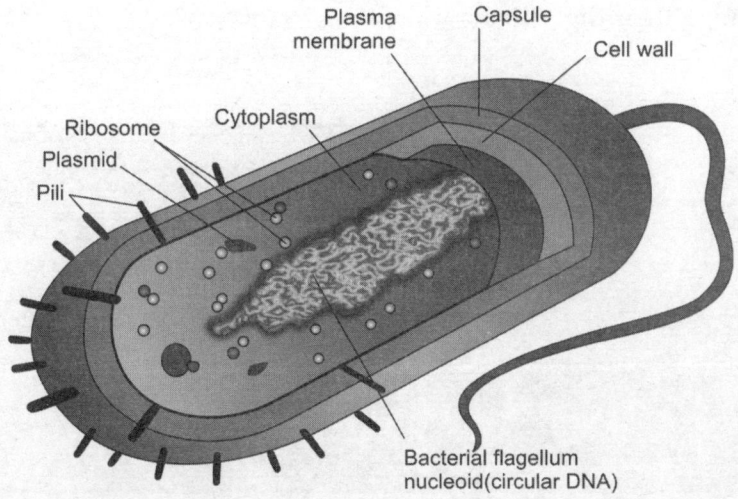

Fig. 9.5: Cross section through a bacterium. The protoplasm inside contains a DNA molecule, but no nucleus. It is surrounded by an inner membrane and a cell wall

molecular weight compounds diffuse into the bacteria (ingestion) through the cell wall. At the same time, some larger complex molecules that have been synthesised within the bacteria, pass outwards. This process is referred to as secretion. The secretions include slimes and gels, that may bond the bacteria together, and also enzymes. The enzymes break down large organic molecules into smaller monomers that are small enough to be ingested. The bacteria use the ingested molecules for the synthesis of new molecules, in the process of growth. When they have reached normal size, the bacterium divides into two, and the process is repeated. If nutrient molecules are not limiting, this results in exponential growth in the numbers of bacteria. The bacteria in a wastewater treatment plant comprise both heterotrophs and autotrophs. The heterotrophic or carbonaceous bacteria are the predominant group of organisms. They are characterised by feeding mainly on organic carbon molecules rather than inorganic ones. By contrast, the autotrophs take in inorganic chemicals, and use these in the synthesis of organic compounds. The nitrifying bacteria that remove ammonia from the wastewater are the most important of this group. There are relatively few species of autotrophs, and since they have low growth rates, they tend to be out-competed by the faster-growing heterotrophs.

9.3.2.4 Bacterial flocs

In a well-maintained aeration tank, the bacteria are concentrated in the flocculent material of the activated sludge, although some always occur free in the wastewater. The flocs are formed from aggregates of non-living organic polymers that are probably secreted by bacteria (Fig. 9.6). They have an open porous structure, and are sufficiently robust to withstand the shear forces created by water movement, during aeration of the tanks. They vary in size from less than 10 µm up to 1mm (1000 µm). The bacteria are adsorbed on to the internal and external surfaces of the floc, and a medium sized floc may harbor several million bacteria. Immediately after the wastewater enters the aeration tank, the fine particulates, colloidal particles and large molecules, become entangled with, and adsorbed to, the floc material. This has the advantage that the enzymes that are secreted by the bacteria into the water, will tend to be confined in the vicinity of the substrate, thereby facilitating their digestion.

Fig. 9.6: Aeration tank with bacterial flocs (a), bacterial flock view in Scanning Electron Microscopy (SEM; b)

However, for the bacteria living on the inside of the floc, oxygen availability may be a problem. This is because oxygen has to diffuse along a concentration gradient from the wastewater through the floc material to the inside. The bacteria of disaggregated flocs may continue to grow when the oxygen concentration of the mixed liquor is only 0.6 mg O_2/L, whereas to ensure this concentration on the inside of a large floc, a mixed liquor oxygen concentration of $1.2-2.0$ mg O_2/L may be required. Quite often, when the aeration tank is operated at below 2.0 mg O_2/L, the centre of the flocs may become oxygen depleted, and colonised by facultative anaerobic bacteria. The outer surface of the activated sludge flocs are frequently colonised by microorganisms of a higher trophic level, including protozoa and rotifers. These feed on bacteria and particulate material in the wastewater. As in all ecosystems, the constituent organisms are in a dynamic steady state. Thus the dominant bacterial species may change, sometimes on a daily basis, in response to changes in the composition of the wastewater. Those species of bacteria that have the ability to secrete the enzymes to break down a novel food source will grow more rapidly, thereby increasing in relative number. This process is known as adaptation or acclimation. In some cases exposure to low levels of potentially toxic chemicals, such as phenol, may result over a period of days in the induction of enzymes that will digest them. These species of bacteria can then exploit the toxicant as a food source.

9.3.2.5 Metabolism of bacteria

Treatment of effluent in the aeration tank involves the removal of organic carbon from the mixed liquor by Ingestion by the bacteria. Once inside, the carbon compounds are metabolised. Metabolism comprises the thousands of simultaneous chemical reactions that are going on at any one time inside the bacterium. In each of these reactions, a substrate, in the presence of an enzyme (which acts as an catalyst), is converted into a product.

$$\text{Substrate} \xrightarrow{\text{Enzyme}} \text{Product}$$

The product then becomes the substrate for the next step in the chain, and is almost immediately converted, in the presence of another specific enzyme, into a different product – and so on. For some of these reactions to take place, chemical energy needs to be provided (endergonic reactions). In other reactions (exergonic reactions), energy is given off, usually in the form of heat. The major divisions of metabolism that concern us here are: Catabolism or Energy Metabolism This comprises a series of reactions in which carbon compounds are broken down to yield cellular energy. This is biological oxidation and involves oxygen uptake by the bacterium. This is also the basis of the process referred to as Respiration. Anabolism, this is a series of biosynthetic reactions in which small molecules are joined together to form large molecular weight macromolecules. This requires an input of energy from catabolism, and is the basis of the process of growth. Although there are many thousands of chemical reactions involved in the metabolism of a bacterium we can identify the three major processes that are relevant to the biological treatment of sewage. These are ingestion, respiration and growth and division. These processes are very highly integrated and the relationship between them in a single bacterial cell can be shown thus: Fig. 9.7 shows the pathway of the ingested organic carbon.

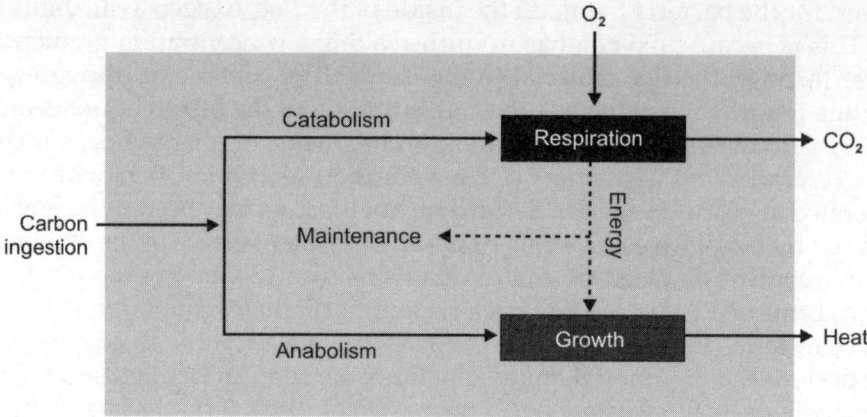

Fig. 9.7: Representation of a single bacterium showing the relationship between the 3 processes

Some goes along the pathway of catabolism or respiration and ends up as carbon dioxide. This carbon is lost to the system. The remaining organic carbon follows the anabolism or growth pathway and ends up in new biomass. This carbon is therefore retained in the system. The purpose of respiration is to provide the energy that is required for growth and for the maintenance of the bacterium. These three processes – Ingestion, Respiration and Growth–are very highly coupled or meshed. No one process can go faster than the other. One implication of this is that, for instance, if you measure the respiration rate, you are indirectly also measuring the rate of growth and the rate of carbon ingestion. Growth is the driver and rate-limiting step. Every bacterium has a genetically programmed maximum rate of growth that will be achieved under ideal conditions. As it grows, it withdraws carbon compounds from the internal pool in its cytoplasm. Carbon flows in (or is ingested) from the mixed liquor in order to keep this pool topped up. At the same time, energy is used for biosynthesis and growth, and hence the catabolism pathways of respiration also withdraw carbon from the internal pool, and this also results in carbon being drawn in by ingestion. It will be noted that the 3 processes correspond to the major processes that we shall see when we examine the operation of the treatment works aeration basin. These include ingestion results into biodegradation, respiration required aeration and growth and division result into biomass production.

(A) Ingestion by bacteria: Ingestion involves the passage of organic carbon compounds, other molecules and ions from the mixed liquor into the bacterium. To do this, they have to pass through the cell wall and the inner membrane. The cell wall does not present much of a barrier, and control over entry is exercised at the inner membrane. Ions such as sodium diffuse in because the concentration in the mixed liquor is higher than inside the bacterium. They then have to be 'pumped' back out again to maintain the internal steady state. Small organic molecules similarly pass in along a concentration gradient, or may be assisted in entry by various mechanisms located in the inner membrane. Most large molecules are excluded. In order to use these for their nutrition and growth, the bacteria secrete enzymes into the water to digest them into small monomers, which can then pass into the cell. Different species of bacteria are specific for what enzymes they secrete, and this determines which chemicals they can exploit as a food source. The ability to secrete a particular enzyme may be latent. In other words, the bacterium requires the presence of the particular chemical compound in the water to switch on the genes for the synthesis of the enzyme required for its digestion.

(B) Respiration: This is a chain of metabolic reactions by which a substrate molecule is oxidised, and the energy made available to do work inside the cell. The energy contained in a substrate such as glucose is rapidly liberated as heat when it is oxidized by burning it in air. When glucose is metabolised in respiration, the same amount of energy is ultimately liberated, but only after some of it has been used to carry out cellular work. During respiration, the energy is initially captured by the molecule adenosine diphosphate (ADP). This adds on another phosphate group to form adenosine triphosphate (ATP). The energy that is captured or transferred is stored in what is sometimes called a 'high energy phosphate bond'.

So when glucose is metabolised the overall reaction is:

$$C_6H_{12}O_6 + 6O_2 + 38 \text{ ADP} + 38P = 6CO_2 + 6H_2O + 38\text{ATP}$$

This is not a perfectly efficient energy capture mechanism, and some of the energy is lost as heat.

The ATP then moves to another site within the cell and releases the energy to do work, as described below. At the same time the phosphate group is released, regenerating ADP again. So overall, we have:

$$38\text{ATP} = 38\text{ADP} + 38P + \text{work} + \text{heat}$$

The ATP is used as quickly as it is produced. The rate-limiting step is in fact the requirement for energy. The faster the cell is using energy, the faster the reactions in respiration proceed. From the equation above, it will be clear that the rate of respiration could be measured by the rate of oxygen uptake, by the rate of CO_2 production or by the rate of heat liberation. Carbon dioxide is difficult to measure in aqueous media. Heat production can be measured in a calorimeter, but the simplest measure of respiration rate is by measuring the oxygen uptake rate with an activated sludge respirometer.

(i) **Why do bacteria need energy?** All living organisms need an input of energy simply to maintain the steady state. For instance, each bacterium is involved in pumping out ions that diffuse through the cell wall, and in various processes of self-repair. All of those processes can be grouped together as maintenance, and all require energy. If the bacterium is mobile (and most are not) energy is used in propulsion. However the main use of energy in bacteria is for biosynthesis for growth. Growth involves the joining together of small molecular weight compounds to form macromolecules, which may then be further modified and assembled to form structures such as membranes, cell walls etc. Thus simple hexose sugars such as glucose, are joined together by glycosidic bonds, amino acids are joined together by peptide bonds to form proteins, and so on. Energy transferred by the ATP molecules is used in doing this work. In the course of this, some of the energy is lost as heat. So as the bacteria grow, they release heat, and this causes the temperature of the aeration tank to be above ambient air temperature.

(ii) **Endogenous respiration:** In a normal growing bacterium, there are a certain number of molecules laid aside as storage products. These are mainly in the form of glycogen and poly-β-hydroxybutyrate (PHB). When all of the biodegradable carbon in the mixed liquor has been used up, as may happen at the end of a plug flow reactor, growth ceases, and the bacterium is then starving. In order to remain alive it still requires energy for maintenance processes. It therefore starts to metabolise its storage products to provide this energy (Fig. 9.8). Although growth has stopped, a low rate of respiration continues in order to provide the energy for maintenance. This is referred to as the Endogenous respiration rate. When the storage products have become exhausted, the bacterium then begins to metabolise cellular proteins and other structural molecules in order to provide the carbon for endogenous respiration. However, this is a like chopping up and burning the furniture to keep the house warm. Eventually the cell dies and splits open, thereby releasing the residual internal molecules, which then become available as potential food source for other bacteria.

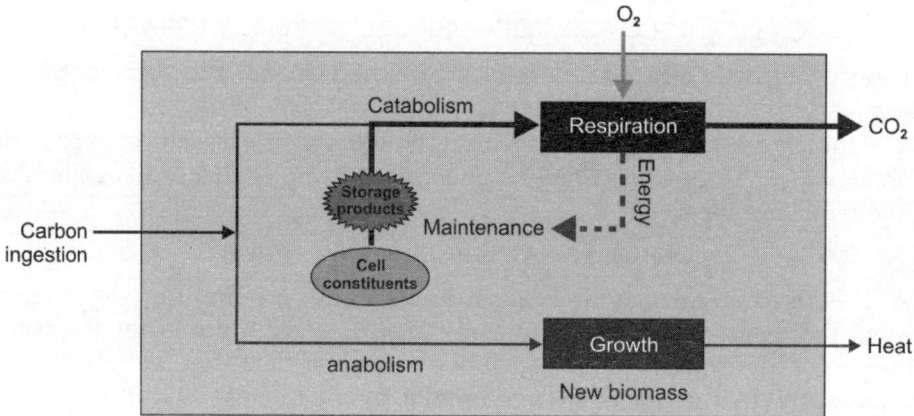

Fig. 9.8: When the bacterium is starving, it has to rely on storage products for the carbon to sustain its endogenous respiration in order to keep itself alive. When these have been exhausted, cell constituents are metabolised

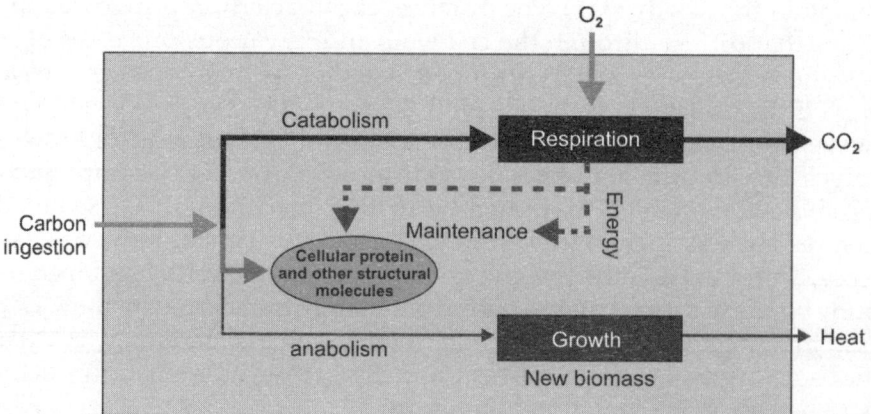

Fig. 9.9: When substrate is provided to a bacterium in Endogenous respiration phase, the carbon is used initially to rebuild that used up during starvation. It also respires some of the carbon to provide the energy for the biosynthesis reactions involved

When a bacterium in the no-growth endogenous phase is presented with feed again, the first process is to rebuild the cell constituents used up. This is followed rapidly by the resumption of growth and the rebuilding of the storage products (Fig. 9.9). It is important to remember that in a growing bacterium in the aeration tanks, part of its respiration is to provide the energy used in biosynthesis and growth, but there is still a small component that is being used in cell maintenance. Because of the tight linking between ingestion, growth and respiration, we find that the rate of respiration is affected by the same factors as that affecting growth rate, viz. substrate concentration, availability of nutrients, oxygen concentration, temperature and toxicity.

(iii) **Effect of substrate concentration on respiration:** When the substrate concentration is at zero, as happens at the end of a plug-flow aeration tank, the bacteria are respiring endogenously, and there is no growth. As the substrate concentration increases, the point is reached at which there is now sufficient carbon intake for growth to take place. As the concentration increases further, the growth rate increases, and the respiration rate also increases. The biosynthetic machinery is switched on very rapidly when bacteria in the endogenous state are presented with food. This can be vividly demonstrated in the lab. If a concentrated feed is added to a flask of endogenous mixed liquor, within seconds it becomes transformed into a foaming and frothing cauldron, as the carbon dioxide bubbles produced in respiration are released.

(iv) **Effect of oxygen on respiration:** If oxygen levels in the mixed liquor are too low, respiration will be inhibited and hence energy will not be available for growth. When saturated with air at 20°C, wastewater will hold approx 9.2 mg O_2/L. Oxygen passes from the water into the bacteria along a concentration gradient (or more accurately a PO_2 gradient). The higher the oxygen concentration in the water, the larger is the gradient to the inside of the bacterial cell, where the oxygen concentration is close to zero. As mentioned previously, oxygen is not limiting above concentrations of about 1.5 – 2.0 mg O_2/L for bacteria in flocs and about 0.6 mg O_2/L for dispersed bacteria. Below these critical concentrations, the respiration rate falls rapidly due to the unavailability of oxygen. Filamentous bacteria have a greater tolerance of low oxygen levels than floc bacteria.

(v) **Effect of Temperature on respiration:** The respiration rate approximately doubles for every 10°C increase in temperature, as noted for growth, above. However, the solubility of oxygen in water decreases with increase in temperature. One consequence of this is increase in the critical oxygen concentration value. Optimum aeration therefore becomes more and more difficult as the temperature in the tanks rises. It is for this reason that most thermophilic plants, operating at 40 – 60°C, have to use pure oxygen for aeration.

(vi) **Toxicity on respiration:** Toxic chemicals can inhibit either the catabolic pathways of respiration or the anabolic pathways of synthesis and growth. Irrespective of which pathway is actually inhibited, all three processes of ingestion, growth and respiration will be similarly inhibited. It is for this reason that toxicity tests conventionally measure the inhibition of respiration, since oxygen uptake rate is easily measured using respirometry. The sequence of events taking place when endogenous activated sludge is presented with feed and then with a slug of toxic waste is shown in Fig. 9.10. Immediately after feeding, the respiration rate rises rapidly to its maximal value. When a toxic wastewater is introduced, the respiration rate falls to a new lower level. The difference between this new rate and the maximal rate is a measure of the inhibition. Inhibition is normally expressed as a percentage of the uninhibited maximum rate of respiration. The percentage inhibition increases with increase in concentration of the toxic chemical in the mixed liquor, as shown in the figure below. Conventionally, toxicity is expressed as the EC_{50}, EC_{20} or EC_{10}, i.e. the concentration causing an inhibition of 50%, 20% or 10% of the respiration rate. We have seen that the rates of both growth and respiration are affected by substrate concentration, availability of other essential nutrients, temperature, and oxygen concentration. Because of the tight coupling of these processes with ingestion, the uptake rate – which corresponds to the rate of biodegradation in the aeration tanks – responds in the same way.

(C) Nitrifying bacteria If the treatment works receives a significant amount of nitrogenous matter in its influents, it will need to be removed by the nitrifiers, in the course of treatment. Nitrifying bacteria are autotrophs, requiring only inorganic chemicals as the starting point for their energy metabolism and growth. Thus ammonia is taken up and oxidised to provide the energy required for growth. Carbon dioxide is used as the carbon source, and this is metabolised into organic carbon compounds inside the bacteria a process which also requires energy. There are relatively few species of nitrifiers and their contribution to the total bacterial biomass is small. The process of ammonia oxidation is referred to as nitrification, and is carried out by two different groups of nitrifiers. The first group oxidise ammonia to form nitrite. The most abundant genus is Nitrosomonas but there are other nitrifiers as well. The overall reaction is:

$$2NH_4 + 3O_2 = 2NO_2 + 2H_2O + 4H^+ + energy$$

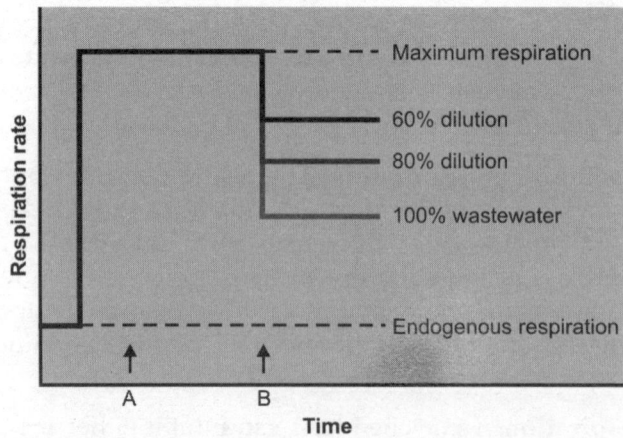

Fig. 9.10: Time course of a laboratory experiment in which samples of activated sludge in endogenous phase are presented (A) with feed in excess, and then (B) with 3 different concentrations of toxic wastewater. The higher the concentration of toxic waste, the greater is the inhibition or the lower is the respiration rate

The oxidation of nitrite to nitrate is carried out by Nitrobacter and by other minor species. The overall reaction here is:

$$2NO_2 + 2O_2 = NO_3 + H_2O + 2H^+ + energy$$

Note that this process should not be referred to as respiration, although oxygen is consumed and the purpose-energy capture is the same. Respiration is the process found in carbonaceous bacteria and in plants and animals. (In mixed activated sludge, the oxygen uptake may comprise both heterotrophic respiration and oxidation by nitrifiers. For convenience only the oxygen uptake rate of activated sludge is referred to as its respiration rate). Chemical oxidation by nitrifiers is not as efficient as the process of respiration in heterotrophs. Relatively large amounts of oxygen are required per unit of energy produced. They therefore have a greater relative oxygen requirement and are slow growing. The major metabolic processes can be represented as if occurring in a single bacterial cell in Fig. 9.11.

Nitrifiers are also characterised by having a low range of temperature tolerance, from 8°C – 30°C and they exhibit a very low metabolic rate below 15 – 20°C. They have a higher critical oxygen concentration of 2.0 – 2.5 mg O_2/L at 20 °C, and their growth is easily disrupted by changes in environmental variables. They are also very much more susceptible to inhibition by toxic chemicals than carbonaceous bacteria, the nitrite oxidising species more so than the ammonia oxidizing species.

(D) Growth of bacteria: Bacteria show prodigious feats of growth. Some bacteria may double their biomass in as little as 20 minutes, provided they have the right conditions of temperature, pH and an abundance of organic carbon, other nutrients, trace elements etc. Note that an individual bacterium has limited capacity for growth, only growing from the size of a daughter cell produced at the time of division to the normal cell size. Growth rate is therefore measured as the increase in number of cells with time. The conditions required for growth vary between species of bacteria. However, there are some general principles. The growth rate observed is a result of both genetic and

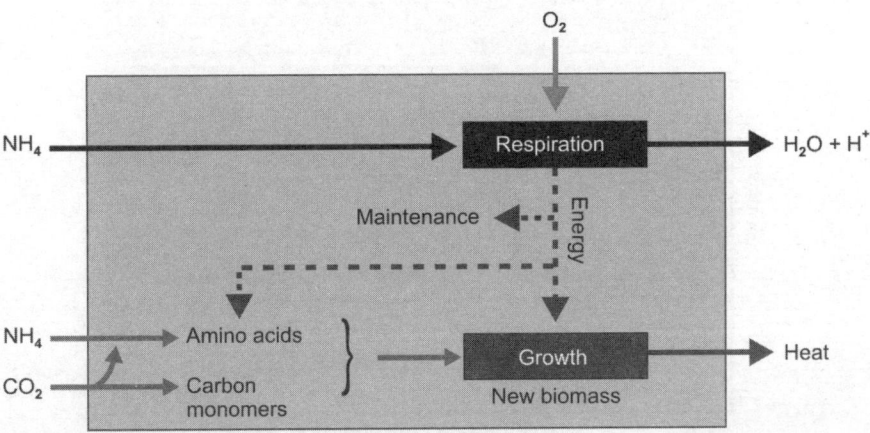

Fig. 9.11: For convenience, the major processes of the two groups of nitrifiers are here combined into a single cell.

environmental factors. The shape of the growth curves and the maximum rate of growth under optimal conditions is genetically determined. The effects of environmental factors are described below.

(i) **Substrate concentration for bacterial growth:** The main substrate for growth is the BOD, or degradable organic carbon in the mixed liquor.

With increase in the concentration of substrate, the growth rate increases exponentially and then levels off. So with further increase in concentration of substrate in the medium, there is no further increase in growth. Note that the curve does not pass through the origin. This is because at very low concentrations, substrate is being used for respiration simply to keep the bacterium alive. At concentrations below A in Fig. 9.12, the bacteria remain alive, but are not growing. The slope of the growth vs substrate concentration curve can be important. A steeper slope indicates a greater affinity for, or ability to use, substrate.

(ii) **Other nutrients for bacterial growth:** Whilst the major substrate requirement is for carbon, growth is also dependent on the intake of nitrogen and phosphorus. The optimum ratio of C : N : P in the mixed liquor is generally thought to be 100 : 5 : 1. The ratio of these nutrients in settled domestic sewage is variously reported as 100 : 17 : 5 or as 100 : 19 : 6. This indicates that nitrogen and phosphorus will not be limiting for growth. Trace components, which include S, Na, Ca, Mg, K and Fe are also required, and are available in abundance in domestic sewage. By contrast, the wastewater from brewing, distilleries, pulp and paper and food-processing industries can be either deficient in nitrogen and phosphorus or non available form. Nutrients therefore need to be added to the mixed liquor to obtain maximum bacterial growth and to optimise carbonaceous treatment. From an operational point of view, lack or an insufficiency of a critical nutrient may result in incomplete treatment, because the bacteria are unable to grow optimally.

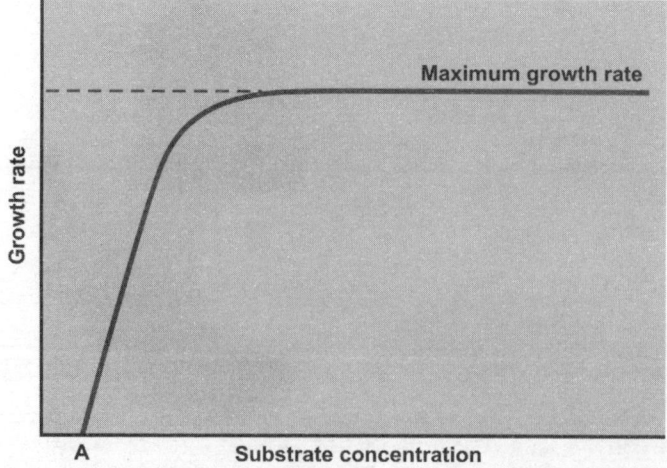

Fig. 9.12: The growth rate increases exponentially with increase in substrate concentration, to a maximum

(iii) **Presence of oxygen for bacterial growth:** Growth can be inhibited if oxygen concentration falls to very low levels in the aeration tank. This is because oxygen becomes limiting for respiration. This is dealt with more fully in the section below, on Respiration.

(iv) **Effect of temperature on bacterial growth:** Bacteria have a genetically determined viable temperature range. For most carbonaceous bacteria of the activated sludge, this is from about 0 to 30°C. However thermophyllic bacteria survive and grow between about 30°C and 60°C. In general, growth rate follows the rule of Arrhenius, that chemical reactions double in rate for a 10°C increase in temperature. Thus as the temperature increases, the rate of growth, and hence requirement for oxygen for respiration, increases.

(v) **Toxicity on bacterial growth**: Toxic chemicals in the wastewater can enter the bacteria and inhibit one or more enzymes of the pathways involved in either anabolism or catabolism (Fig. 9.13). If the catabolic reactions of respiration are affected, the rate of respiration and energy production is reduced and the rate of growth is therefore reduced. On the other hand, if the anabolic pathways of biosynthesis are inhibited, the rate of growth is reduced, and this is accompanied by a fall in the rate of respiration, as the requirement for energy is reduced.

The tight coupling between ingestion, respiration and growth was mentioned previously. It follows that irrespective of where the toxic chemical exerts its inhibitory effect, growth, respiration and ingestion will be equally inhibited. In the aeration tank therefore, toxicity will have the effect of reducing the rate at which organic carbon is degraded. This can be easily monitored by observing changes in the rate of respiration of the activated sludge.

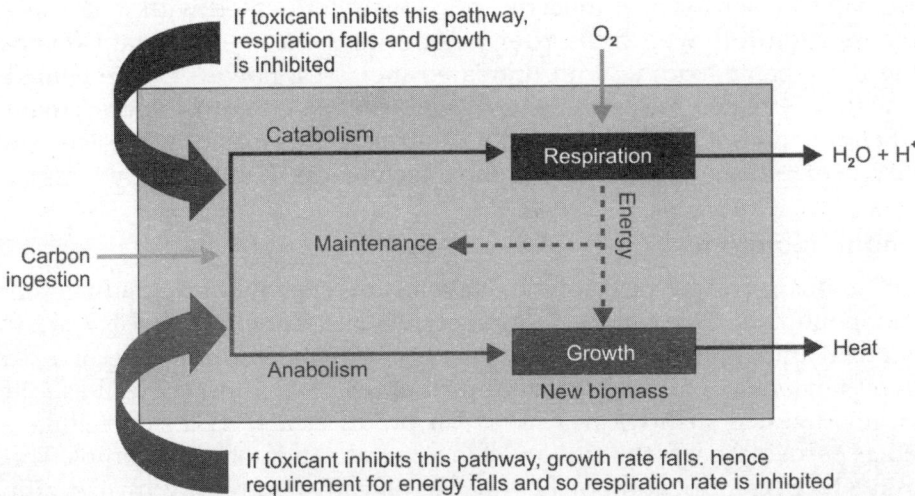

Fig. 9.13: Toxic chemicals can inhibit either the respiration metabolic pathway or the growth metabolic pathway. In both cases the influx of organic carbon into the bacterium, which corresponds to the rate of biodegradation in the aeration tank, is inhibited as a result

9.4 AEROBIC TREATMENT

9.4.1 Biocomposting

The spentwash, either directly, or after biomethanation is sprayed in a controlled manner on sugarcane pressmud. The latter is the filter cake obtained during juice clarification in the manufacture of sugar. Biocomposting is an aerobic, thermophilic process resulting in a product rich in humus which is thus used as a fertilizer. This is a popular option adopted by several Indian distilleries attached to sugar mills with adequate land availability. The most common post-biomethanation step is the activated sludge process wherein research efforts are targeted at improvements in the reactor configuration and performance. For instance, aerobic sequencing batch reactor (SBR) is reported to be a promising solution for the treatment of effluents originating from small wineries. The treatment system consisted of a primary settling tank, an intermediate retention trough, two storage tanks and an aerobic treatment tank. A start up period of 7 d was given to the aerobic reactor and the system resulted in 93% COD and 97.5% BOD removal. Another configuration that has been examined is the rotating biological reactor (RBR). The system consisted of anaerobic fluidized bed reactor coupled with a RBR. Both the reactors were tested with diluted raw vinasse with a COD of 60–70 g/L. At a HRT of 2 d and COD loading up to 20 kg COD m^3/day, the anaerobic unit resulted in 70% COD removal while a lower 46% COD removal was obtained in the aerobic step. Further, testing in the aerobic system was planned with the effluent from anaerobic reactor but no results were reported. The activated sludge process and its variations utilize mixed cultures. To enhance the efficiency of aerobic systems, several workers have focused on treatment by pure cultures. Further, aerobic treatment has also been examined as a precursor to anaerobic treatment. In studies on both beet spentwash and molasses, aerobic pretreatment of beet spentwash with Penicillium decumbens resulted in about 74% reduction in phenolics content and 40% reduction in colour. Anaerobic digestion without aerobic pre-treatment resulted in a sharp drop in COD removal efficiencies with decreasing HRT. The organic matter removal was marginally higher for beet molasses previously fermented with *P. decumbens*. The anaerobic reaction followed first-order kinetics and the rate constant decreased on increasing the organic loading with untreated molasses; however, it remained almost constant with pre-treated molasses. *Geotrichum candidum* is another species that resulted in partial elimination of phenolic inhibitors such as gentisic acid, gallic acid, quercetin, p-coumaric acid, etc. thereby enhancing the effectiveness of anaerobic process.

9.4.2 Fungal Treatment

White rot fungus secreting ligninolytic enzymes are capable of degrading xenobiotics and organopollutants. *Phanerochaete chrysosporium* and *Trametes versicolor* are the most widely studied among these. *P. chrysosporium* JAG 40 resulted in 80% decolourization of diluted synthetic melanoidin (absorbance unit of 3.5 at 475 nm), as well as with 6.25% anaerobically digested spentwash. *T. versicolor* produces a 47 kDa extracellular enzyme identified as peroxidase which is involved in mineralization of melanoidins. The fungus resulted in 82% decolourization of 12.5% anaerobically-aerobically treated effluent. Of this, 90% colour was removed biologically and the rest by adsorption on the mycelium. In addition, treatment by *Trametes species* I-62 (CECT 20197) detoxifies the effluent by degrading furan derivatives as observed by gas chromatography analysis. Decolourization of melanoidin pigment has also been reported by extracellular H_2O_2 and peroxidase produced by *Coriolus hirsutus*. For 6.25% anaerobically digested spentwash, this species showed 71–75% reduction in colour and 90% reduction in COD. Further, *Coriolus versicolor* Ps4a decolourizes the effluent by decomposition of melanoidins and not by the partial transformation of chromophores. The decolourizing activity was attributed to an intra-cellular enzyme which is induced in the presence of melanoidin pigment. Treatment of biodigested distillery wastewater by *C. versicolor* has also been investigated by Chopra et al. (2004). The fungus was able to reduce both COD and colour up to 53% in 8 d; however, glucose and peptone were required as additional nutrient sources. It was suspected that Maillard reaction also resulted in the formation of pyrogenic compounds like polycyclic aromatic hydrocarbons (PAHs) that are toxic to estuarine fish. Treatment by *F. flavus* detoxified the effluent by 68% reduction in PAH and resulted in 73% colour removal. The fungus was more effective in decolourizing raw molasses spentwash than the anaerobically and aerobically treated streams. This was possibly due to changes in the chemical structure of the melanoidin pigments during anaerobic and aerobic treatment. However, the oxygen demand of the fungus was reportedly high. The effects of filamen-tous fungi have also been studied on distillery wastewater. These are comparatively slow growing species and more susceptible to infection but the production of a series of extra-cellular hydrolytic enzymes makes it easier for them to grow on starch and cellulose substrates.

Table 9.1: Fungi employed for the decolourisation of distillery effluent

S. No.	Name	Comments	Colour removal (%)	Reference
1.	*Coriolus versicolor* Ps4a	Two types of enzymes, sugar-dependent and sugar-independent, were found to be responsible for melanoidin decolourizing activity	80	Ohmomo et al. (1985)
2.	*Aspergillus fumigatus* G-2-6	Thermophilic strain tried for molasses wastewater decolourization but colouring compounds hardly degraded	56	Ohmomo et al. (1987)
3.	*Mycelia sterilia*	Organism required glucose for the decolou-rizing activity	93	Sirianuntapi-boon et al. (1988)
4.	*Aspergillus oryzae* Y-2-32	The thermophilic strain adsorbed lower molecular weight fractions of melanoidin and required sugars for growth	75	Ohmomo et al. (1988a)
5.	*Rhizoctonia sp.* D-90	Mechanism of decolourization of melanoidin involved absorption of the melanoidin pig-ment by the cells as a macromolecule and its intracellular accumulation in the cytoplasm and around the cell membrane as a melanoi-din complex, which was then gradually decolourized by intracellular enzymes	90	Sirianuntapi-boon et al. (1995) *(Contd.)*

Table 9.1: Fungi employed for the decolourisation of distillery effluent (*Contd.*)

S. No.	Name	Comments	Colour removal (%)	Reference
6.	Phanerochaete chrysosporium	Both the fungi required a readily available carbon source for melanoidin decolourization while N source had no effect. Maximum decolourization was observed in 6.25% (v/v) spent wash	53.5	Kumar et al. (1998)
7.	Coriolus versicolor		71.5	
8.	Geotrichum candidum	Fungus immobilized on polyurethane foam showed stable decolourization of molasses in repeated-batch cultivation	80	Kim and Shoda, (1999)
9.	Trametes sp. I-62	No colour observed associated with either fungal mycelium or polysaccharides secreted by the fungus and therefore colour removal was attributed to fungal degradation and not to a simple physical binding	73	Gonzalez et al. (2000)
10.	Coriolus hirsutus	A large amount of glucose was required for colour removal but addition of peptone reduced the decolourizing ability of the fungus	80	Miyata et al. (2000)
11.	Phanerochaete chrysosporium JAG-40	This organism decolourized synthetic and natural melanoidins when the medium was supplemented with glucose and peptone	80	Dahiya et al. (2001a)
12.	Aspergillus niger UM2	Decolourization was more by immobilized fungus and it was able to decolourize up to 50% of initial effluent concentrations	80	Patil et al. (2003)
13.	Citeromyces sp. WR-43-6	Organism required glucose, Sodium nitrate and KH_2PO_4 for maximal decolourization	68.91	Sirianuntapiboon et al. (2003)
14.	Coriolus versi-color sp no. 20	10% diluted spent wash was used with glucose @ 2% added as carbon source	34.5	Chopra et al. (2004)
15.	Flavodon flavus	MSW was decolourized using a marine basidiomycete fungus. It also removed 68% benzo(a)pyrene, a PAH found in MSW	80	Raghukumar et al. (2004)
16.	Phanerochaete chrysosporium	Sugar refinery effluent was treated in a RBC using polyurethane foam and scouring web as support	55	Guimaraes et al. (2005)
17.	Marine Basidiomycete NIOCC # 2a	Experiment was carried out at 10% diluted spent wash	100	D'souza et al. (2006)
18.	Phanerochaete chrysosporium NCIM 1073 NCIM 1106 NCIM 1197	Molasses medium decolourization was checked in stationary and submerged cultivation conditions	0 82 76	Thakkar et al. (2006)
19.	Pleurotus florida Eger EM1303	Hydroponically treated distillery effluent was subjected for treatment by fungus	86.3	Pant and Adholeya, (2009)

Among the filamentous species, *Aspergillus sp.* is the most popular. *A. niveus* and *A. niger* resulted in 60–69% reduction in colour and 75–95% COD removal; also, the treated effluent enhanced the seedling growth in Zea mays. Immobilized fungal isolate of *A. niger* UM2 resulted in a 72% decolourization of diluted synthetic melanoidin (absorbance unit of 3.5 at 475 nm) and 80% decolourization of 50% biodigested effluent. Similarly, a thermophilic strain of *A. fumigatus* G-2-6 decolourized 75% of melanoidin pigment solution at 45°C. Gel filtration chromatography revealed that large molecular weight fractions of melanoidins, in particular, were degraded rapidly. The dye decolourizing fungus *G. candidum* Dec 1 immobilized on polyurethane foam resulted in 80% removal in colour in diluted molasses solution (40–50 g/L). In case of pure culture experiments, *Candida utilis* and *Trichoderma viridiae* each showed less than 65% reduction in COD whereas *C. utilis* and *A. niger* together resulted in 89% COD removal. This reduction was from sugarcane stillage-based media with an initial COD of 40–75 g/kg. The fungus *Mycelia sterilia* D90 resulted in 91% decolourization of raw spentwash. The colour intensity (in terms of absorbance unit at 475 nm) was originally 47 for the raw spentwash but it was diluted to a value of 3.5 before use. However, a lower colour removal of 60–65% was obtained for the wastewater from anaerobic and aerobic ponds. This was possibly due to either the formation of some toxic compounds during anaerobic and aerobic treatment or the inability of the strain to attack the colour causing compound due to a change in their structure during anaerobic and aerobic treatment.

9.4.3 Bacterial treatment

Treatment of distillery wastewater by the use of Pseudomonas putida followed by *Aeromonas sp.* in a two-stage bioreactor resulted in COD as well as colour reduction. *P. putida* produces hydrogen peroxide which is a strong decolourizing agent. Since the organism cannot use spentwash as a source of carbon, 1% w/v glucose supplement was provided along with 12.5% spentwash. *Aeromonas sp.* utilizes the carbonaceous compounds present in spentwash as the sole carbon source, thereby eventually reducing the effluent COD by 66% in a 24 h period. *P. putida* also resulted in 44% COD removal accompanied by 60% colour reduction. In another study on predigested distillery effluent with *Aeromonas formicans*, 57% COD reduction and 55% decrease in colour was observed after 72 h. The colour removal efficiency increased up to 68% using the bacteria immobilized on calcium alginate beads; however, the COD reduction remained unchanged with longer incubation period of up to 96 h. *P. fluorescence* immobilized on porous cellulose carrier resulted in 66% colour removal with non-sterile diluted spentwash (absorbance unit of 3.5 at 475 nm) and 90% decolourization with sterile samples at 30°C over a 4 d period. The decolourization efficiency was further increased to 94% with cellulose carrier coated with collagen. These immobilized cells could be reused but the efficacy of colour removal was reduced. In another study, three different bacterial strains *Xanthomonas fragariae*, *Bacillus megaterium* and *Bacillus cereus* were used both in free form as well as after immobilization on calcium alginate beads for the treatment of 33% predigested distillery effluent. *B. cereus* resulted in maximum COD (81%) and colour (75%) reduction in free form. The reduction efficiencies increased marginally with immobilization. The decolourization activity of acetogenic bacteria has been reported for the first time by Sirianuntapiboon et al. (2004). Acetogenic bacteria is capable of oxidative decomposition of melanoidins thereby removing low molecular weight compounds in untreated molasses spentwash and almost all the low and high molecular weight compounds in anaerobically treated molasses spentwash. Nearly 76% decolourization, which is possibly due to a sugar oxidase, has been observed. The nitrifying bacteria *Nitrosococcus oceanus* is capable of detoxifying the spentwash accompanied by a reduction in the chloride content. However, no explanation was provided for this observation. The treated wastewater leads to better growth of rice plant due to adequate nitrogen content and can therefore be used as a low cost fertilizer.

Table 9.2: Bacteria employed for the decolouristion of distillery effluent

S. No.	Name	Comments	Colour removal (%)	Reference
1.	*Lactobacillus hilgardii*	Immobilized cells of the heterofermentative lactic acid bacterium decolourized 40% of the melanoidins solution within 4 days aerobically	40	Ohmomo et al. (1988b)
2.	*Bacillus smithii*	Decolourization occurred at 55°C in 20 days under anaerobic conditions in presence of peptone or yeast extract as supplemental nutrient. Strain could not use MWW as sole carbon source	35.5	Kambe et al. (1999)
3.	*Acinetobacter sp.*	All these organisms were isolated from an air bubble column reactor treating winery waste-water after 6 months of operation. Most isolates from the colonized carriers belonged to species of the genus Bacillus	Not checked in this study	Petruccioli et al. (2000)
4.	*Aeromonas sp.*			
5.	*Alcaligens faecalis*			
6.	*Bacillus sp.*			
7.	*Flavobacterium sp.*			
8.	*Pseudomonas sp.*			
9.	*P. paucimobilis*			
10.	*Pseudomonas fluorescens*	This decolourization was obtained with cellulose carrier coated with collagen. Reuse of decolourized cells reduced the decolouriza-tion efficiency	94	Dahiya et al. (2001b)
11.	*Pseudomonas putida*	The organism needed glucose as a carbon source, to produce hydrogen peroxide which reduced the colour	60	Ghosh et al. (2002)
12.	*Xanthomonas fragariae*	All the three strains needed glucose as carbon source and NH4Cl as nitrogen source. The decolourization efficiency of free cells was better than immobilized cells	76	Jain et al. (2002)
13.	*Bacillus megaterium*		76	

(Contd.)

Table 9.2: Bacteria employed for the decolouristion of distillery effluent (*Contd.*)

S. No.	Name	Comments	Colour removal (%)	Reference
14.	*Bacillus cereus*		82	
15.	*Acetobacter acetii*	The organism required sugar especially, glucose and fructose for decolourization of MWWs	76.4	Sirianuntapi- poon et al. (2004)
16.	*Bacillus thuringiensis*	Addition of 1% glucose as a supplementary carbon source was necessary	22	Kumar and Chandra (2006)
17.	*Bacillus brevis*		27.4	
18.	*Bacillus sp.*		27.4	
19.	*Pseudomonas Aeruginosa,*	The three strains were part of a consortium which decolourized the anaerobically digested spent wash in presence of basal salts and glucose	67	Mohana et al. (2007)
20.	*Stenotrophomonas maltophila,*			
21.	*Proteus mirabilis*			

In yet another investigation, two aerobic bacterial strains TA 2 and TA 4 have been isolated from sites contaminated with anaerobically treated distillery effluent. These bacteria, which were identified to be Gram negative and Gram-positive, respectively, resulted in 66% and 62% BOD reduction in anaerobically treated spentwash. However the reduction in BOD was found to be higher (80%) when the two were used together; further the combination resulted in 76% colour removal after 72 h. More recently, the decolourization of four types of synthetic melanoidins, i.e. glucose-glutamic-acid (GGA), glucose-aspartic-acid (GAA), sucrose-glutamic acid (SGA), and sucrose-aspartic-acid (SAA), were investigated using three different isolates, viz. *Bacillus thuringiensis*, *Bacillus brevis* and *Bacillus sp.* The degree of decolourization of the melanoidins separately by each isolate was in the 1–31% range; however, when used collectively, these isolates resulted in up to 50% decolourization due to the enhanced effect of coordinated metabolic interactions. The results also indicated that the GAA polymer was the most recalcitrant among the melanoidins tested. The biodegradability of spentwash can be enhanced by enzymatic pre-treatment prior to the aerobic step. After 24 h of treatment with Gram positive bacterium ASN6, the COD reduction of cellulose pre-treated spentwash was 28.8% in comparison to 18.3% for untreated effluent. This was explained by the fact that pre-treatment affected the metabolic value (microbial acceptability) by generating intermediate hydrolysis products from the parent cellulosic compounds present in the spentwash. The biodegradability was further enhanced by combined ultrasound and enzymatic pre-treatment resulting in 62.2% COD reduction after 36 h as compared to 39.4% COD removal for the untreated effluent. The enhancement in biodegradability was attributed to molecular transformation of effluent constituents by ultrasound pre-treatment. A detailed list of bacteria tried by different researchers for decolourisation of distillery effluent is given in Table 9.2.

9.4.4 Algal Treatment

The treatment of anaerobically treated 10% distillery effluent using the microalga *Chlorella vulgaris* followed by *Lemna minuscula* resulted in 52% reduction in colour. In another study, Kalavathi et al. (2001) examined the degradation of 5% melanoidin by the marine cyanobacterium *Oscillatoria boryana* BDU 92181. The organism was found to release hydrogen peroxide, hydroxyl ions and molecular oxygen during photosynthesis resulting in 60% decolourization of distillery effluent. In addition, this study suggested that cyanobacteria could use melanoidin as a better nitrogen source than carbon. Further, cyanobacteria also excrete colloidal substances like lipo-polysaccharides, proteins, polyhydroxybutyrate (PHB), polyhydroxy-alkanoates (PHA), etc. These compounds possess COO^- and ester sulphate (OSO_3^-) groups that can form complexes with cationic sites thereby resulting in flocculation of organic matter in the effluent. It was observed that the strain Oscillatoria resulted in almost complete colour removal (96%) whereas Lyngbya and Synechocystis were less effective resulting in 81 and 26% colour reduction, respectively. The consortium of the three strains showed a maximum decolourization of 98%. This was attributed to adsorption in the initial stages followed by degradation of organic compounds which dominated in the subsequent stages.

9.4.5 Decolourization of Distillery Effluent by Yeast

Apart from bacteria, white-rot and filamentous fungi, yeast has also been investigated for distillery wastewater treatment. Yeast is characterized by quick growth and is less susceptible to contamination by other microorganisms; further, yeast produces biomass with high nutritive value. The yeast Citeromyces WR-43-6 resulted in high and stable removal efficiency in both colour intensity and organic matter. The removal efficiencies for diluted spentwash (absorbance unit of 3.5 at 475 nm) were 75% for colour intensity and 76% for BOD. Besides, the yeast was also found to utilize lactate and acetate that are inhibitory to ethanol production. As a result, the treated effluent could be used as dilution water for fermentation thereby reducing the residual stillage volume by 70%. Another strain of *Hansenula anomala* J 45-N-5 and I-44 isolated from soil, resulted in 74% reduction in total organic carbon (TOC). *S. cerevisiae* also provides promising results on a larger scale. The use of pure culture of *S. cerevisiae* resulted in 82.7% decolourization in the 10% anaerobically treated distillery effluent along with 84% reduction in COD. It was also reported that the nitrogen present in the distillery effluent was sufficient for the growth of yeast. Yeast, Citeromyces was used for treating MWW and high and stable removal efficiencies in both colour intensity and organic matter were obtained. However, the semi-pilot and pilot-scale experiments are to be tested for checking the stability of *Citeromyces sp.* Microorganisms associated with a rotating biological contactor (RBC) treating winery wastewater was studied. One of the yeast isolates was able to reduce the COD of synthetic wastewater by 95% and 46% within 24 h under aerated and non-aerated conditions, respectively. Two flocculant strains of yeast, *Hansenula fabianii* and *Hansenula anomala* was used for the treatment of wastewater from beet molasses-spirits production and achieved 25.9% and 28.5% removal of TOC respectively from wastewater without dilution. Dilution of wastewater was not favourable for practical treatment of wastewater due to the longer treatment time and higher energy cost.

9.5 NOVEL BIOLOGICAL TREATMENT TECHNIQUES

9.5.1 Phytoremediation

Algal growth potential bioassay is a standard assay to determine the potential of water bodies, natural waters and wastewaters, to support or inhibit the microalgae growth. Algae growth potential was determined in distillery wastewater pretreated by anaerobic processes and by a combined anaerobic-aerobic system. The biologically treated distillery wastewater provided satisfactory conditions for microalgae growth. Billore et al. (2001) used *Phragmites kharka* in a constructed wetland for treatment of wastewater from the same industry and obtained 36% removal of total Kjeldahl nitrogen and 48% removal of total suspended solids (TSS). Enhanced decolourization was achieved by phytoremediation of distillery effluent by a macrophyte, *Spirodela polyrrhiza* (L.) Schliden pretreated with *Bacillus thuringiensis* (Kumar and Chandra, 2004). Recently, macrophyte *Potamogeton pectinatus* was used for bioaccumulating heavy metals from distillery effluent. Increasing concentration of the effluent greatly reduced the biomass of the plant with maximum accumulation of Fe being recorded in plants growing in 100% effluent.

It is possible that constructed wetlands may allow for effective, low-maintenance purification of distillery and winery wastewaters. Although they are land intensive treatment systems, wetlands provide many benefits, which include water quality improvement, food and habitat for wildlife, flood protection, and shoreline erosion control. Constructed wetlands are engineered systems that have been designed to make use of macrophytic vegetation, soils, and a variety of aerobic and anaerobic bacteria populations and fungi and yeasts to remediate wastewaters. Wetland treatment is advantageous to lagooning in that subsurface flow (through a permeable medium) allows for odor control. However, surface flow simulates wetlands and is sometimes more economical than subsurface flow. A variety of processes such as sedimentation, filtration, precipitation, sorption, and plant and microbial nutrient uptake allow for removal of organic materials from wastewaters. Constructed wetland systems can vary from a single cell to multiple cells. Appropriately designed constructed wetlands have a large potential application in the treatment of wastewaters. Billore et al. (2001) studied a field-scale, four-celled, horizontal subsurface flow constructed wetland for the treatment of a molasses-based distillery wastewater. Although the COD was reduced by 64%, the wastewater strongly impacted plant morphology, aeration anatomy in the chiselled plant tissues, reed growth, and composition of the biofilm in the specialized substratum. However, when new plant growth emerged 8 months later, it appeared healthy. Shepherd et al. (2001) combined a subsurface-flow constructed wetland with up flowing sand prefilter and measured average removal efficiencies of 98% for COD, 97% for total suspended solids, and 78.2% for nitrogen in winery wastewater. In addition, removals of sulphide (98.5%), o-phosphate (63.3%), VFAs (99.9%), phenols (100%), tannins and lignins (77.9%), and settleable solids were observed and the acidic pH was neutralized. Although wetlands do allow for a relatively low maintenance and good removal of COD and phenolic compounds, they are not without their problems. The most obvious is the cost of land associated with the spatial requirement for such a treatment system. Another is the gradual sedimentation of the wetland. Black sediment formed from the precipitate of polymerized phenolic compounds is recalcitrant, and during peak season forms far quicker than it can be degraded. It would have to be physically removed to prevent clogging of the artificial wetland. Direct land disposal has relatively low running costs but is dependent on the volume and characteristics of the wastewater produced and land availability. In a study *Typha latipholia* was used for distillery effluent treatment in a constructed wetland. The system resulted in 78% and 47% reduction in COD and BOD respectively in a period of 10 days. Combined treatment with *Lemna minuscula* and *Chlorella vulgaris* showed 52% colour removal from distillery effluent. The microalgal treatment removed nutrients and organic matter from wastewater and produced oxygen for other organisms. The macrophyte removed organic matter and eliminated the microalgae form treated wastewater. However, despite the potential of aquatic macrophytes in cleaning wastewaters the use of these plants in designing a low cost treatment system is still at experimental stage and is considered to be a potentially important area of environmental management. Billore et al. (2001) have demonstrated a four-celled horizontal subsurface flow (HSF) constructed wetland (CW) for the treatment of distillery effluent after anaerobic treatment. The post-anaerobic treated effluent had a BOD of about 2500 mg/L and a COD of nearly 14,000 mg/L. A pre-treatment chamber filled with gravel was used to capture the suspended solids. All the cells were filled with gravel up to varying heights and cells three and four supported the plants *Typha latipholia* and *Phragmites karka* respectively. The overall retention time was 14.4 d and the treatment resulted in 64% COD, 85% BOD, 42% total solids and 79% phosphorus content reduction. In another study, a laboratory scale CW employing *T. latipholia* was used to treat diluted distillery effluent. A root zone of $1.5 \times 0.3 \times 0.3$ m, filled with 75% sand and gravel and 25% soil was used and the diluted effluent was applied after 4 weeks of planting. The system resulted in 76% COD reduction in 7 d which increased marginally to 78% COD reduction in 10 d. The BOD reduction was 22% and 47% on days 7 and 10, respectively. In yet another instance, a distillery in northern India is presently employing CW for polishing the effluent prior to land discharge for irrigation in the surrounding paddy fields. The effluent is initially subjected to primary treatment which includes settling and anaerobic digestion in a structured media attached growth (SMAG)-type anaerobic reactor. The primary treated effluent, with a COD of 28,000–35,000 mg/L, is subjected to two-stage aeration to bring down the COD to 400 mg/L. Thereafter, it is directed to a CW before final discharge.

9.5.2 Extensive/ Recycling Process

Land application was considered to be the most effective and practical disposal method for water, nutrient and organic matter contents of winery wastewater. Tano et al. (2005) compared increasing doses of distillery vinasses factorially combined with three levels of urea on vegetative growth, leaf mineral levels, and grape yield and quality over a 4-year period. The improved the ripening levels of grapes, increased sugars, and reduced grape juice acidity all indicated the possibility for the reuse of vinasse within the vineyard. However, raw distillery wastewater highly toxic to the soil microorganisms, which are important in the soil ecosystem. A pot culture experiment found raw wastewater application decreased the population of bacteria, fungi, and actinomycetes, whereas the growth rates of Rhizobium and Azotobacter were reduced. The toxic effects were minimized by mixing (1:1) the raw wastewater with stabilization pond wastewater. Ramana et al. (2002a) found the effect of the distillery wastewater on seed germination crop-specific and care was advised when using distillery wastewater for pre-sowing irrigation purposes. They observed complete failure of germination when the wastewater concentration was greater than 50%. Other work done with maize (Ramana et al., 2002b) and groundnut (Ramana et al., 2002c) suggested that distillery wastes should not be used in isolation, but could serve well as a supplement to lower fertilizer requirements. Similarly, adverse effect of distillery sludge on *Phaseolus mungo* at higher concentration (>10%) was also noted but 10% concentration are found supporting for seed germination, biomass production (Chandra et al., 2008a). Winery wastewaters pose severe pollution problems, mainly because the organic component leaches down to the water table where decomposition is slow, due to the shortage of oxygen. Unpleasant odors may also be released if the soil is disturbed. The organic material of effluent is mobile in the groundwater and constitutes a major off-site pollution hazard. The major potential problem with land application is that of inorganic ions. These can lead to the degradation of soil structure and fertility in the longer term. Root zone accumulation of salt must also be avoided as it may negatively affect plant growth. Although land application has been successfully used to dispose of these wastewaters, it must be adequately monitored in order to protect water resources from the cumulative effect of many small additions. In *California multistage*, facultative aerobic ponds have been used successfully for treatment and storage of winery wastewater for more than 40 years. Ponds are lined to prevent seepage into the water table and aeration prevents the generation of malodorous compounds. The treated water is utilized for irrigation during periods of low rainfall. PMDE has been used for pisciculture near Chennai city in southern India. The biodigested effluent, which is a rich growth medium, is directed to bioconversion ponds after which it is spread in about 6 ha of fishponds. The BOD is reduced to nearly zero and the initiative yields about 50 tons per hectare per year of fish.

9.5.3 Role of Enzymes in Effluent Decolourization

Although the enzymatic system related with decolourization of melanoidins is yet to be completely understood, it seems greatly connected with fungal ligninolytic mechanisms. The white-rot fungi have a complex enzymatic system which is extracellular and non-specific, and under nutrient-limiting conditions is capable of degrading lignolytic compounds, melanoidins, and polyaromatic compounds that cannot be degraded by other microorganisms. A large number of enzymes from a variety of different plants and microorganisms have been reported to play an important role in an array of waste treatment applications. Several studies regarding degradation of melanoidins, humic acids and related compounds using basidiomycetes have also suggested a participation of at least one laccase enzyme in fungi belonging to Trametes (Coriolus) genus. The role of enzymes other than laccase or peroxidases in the decolourization of melanoidins by Trametes (Coriolus) strain was reported during the 1980s. Several reports claimed that intracellular sugar-oxidase- type enzymes (sorbose-oxidase or glucose-oxidase) had melanoidin-decolourizing activities. It was suggested that melanoidins were decolourized by the active oxygen (O_2; H_2O_2) produced by the reaction with sugar oxidases. Decolourization by microbial methods includes the enzymatic breakdown of melanoidin and flocculation by microbially secreted substances. Ohmomo et al. (1985) used C. *versicolour* Ps4a, which decolourized molasses wastewater 80% in darkness under optimum conditions. Decolourization activity involved two types of intracellular enzymes, sugar-dependent and sugar-independent. One of these enzymes required no sugar and oxygen for appearance of the activity and could decolourize MWW up to 20% in darkness and 11–17% of synthetic melanoidins. Thus, the participation of these H_2O_2 producing enzymes as a part of the complex enzymatic system for melanoidin degradation by fungi should be taken into account while designing any treatment strategy. Colour removal of synthetic melanoidin by *C. hirsutus* involved the participation of peroxidases (MnP and MIP) and the extracellular H_2O_2 produced by glucose-oxidase, without disregard of a partial participation of fungal laccase. Mansur et al. (1997) obtained a maximum decolourization of around 60% on day 8 after inoculating with fungus *Trametes sp.* I-62. Here effluent was added at a final concentration of 20% (v/v) after 5 days of fungal growth, the time at which high levels of laccase activity were detected in the extracellular mycelium. The white-rot basidiomycete *T. versicolour* is an active degrader of humic acids as well as of melanoidins. A melanoidin mineralizing 47 kDa extracellular protein corresponding to the major mineralizing enzyme system from *T. versicolour* was isolated. This Mn^{2+} dependent enzyme system required oxygen and was described to be as peroxidase. Uniform, small and spongy pellets of the fungus *T. versicolour* were used as inoculum for colour removal using different nutrients such as ammonium nitrate, manganese phosphate, magnesium sulphate and potassium phosphate and also sucrose as carbon source. Maximum colour removal of 82% and 36% removal of N-NH4 was obtained on using low sucrose concentration and KH_2PO_4 as the only nutrient. Some studies have identified the lignin degradation related enzymes participating in the melanoidin decolourization. Intracellular H_2O_2 producing sugar oxidases have been isolated from Coriolus strains. Also, *C. hirsustus* have been reported to produce enzymes that catalyze melanoidin decolourization directly without additions of sugar and O_2. Miyata et al. (1998) used *C. hirsutus* pellets to decolourize a melanoidin-containing medium. It was elucidated that extracellular H_2O_2 and two extracellular peroxidases, a manganese-independent peroxidase (MIP) and manganese peroxidase (MnP) were involved in decolourization activity. Lee et al. (2000) investigated the dye-decolourizing peroxidase by cultivating *Geotrichum candidum* Dec1 using molasses as a carbon source. Components in the molasses medium stimulated the production of decolourizing peroxidase but inhibited the decolourizing activity of the purified enzyme. It was found that the inhibitory effect of molasses can be eliminated at dilution ratios of more than 25. Recently D'souza et al. (2006) reported 100% decolourization of 10% spentwash by a marine fungal isolate whose laccase production increased several folds in the presence of phenolic and non-phenolic inducers. A combined treatment technique consisting of enzyme catalyzed in situ transformation of pollutants followed by aerobic biological oxidation was investigated by Sangave and Pandit (2006a) for the treatment of alcohol distillery spentwash. It was suggested that enzymatic pretreatment of the distillery effluent leads to in situ formation of the hydrolysis products, which have different physical properties and are easier to assimilate than the parent pollutant molecules by the microorganisms, leading to faster initial rates of aerobic oxidation even at lower biomass levels. In another study, Sangave and Pandit (2006b) used irradiation and ultrasound combined with the use of an enzyme as pretreatment technique for treatment of distillery wastewater. The combination of the ultrasound and enzyme yielded the best COD removal efficiencies as compared to the processes when they were used as stand-alone treatment techniques. Enzymatic decolourization of molasses medium has also been tried using *P. chrysosporium*. Under stationary cultivation conditions, none of the strains could decolourize molasses nor produce enzymes lignin peroxidase, manganese peroxidase and laccase. All of them could produce lignin peroxidase and manganese peroxidase when cultivated in flat bottom glass bottles under stationary cultivation conditions.

9.5.4 Biphasic Treatment System of PMDE using Bacteria and Wetland Plants

Complex industrial wastewater i.e. distillery wastewater that are treated by biological means are good candidate to be addressed by constructed wetland treatment system. Wetlands are considered sink for contaminates due to its remediation potential for wide varieties of pollutants. Wetland plants which accumulate different environmental pollutants included heavy metals either through phytostabilization or phytoextraction process. The proper functioning of the wetland system is dependent on the interactions between plants, soil, wastewater characteristics, microorganisms and operational conditions. Several designed features may affect the processes occurring in a constructed wetland; for instance, changes in design and loading regime may improve the oxidation in constructed wetland system. It has been suggested that macrophyte species also affect the pollutant removal efficiency in constructed wetland system. Recently, the application of constructed wetland treatment has been reported promising for the industrial wastewater remediation/treatment. Plant roots and rhizomes are important for the microbial transformation processes of wastewater purification. Microbial assemblages can be found as a biofilm on substrate and root surfaces. Many parameters affect biofilm structure, especially nutrient availability or other environmental conditions. Use of constructed wetland treatment process is accepted as eco-friendly, cost-effective and energetically sustainable technique compared with the conventional wastewater treatment system. Constructed wetland for wastewater treatment may be classified according to the life form of the dominating macrophyte, into systems with free-floating, floating leaved, rooted emergent and submerged macrophytes. Further, division could be made according to the wetland hydrology (free water surface and subsurface systems) and subsurface flow constructed wetland could be classified according to the flow direction (horizontal and vertical). A simple scheme for various types of constructed wetlands is shown in Figure 9.14.

Constructed wetland treatment system generally fall in to two categories: (a) horizontal flow system (HFS) and (b) vertical flow system (VFS). HFS may be free water surface system or subsurface system depending up on their designing. Free water surface system

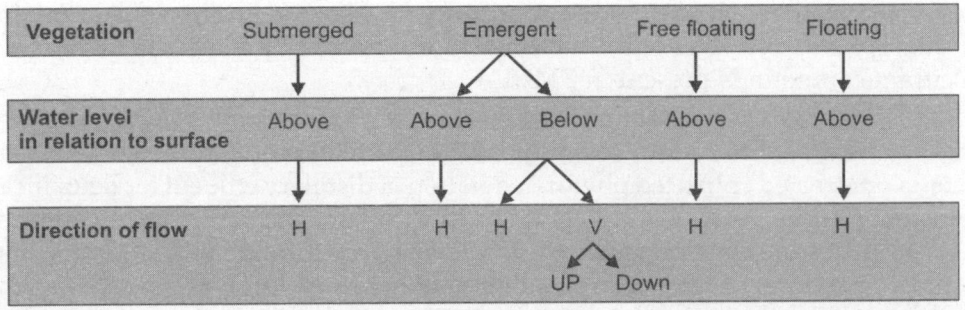

Fig. 9.14: The major characteristics of various types of constructed wetlands for wastewater treatment. H = horizontal, V = vertical

can mimicked the pollutants by flow over the bed surface and it filter through a dense stand of aquatic plants, while in subsurface flow water system consisting of an excavated, but usually lined shallow basin containing pebbles, gravels and uppermost layer is sand (EPA, 1998). Lower most zone containing pebbles gives ample opportunity for growing network of fibrous root, where rhizospheric bacteria makes biofilm and it act as biofilter. Thus, during the treatment of wastewater, water flows under subsurface of designed layer. In contrast, VFS, wastewater during treatment moves either down flow or up flow. But, it required comparatively higher operational maintenance cost compared to HFS due to the necessity to pump wastewater intermittently on the wetland surface containing plants. The wastewater is fed into large batches and then it percolates down through sand medium. The new batch is fed only after all the water percolates and the bed is free of water. This enables diffusion of oxygen from the air into the bed. As a result, VFS are for more aerobic than HFS wetland system and provide suitable condition for nitrification, while HFS provides more suitable condition for denitrification due to semi-anaerobic condition. VFS is very effective in removing organic and suspended solids. Moreover, it required more land area. Therefore, VFS are very often using to treat domestic and municipal wastewater or wastewater in which discharge limit of ammonia-nitrogen is fixed. Hence, in this study constructed system has been designed with subsurface horizontal flow wastewater treatment system, which can degrade variety of pollutants in semi-aerobic condition due to specific microbial biofilm on the root zone of wetland plants.

Since, PMDE containing different types of pollutants which do not easily decolourized by single step treatment. Therefore, biphasic treatment of complex wastewater using bacteria and constructed wetland might be more promising. Hence, bacterial pre-treated effluent was integrated with wetland treatment system consisting different group of potential wetland plants (*Phragmites cummunis, Typha anguistifolia and Cyperus esculentus*) (Fig. 9.15). The commonly growing wetland plants in industrial contaminated wetland has been selected as described in previous chapter seven and acclimatized in constructed wetland system for the improvement of decolourisation and degradation of PMDE in integrated bacteria and wetland treatment system. *T. anguistifolia* and *C. esculentus* is faster growing plants than the *P. cummunis*. However, *P. cummunis* has double advantage; first it has bigger vegetable growth and second has strong root system. Moreover, advantage of mixed wetland plants has differential growth and tolerance potential. This increases the tolerance of organic and hydraulic load. Furthermore, structural and physiological variation will offer wide opportunity for different group of micro-organism in their rhizospheric zone. This will established more effective biofilter system for removal of recalcitrant compounds present in PMDE.

The sequential bioreactor treatment of wastewater has become promising over the conventional treatment technique to remove the mixture of pollutants from wastewater. The bacteria pre-treated enhanced phytoremediation of distillery effluent for detoxification has been reported by Kumar and Chandra (2004) where the recalcitrant compounds present in effluent were ameliorated by *Bacillus thurenginsis* through utilization of complex compounds. Therefore, foster phytoremediation of heavy metals was observed from the PMDE. Enhancement in biodegradability of distillery wastewater using enzymatic pre-treatment has been reported by researcher. A combined treatment technique consisting of enzymatic hydrolysis followed by aerobic oxidation was investigated for the treatment

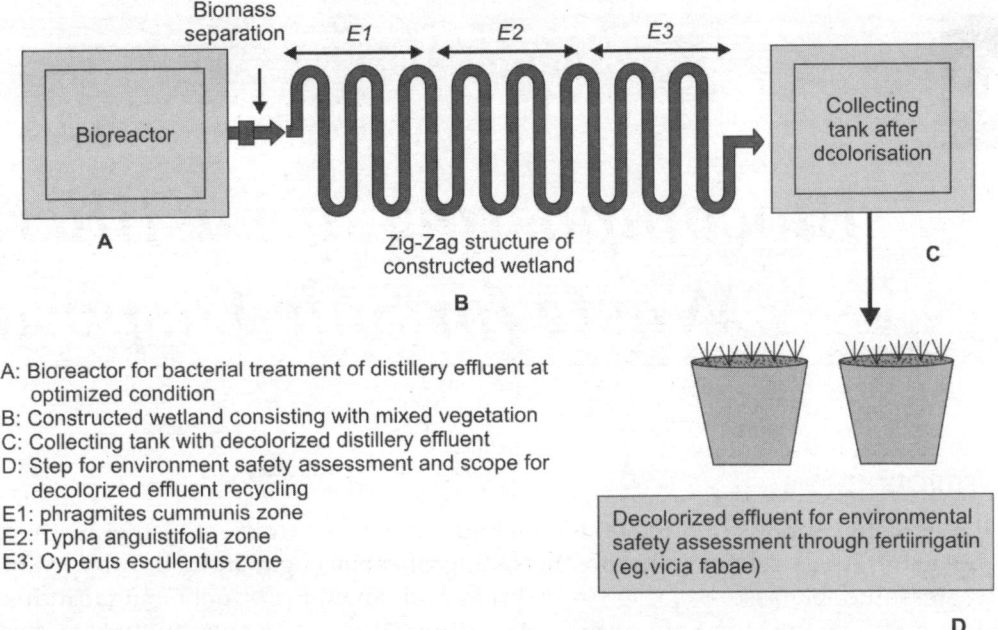

A: Bioreactor for bacterial treatment of distillery effluent at
   optimized condition
B: Constructed wetland consisting with mixed vegetation
C: Collecting tank with decolorized distillery effluent
D: Step for environment safety assessment and scope for
   decolorized effluent recycling
E1: phragmites cummunis zone
E2: Typha anguistifolia zone
E3: Cyperus esculentus zone

Fig. 9.15: Schematic diagram showing biphasic treatment of PMDE for environmental safety

of alcohol distillery spentwash. The enzyme cellulase was used for the pre-treatment step with an intention of transforming the complex and large pollutant molecules into simpler biologically assimilable smaller molecules. Batch experiments were performed in order to analyze the influence of various parameters like pre-treatment time, enzyme concentration and pH during the pre-treatment step on the subsequent aerobic oxidation kinetics. The rate of anaerobic oxidation enhanced by 2.3 fold in the pre-treated sample as compared to the untreated sample when the pH during the pre-treatment step was maintained at a value of 4.8. Similarly, two-fold increase in the aerobic oxidation rate was observed when the effluent was pre-treated with the enzyme without any pH control (effluent pH 3.8).

10

Biocomposting of Distillery Waste for Safe Disposal

10.1 INTRODUCTION

Since the effluent contains organic materials and many plant nutrient elements, there is scope for using it advantageously in composting of other sugar industrial byproducts namely pressmud, bagasse, etc. (Fig. 10.1). Pressmud, an end product of sugar industry is used as a one of the substrate in bio-composting. It is a soft spongy, brown colour fibrous material.

Fig. 10.1: Showing some common products uses in distillery effluent composting pressmud (a), molasses (b), coir pith (c) and rice straw (d)

The composting enables the degradation of coloured organics in the distillery effluent and evaporation of water rapidly and also reduction in BOD: Distillery effluent will also enrich the compost with plant nutrients especially potassium. Suitable compost can be made in three, months from vinasse in different, combinations e.g. vinasse + pressmud cake or from vinasse + sugarcane bagasse + rice straw. The distillery effluent based compost can be prepared by using pressmud and the compost could be enriched with the use of rock phosphate, gypsum, yeast sludge, bagasse, sugarcane trash, boiler ash, coir pith and water hyacinth. First, the pressmud is spread in the compost yard to form a heap of 1.5 m height, 3.5 m width and 300 m length. Ten Iitres of bacterial culture, diluted with water, in the ratio of 1: 10 is sufficient for a tonne of pressmud. A consortium of efficient microbial decomposers viz., *Phanerocheate chrysosporium, Trichurus spiralis, Pacelomyces fusisporus, Trichoderma spp.*, etc. are sprayed on the pressmud and mixed thoroughly using aerotiller which makes the pressmud aerable and hastens the process of decomposition. After 3 days, distillery effluent is sprayed on the heaps to a moisture level of 60 percent and the pressmud heaps are allowed for 4–5 hours to absorb the effluent. The heaps are then thoroughly mixed by aerotiller (Fig. 10.2).

When the moisture level drops below 30–40 percent, again the effluent is sprayed, mixed with pressmud and heaps are again formed. Effluent can be sprayed once or twice in a week depending on the moisture content of pressmud heaps. Mixing of effluent and heap formation will be repeated for 8 weeks so that the pressmud and effluent proportion reaches an optimum ratio of 1 : 3, Then the heaps are allowed for curing for a month. The compost obtained from this process is neutral in pH with an EC of 3.12 to 6.40 dS/m. It contains 1.53% N, 1.50% P, 3.10% K, 300 ppm Fe, 130 ppm Cu, 180 ppm Mn and 220 ppm Zn. The organic carbon and C: N ratio reduced from 36 to 18 % and from 28.12 to 16.3% respectively. The technology of using distillery effluent for composting of pressmud, pressmud along with sugarcane trash and coir waste, pressmud plus bagasse ash and city garbage have been successfully treated in so many places. Biocomposting is an eco-friendly approach for bioconversion into value added products which may be utilized as plant nutrients. It also reduces the disposal and pollution problems arising from spent wash. Davamani et al. (2006) also reported that biocompost significantly enhanced the yield and yield components and juice quality of sugarcane. Bhalerao et al. (2006) and Jadhav et al. (1992) observed that increased nutrient uptake by sugarcane is due to use of spent wash and pressmud compost. The application of biomethanated distillery

Fig. 10.2: Mixing of effluent and pressmud with the help of aerotiller

spentwash and pressmud biocompost substantially increased the microflora and enzyme activities of soil throughout the crop growth period of sugarcane. The increased microbial biomass and enzymatic activities in sugarcane grown soil expedited mineralization of biomethanated distillery spentwash and biocompost, nutrient cycling and formation of organic matter and soil structure. Thus, bio-composting not only solves the disposal problems but also helps in saving the cost on chemical fertilizers. Many workers have studied the effect of spent wash as a source of plant nutrients; however, very little information is available on the use of BMDE and bio-compost in sugarcane.

Composting involves the conversion of organic residues of plant and animal origin into manure. It is largely a microbiological process based upon the activities of several bacteria, actinomycetes and fungi. The main product is rich in humus and plant nutrients; the by-products are carbon dioxide, water and heat. In the composting process, aerobic microorganisms use organic matter as a substrate. The microorganisms decompose the substrate, breaking it down from complex to intermediate and then to simpler compounds. During composting, compounds containing carbon and nitrogen are transformed through successive activities of different microbes to more stable organic matter, which chemically and biologically resembles humic substances. The rate and extent of these transformations depend on available substrates and the process variables used to control composting.

$$\text{Fresh Organic Waste} + O_2 \xrightarrow{\text{Microbial Metabolism}} \text{Stabilized Organic Waste Material} + CO_2 + H_2O + \text{Heat}$$

In nature, composting takes place when leaves pile up and begin to decay. Eventually the decayed leaves are returned to the soil, where living roots reclaim the nutrients from the remains of the leaves. Ancient people dumped food wastes in piles near their camps and found that the wastes rotted and formed habitat for the seeds of many food plants that sprouted there. Perhaps this led to the realization that dump heaps were good places for food crops to grow and humans began to put seeds there intentionally. By all accounts, recycling of organic residues through composting appears to be an ancient practice. It has acquired ever greater relevance and in the present times the use of composting to turn organic wastes into resource should be practiced with a sense of urgency as landfill space becomes increasingly more scarce and expensive. The first microorganisms to colonize a heap of biodegradable solid waste are mesophilic bacteria, actinomycetes, fungi and protozoa. They grow between 10 and 45°C and break down easily degradable components such as sugars and amino acids. The degradation of fresh matter starts as soon as it is piled into heap. Due to the oxidative action of microorganisms the temperature increases. Even though there is a drop in pH at the very beginning of composting, caused by the formation of volatile fatty acids, the subsequent degradation of acids brings about an increase in pH. When the temperature of a waste heap reaches 45–50°C, thermophilic microorganisms replace mesophilic ones. The second phase is called the thermophilic phase and can last several weeks. It is the active phase of composting: Most of the organic matter is degraded and consequently most oxygen is consumed in this phase. Lignin degradation also starts during this phase. Indeed, the optimum temperature for thermophilic micro-fungi and actinomycetes which mainly degrade lignin is 40–50°C. Above 60°C, these microorganisms cannot grow and lignin degradation is slowed down. After the thermophilic phase, the peak of degradation of fresh organic matter, the microbial activity decreases, as does the temperature. This is termed the cooling phase. The compost maturation phase then begins when the compost temperature falls to that of the ambient air. During this phase, mesophilic microorganisms colonize the compost heap and slowly degrade complex organic compounds such as lignin. This last phase is important because humus like substances is produced in this phase to form mature compost. The microbial succession is show in figure 10.3.

During composting, mineralization and humification occur simultaneously and are the main processes causing the degradation of the fresh organic matter. During mineralization, transformations of nitrogenous compounds occur involving several biochemical reactions. Degradation of protein, urea or uric acid produces ammonium ion (NH_4^+). During this process, high pH, high temperature and moisture determine the NH_3/NH_4^+ balance and the NH_3 emission. The solubility of NH_3 is reduced by about 30% when temperature increases from 40 to 50°C, and when pH also increases. Another step of degradation is the nitrification, which transforms NH_4^+ into NO_3^+ (nitrate) by oxidation under aerobic conditions.

One of the byproducts of nitrification is N_2O (Nitrous oxide, commonly known as laughing gas). Although composting is essentially an aerobic transformation of organic matter, anaerobic conditions can occur in pockets of the waste heap where free oxygen is exhausted. It may lead to formation of volatile fatty acids, which lower the pH of the anaerobic zone. Under these conditions NO_3 is reduced to N_2O and then to N_2. In addition, N_2O, NO (Nitric oxide), and NO_2 may be produced in a compost heap that is not completely aerobic. Due to these reasons steps must be taken to avoid anaerobic zones from developing in a compost heap. During composting, carbon is transformed into CO_2 and is integrated into humus like substances as a result of humification. If anaerobic zones form in a compost heap, methane can be released from such zones. Low redox potential and high temperature provide suitable conditions for the development of thermophilic methanogenic bacteria. Moreover, during the thermohilic phase, oxygen is liberally consumed by aerobic microorganisms; the subsequent reduction of oxygen concentration in the heap favors anaerobic conditions for methane production.

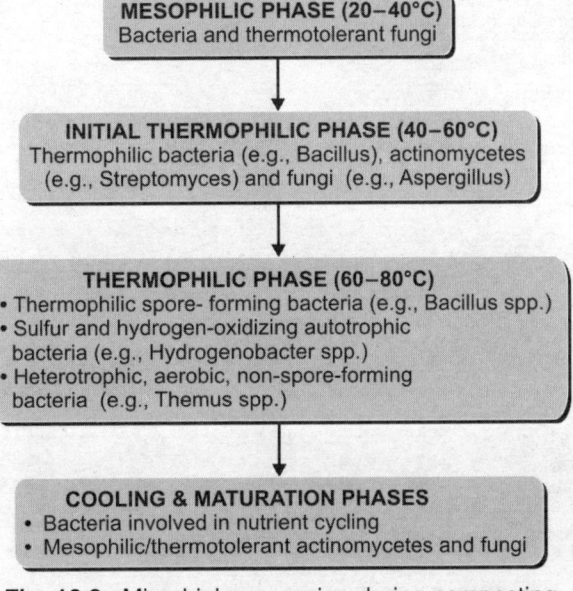

Fig. 10.3: Microbial succession during composting

10.2 PROCESS DESCRIPTION

Compositing consists essentially of mixing sludge with a bulking agent, stabilizing the mixture in the presence of air, curing, screening to recover the bulking agent, and storing the resulting compost material. Different views of composting are shown in Figure 10.4. There are three main types of composting systems

10.2.1 Aerated static pile process

This process is the most widely used in the United States and consists of mixing dewatered sludge (raw, anaerobically or aerobically digested sludge) with a bulking agent (new or recycled) such as wood chips, leaves, corncobs, bark, peanut and rice hulls, or dried sludge. Wood chips are the most commonly used bulking agent in composting (they have a high C/N ratio). Bulking materials offer structural support and favor aeration during composting. The pile is covered with screened compost to reduce or remove odor and to maintain high temperatures inside the pile.

Aeration is provided by blowers and air diffusers during a 21- day active composting period in the pile. Afterward, the compost is cured for at least 30 days, dried, and screened to recycle the bulking agent.

Fig. 10.4: Different view of composting

10.2.2 Windrow process

Dewatered sludge is mixed with the bulking agent and stacked in 1 to 2 m high rows called *windrows*. The composting period lasts approximately 30–60 days. Aeration is provided by turning the windrows two or three times per week. Additional induced aeration may be provided in the aerated windrow process.

10.2.3 Enclosed systems

These systems are enclosed to ensure a better control of temperature, oxygen concentration, and odors during composting. They require little space and minimize odor problems. Their cost is, however, higher than open systems.

10.3 FACTORS CONTROLLING COMPOSTING

In order to control and optimize the composting process toward achieving a product of desired quality, it is necessary to understand the factors that influence the process in one way or the other. A compost heap is a miniature ecosystem where interactions between biotic and abiotic factors bring about the desired changes. By providing a favorable environment for the growth and activities of the desired biota in the system, good quality compost can be produced. The criteria used in the evaluation of the composting process, compost stability (maturity), and quality are based on the physical and chemical characteristics of the organic material. These parameters include a drop in temperature, degree of self heating capacity, oxygen consumption, cation-exchange capacity, organic matter, nutrient contents, and C : N ratio.

10.3.1 Abiotic Factors

10.3.1.1 Nature of the substrate

All kinds of organic residues amenable to the enzymatic activities of the microorganisms can be converted into compost if suitable conditions for biodegradation are provided. As the substrate becomes the only source of food to the microorganisms in a compost heap, the nature of the substrate is the most basic controlling factor in any composting process. Most of the substrates are largely made up of polymers, which are insoluble in water. The extracellular enzymes released by the microbes hydrolyze these polymers into monomers, which then dissolve into water and enter the microbial cell where further decomposition takes place. The maturity of the compost also depends upon the nature of the substrate. Use of compost agronomic or horticultural is based on the compost's chemical composition. If the substrate is of plant origin, then the main constituents are the carbonaceous compounds such as cellulose, hemicellulose, and lignin. Nitrogenous constituents (proteins) occur to a lesser extent. Protein constituents, cellulose and hemicellulose decompose easily. Although cellulosic substrates form good raw material for composting, lignin, being a complex aromatic polymer, is resistant to microbial attack to a considerable extent. However, it is not entirely recalcitrant to microbial decomposition; it undergoes slow degradation. The elevated temperature found during the thermophilic phase is essential for rapid degradation of lignocellulose. A number of fungi, particularly those belonging to the Basidiomycetes group, are well known for their ability to decompose lignin. Some bacteria and actinomycetes also have lignolytic characteristics. The organic compounds in the biowaste could be divided into three main fractions: (1)

carbohydrates (polymers and simple sugars), (2) lignin and (3) nitrogen compounds. In the beginning of the composting process, simple carbohydrates are converted to carbon dioxide and water and degradation of nitrogenous compounds results mainly in the production of ammonia. In the later stages of composting, cellulose and hemicellulose are utilized by the compost microflora and finally lignin is also subjected to degradation. Besides mineralization, organic matter is converted to humic substances. The porosity of the substrate plays a major role in the composting process. Porosity facilitates gas exchange with the atmosphere, enabling the aerobic metabolism to become dominant, liberating heat profusely. Materials that should not be included while setting up a composting pile include soil, ashes from a stove or fireplace and manure from carnivorous (meat-eating) animals. Manure from herbivorous animals such as rabbits, goats, cattle, horses, elephants, or fowl can be used, as it is much leaner in proteins than the manure from carnivores. Once a pile is started, no further substrate should be added; the reason is that it takes a certain length of time for the substrate to break down and anything added has to start at the beginning, thus lengthening the decomposition time for the whole pile.

10.3.1.2 Carbon/nitrogen ratio

The relative proportion of carbon and nitrogen is a major controlling factor in the composting process. Carbon serves primarily as an energy source for the microorganisms, while a small fraction of the carbon is incorporated to the microbial cells. Nitrogen is critical for microbial population growth, as it is a constituent of protein that forms over 50% of dry bacterial cell mass. If nitrogen is limiting, microbial populations will remain small and it will take longer to decompose the available carbon. Excess nitrogen, beyond the microbial requirements, is often lost from the system as ammonia gas. In the composting process, the substrate should achieve a C/N ratio of 30 : 1 for stimulating degradation and immobilization of nitrogen. A balanced carbon to nitrogen (C : N) ratio of 25 : 1 to 30 : 1 is ideal for an active compost pile. C : N ratios of as low as 20 : 1 or as high as 40:1 also produce good quality finished compost.

If $\qquad$ C : N < 20 : 1

Excess nitrogen will off gas to the atmosphere as NH_3 or N_2O, resulting in an undesirable odor.

If $\qquad$ C : N > 40:1

Nitrogen mineralization generally occurs in two phases, a rapid exponential immobilization or mineralization phase, followed by a slow linear mineralization phase. Nitrogen mineralization is the process by which organic nitrogen is converted to plant available inorganic form like ammonium and nitrate. The C/N ratio of the substrate determines whether immobilization or mineralization will dominate in the early stages of composting. The rate of inorganic N release to the soil from composted manure depends on the rate of decomposition of the organic matter and on subsequent turnover of the decomposed C and N in soil. Release of plant available N from manure in the soil is controlled by the balance of N immobilization and mineralization, which in turn is controlled, to a large extent, by the C/N ratio of the decomposing organic material. Decomposition rate (i.e. composting process) slows down.

10.3.1.3 Moisture

Moisture is one of the composting variables that affect microbial activities, as it provides a medium for the transport of dissolved nutrients required for the metabolic and physiological activities of microorganisms. It is essential for the decomposition process, as most of the decomposition occurs in the thin liquid films on the surfaces of particles. Moisture content of 60–70% is generally considered ideal to start with. At later stages of decomposition, the ideal moisture content may be 50–60%. Moisture management requires a balance between microbial activity and oxygen supply. Very low (<30%) or high moisture content (>75%) inhibits microbial activities due to early dehydration or anaerobiosis. Excess moisture will fill many of the pores between particles with water, thereby limiting oxygen transport. This in turn would create anaerobic conditions and brings about putrefaction, resulting in disagreeable odor and undesirable products. On the other hand, if the composting substrate is supplied with insufficient water, the growth and proliferation of microorganisms as well as the rate of decomposition of the organic material would be slowed down or even stopped. It is important, therefore, to ensure adequate moisture in each layer of the compost heap.

10.3.1.4 Oxygen and temperature

The decomposition process enhances the interplay between two of the key environmental parameters, oxygen and temperature. The temperature within a composting mass determines the rate at which many of the biological processes take place and plays a selective role in the development and the succession of the microbiological communities. Temperature and oxygen fluctuate in response to microbial activity, which consumes oxygen and generates heat. Both are linked by a common mechanism of control: aeration. Aeration is one of the components of the controlling process, as it ensures the growth of adequate aerobic microbe populations and the development of stabilizing temperature. Aeration supplies the depleted oxygen to the composting mixture and carries away excess heat from the system. Inadequate oxygen may lead to the growth of anaerobic microorganisms, which can produce odorous compounds. Usually, in an aerobic system, the temperature rises to 50–60°C in just a few days and can even go up to 70°C in some cases. If done correctly, a compost pile will heat to high temperatures within 24 to 48 hours. If it doesn't, the pile is too wet or too dry or there is not enough green material (or nitrogen) present. The high temperature rise in the compost heap destroys weed seeds, pathogenic microorganisms, maggots, and worms and prevents fly breeding. This happening and the generation of antibiotics during composting drastically reduce pathogens in the final compost. A temperature in the range of 55 to 65°C ensures destruction of pathogenic organisms. A temperature of 65°C for at least 30 minutes is considered a critical threshold for plant pathogens. Human pathogens are also inactivated at high temperatures. The temperature and the time interval required destroying most common types of pathogenic microorganisms and parasites are given in Table 10.1. The heat resistance of human pathogens increases markedly under dry conditions. Therefore, wet conditions must prevail in the compost pile. The maximum temperature of the composting process reaches 60–70°C, the temperature level where many microorganisms become less active. At the top of the pile, the temperature is slightly lower due to conductive heat loss

Table 10.1: Temperature and the time interval required to destroy most common types of pathogenic microorganisms and parasites

Pathogen	Temperature and time
Salmonella typhosa	Further growth is stopped above 46°C dies within 20–30 minutes at temperature of 55–60°C
Salmonella sp.	Dies within 60 and 20 minutes at a temperature of 55 and 60°C
Sbigella sp.	Dies within 60 and 20 minutes at a temperature of 60°C

from the top to the surroundings. Over time, the temperature gradually drops off as the degradation rate of organic matter becomes less. This course in composting will result in adequate stabilization of organic matter, drying of the compost and killing of pathogens and weeds.

Low temperature typically indicates low aerobic activity in the composting pile. Temperature alone is not a fool proof indicator of aerobic activity, as it is a result of heat production and heat removal. Lack of aerobic activity can only be confirmed by measuring the oxygen content within the compost bed. To attain temperatures high enough for heat activation throughout in the compost, the vessel has to be insulated to retain the heat produced. High temperature combined with high exchange rates of the air will increase the ammonia losses. In a composting pile, however, the rate of degradation is a result of metabolic activity of a mixed microbial population that may originally include microorganisms with different temperature optima. These microorganisms adapt to the environmental temperature during composting and have a collective temperature optimum at which respiration from the microbial community is highest. Not only is microbial metabolism highly temperature dependent, but it also dramatically influences the population dynamics (e.g. composition and density) of microbes are dramatically influenced by temperature. Temperature increase within composting materials is a function of initial temperature, metabolic heat evolution and heat conservation. Indeed, temperatures of composting material below 20°C have been demonstrated to significantly slow or even stop the composting process. Temperature in excess of 60°C has also been shown to reduce the activity of the microbial community and above this temperature, microbial activity declines as the thermophilic optimum of microorganisms is surpassed. If the temperatures reach 82°C, the microbial community is severely impeded.

10.3.1.5 Aeration

Aerobic organisms need to breathe air to survive. Aeration is necessary in high temperature aerobic composting for rapid odor free decomposition. Aeration is also useful in reducing high initial moisture content in composting materials. Several different aeration techniques can be used. Turning material is the most common method of aeration when composting is done in stacks. Hand turning of the compost piles or in units is most commonly used for small garden operations. Mechanical turning or static piles with a forced air system are most economical in large municipal or commercial operations. The most important consideration in turning compost, apart from aeration is to ensure that material on the outside of the pile of units is turned into the center where it will be subject to high temperatures. In hand turning with forks, this can be easily accomplished.

For piles or windrows on top of the ground material from the outer layers can be placed on the inside of the new pile. For static piles with a forced air system finished compost or a physical "cover" can be placed on the composting material, ensuring it reaches high temperatures uniformly. Volume reduces during the compost process. Piles or windrows can eventually be combined when turned, particularly if long composting periods are used.

10.3.1.6 pH

The pH is another parameter that greatly affects the composting process. The range of pH values suitable for bacterial development is 6.0–7.5, while fungi prefer an environment in the range of pH 5.5–8.0. An initial phase characterized by a low pH is often observed during composting of organic wastes and perhaps especially of easily degradable energy rich materials like waste. This is due to the formation of carbon dioxide and volatile fatty acids. With the subsequent evolution of CO_2 and utilization of VFAs, the pH begins to rise and may reach even values exceeding 8.0. Organic acids are produced during decomposition of the organic matter, but their existence is only transitory. Problems may arise if the material obtained undergoes putrefaction, as appreciable amounts of troublesome organic acids are produced during anaerobic decomposition and may produce malodour. However, a rise in pH beyond 7.5 could make the environment alkaline which may cause loss of nitrogen as ammonia. The growth of active microorganisms is inhibited by temperature above about 40°C if short chain fatty acids and low pH are present. Microbial tolerance to thermophilic temperature is reduced by the combination of low pH and increasing concentrations of fatty acids. The optimum pH range for decomposition is between 6.5 and 8.5. The pH affects the potential for beneficial bacteria to colonize composts; below pH 5.0, bacterial biocontrol agents are inhibited. To curtail excessive ammonia loss, Hoitink and Kuter (1986) suggest that pH should be below 7.4 in aerated composting systems. The pH is an indicator of aeration levels within a composting pile. Well-aerated compost piles generally have a high pH, whereas piles with anaerobic conditions have decreased pH values. The decrease in pH during the initial period of composting is expected because of the acids formed during the metabolism of readily available carbohydrates. After the initial stage, the pH is expected to rise, with evolution of free ammonia and to stabilize or drop slightly again to near neutral as a result of humus formation with its pH buffering capacity at the termination of composting activity.

A list of the important abiotic parameters associated with the success of composting process and the range in which they should preferably remain, is presented in Table 10.2. The relationship between the degree of maturity of the compost, the temperature, and the O_2 consumption is presented Table 10.3.

In composting, carbohydrates are also broken down into humic and fulvic acids. However, the fulvic acid is subsequently degraded. This, together with ammonification of inorganic nitrogen, accounts for neutral pH, which is generally attained at the end of the process.

10.3.1.7 Electrical Conductivity (EC)

Generally, it is found that EC increases during composting as volatile solids (VS) are degraded and the amount of water-soluble salts increases on a total solids (TS) basis. At lower pH values, negatively charged surface sites of organic matter are occupied by protons, which thus lowers CEC. A decrease in CEC results in a lower adsorption of cations to organic matter and thus an increase in EC.

Table 10.2: Key parameters that influence the composting process and their optimum values

Parameter	Optimum value for composting
C/N ratio of the feed	25 to 35
Particle size	10 mm for agitated systems and forced aeration, 50 mm for long heaps and natural aeration
Moisture content	50 to 60% (higher values when bulking agents are used)
Air flow	0.6 to 1.8 m^3 air/day/kg volatile solids during thermophilic stage, or maintain oxygen level at 10% or higher
Temperature	55 to 60°C held for 3 days
Agitation	No agitation to periodic turning in simple systems and short bursts of vigorous agitation in mechanized systems
pH control	Normally not necessary
Heap size	Any length, 1.5 m and 2.5 m wide for heaps using natural aeration. With forced aeration, heap size depends on need to avoid overheating
Activators	Use of efficient cellulolytic fungi and biofertilizers

Table 10.3: The relation between the degree of maturity, temperature and O_2 consumption in a compost system

Degree of maturity	Maximum temperature (°C)	O_2 consumption (mg/g OS) (according to Jourdan)	O_2 consumption (mg/g OS) (according to Becker)	Material Status
I	> 60	> 40	> 80	Raw material
II	60–50.1	40–8.1	80–50	Fresh compost
III	50–40.1	28–16.1	50–30	Fresh compost
IV	40–30.1	16–6.1	30–20	Matured compost
V	≤ 30	≤ 6	≤ 20	Matured compost

10.3.2 Biotic Factors

Composting involves a myriad of microorganisms. The composition and magnitude of these microorganisms are important components of the composting process. The microbes decompose the organic matter, and transform the nitrogen component through oxidation, nitrification, and denitrification. During the process, there may be depletion of nutrients if the microbes incorporate minerals from the waste into biomass. However, composting may also involve sequential growth and degradation of subpopulations. Hence, there may be no significant change in overall levels of microorganisms or inorganic nutrient requirement. Microbes cannot directly metabolize the insoluble particles of organic matter. All biochemical reactions during composting are catalyzed by enzymes. The microbes produce hydrolytic extracellular enzymes to depolymerize the larger compounds (i.e. plant polymers, cellulose, hemicellulose, and lignin) to smaller fragments that are water-soluble. Common microorganism growing in compost are shown in Table 10.4.

Table. 10.4: Common microorganisms associated with composting

Organisms	Mesophilic phase	Thermophilic phase
Bacteria	Pseudomonas, Bacillus, Bacillus licheniformis, Bacillus cereus, Serratia marcescens, Alcaligenes sp., Flavobacterium, Clostridium, Leuconostac, Pediococcus	**(i) Thermophilic spore-forming bacteria** Bacillus stearothermophilus, B. thermoglucosidasius, B. pallidus, B. thermodenitrificans, B. coagulans, Bacillus schlegelii, **(ii) Sulfur and hydrogen-oxidizing autotrophic bacteria** i.e. Hydrogenobacter spp **(iii) Heterotrophic, aerobic, non-spore forming bacteria** Thermus thermophilus, T. aquaticus,
Actinomycetes	Streptomyces	Micromonospora, Streptomyces sp., Micropolyspora, Thermoactinomyces sp., Thermomonospora
Fungi	Alternaria, Cladosporium, Aspergillus, Mucor, Humicola, Penicillium, Absidia corymbifera, Absidia orchidis, Absidia sp., Acremonium atrogriseum, Acremonium chrysogenum, Acremonium furcatum, Acremonium kiliense, Acremonium murorum (Corda), Acremonium sp., Acremonium strictum , Acremonium thermophilum , Actinomucor elegans, Actinomucorsp. Aleurisma sp., Alternaria alternate, Alternaria sp., Arthrobotrys amerospora, Arthrobotrys oligospora, Ascotricha sp., Aspergillus clavatus, Aspergillus erythrocephalus, Aspergillus flavipes, Aspergillus fumigates, Aspergillus orchraceous, Aspergillus parasiticus, Aspergillus sp., Aspergillus versicolor, Aspergillus wentii, Aureobasidium pullulans, Aureobasidium sp., Botryosporium sp., Cephaliophora irregularis, Cephaliophora sp., Cephalosporium sp., Coprinus cinereus , Emericella nidulans (Eidam), Hormiscium sp., Paecilomyces sp., Penicillium sp., Trichothecium roseum	Torula (yeast), Thermoascus Absidia corymbifera, Absidia ramosa (Lindt), Acremoniella sp., Acremonium thermophilum, Aspergillus fumigates, Aspergillus sp., Chaetomium sp., Chaetomium thermophile var. coprophile, Chaetomium thermophile var. dissetum, Chaetomium thermophilum, Coonemeria crustacean, Coprinus cinereus, Emericella nidulans (Eidam), Fomes sp., Geosmithia sp, Hormiscium sp., Humicola grisea var. thermoidea, Humicola sp., Lenzites sp., Malbranchea cinnamomea, Mollisia sp., Monotospora sp., Mucor miehei, Mucor pusillus, Myceliophthora thermophila, Oïdium sp., Paecilomyces sp., Paecilomyces variotii, Penicilium dupontii, Penicillium sp., Rhizomucor sp., Rhizopus chinensis, Rhizopus microspores, Scytalidium thermophilum, Sporotrichum sp., Stibella thermophila, Talaromyces thermophilus, Thermomycessp., Thielavia heterothallica, Thielavia terrestris (Apinis), Trichothecium roseum

10.3.2.1 Bacteria

Bacteria play by far the most dominant role during the most active stages of composting process because of their ability to grow rapidly on soluble proteins and other readily available substrates. They may also attack more complex materials, or may exploit substances released from the less degradable materials due to extracellular enzyme activities of other organisms. Among bacteria that occur commonly in aerobically decomposing substrate are species of *Bacillus, Cellulomonas, Pseudomonas, Klebsiella* and *Clostridium* occurs substantially in anaerobic conditions. Typical bacteria of the thermophilic phase are species of *Bacillus*, e.g. *B. subtilis, B. licheniformis,* and *B. circulans.* About 87% of the randomly selected colonies during the thermophilic phase of composting belong to the genus *Bacillus.*

Many thermophilic species of *Thermus* have been isolated from compost at temperatures as high as 65°C and even 82°C. *Nitrosomonas spp.* and *Nitrobacter spp.* are the ammonium oxidizing and nitrite oxidizing bacteria respectively present in the compost heap. Establishment of a large population of denitrifying bacteria suggests that some anaerobic microhabitat exists within the compost piles. These microhabitats could have been developed within the piles partially due to the initial high water content (65%) of the piles and partially because of the rich contents of organic matter and nitrogen present in the substrate, which promote microbial activity to the extent of causing depletion in O_2 content in isolated pockets within the piles. Moreover, some species of denitrifying bacteria may be facultative and grow aerobically. Some microbial genera capable of denitrification are Bacillus, Flavobacterium and Pseudomonas. Mesophilic microorganisms are partially killed or poorly active during the thermogenic stage (40–60°C). The diversity decreased as temperature increased, with a shift from *Pseudomonas, Achromobacter, Flavobacterium, Micrococcus* and *Bacillus* to one dominated by *Bacillus.* Bacteria related to *B. schlegelii, Hydrogenobacter spp.* and particularly to the genus Thermus (*T. thermophilus, T. aquaticus*) appear to be the main active microbes in hot compost (65–80°C). Bacterial survival in high temperature composting material is possible through formation of microcolonies. Mesophiles are likely to contribute little to compost degradation at these temperatures. Microbial fermentation of carbohydrates generally results in an increase in acidity. Clostridium species commonly ferment glucose to yield butyl and ethyl alcohols and certain acids. *Lactobacillus lactis* yields almost entirely lactic acid, while *Lactobacillus bevis* yields lactic and acetic acids, ethyl alcohol and carbon dioxide.

10.3.2.2 Fungi

The role of fungi starts when simple, easily degradable substances such as sugar, starch and protein are acted upon by bacteria and the substrate is predominated by cellulose and lignin, which normally occurs toward the later stages of composting (curing process). Most fungi are eliminated by high temperatures, but they commonly recover when temperatures are moderate and the remaining substrates are predominantly cellulose or lignin. Being efficient consumers of carbon, fungi build up much higher biomass than other microorganisms. The most commonly observed species of celluloytic fungi in composting materials are *Aspergillus, Penicillium, Rhizopus, Fusarium, Chaetomonium, Trichoderma, Alternaria* and *Cladiosporium* (Fig. 10.5). Some of the species of *Paecilomyces* and *Sporotrichum* have also been named as efficient degraders of lignocellulosic wastes.

White rot fungi are known as the most efficient lignolytic microorganisms. *Phanerochaete chrysosporium* is probably the best suited microorganism with this activity and it is often used as a reference. Among other well known white rot fungi, *Coriolus versicolor* show even higher efficiency and a wider range of lignolytic activities together with an important celluloytic activity. *Phanerochaete flavidoalba* causes preferential loss of lignin rather than

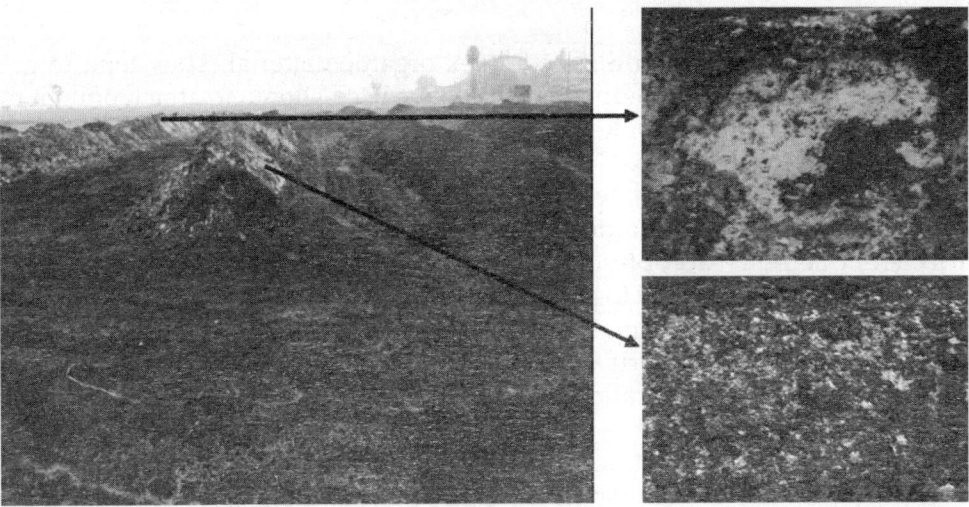

Fig. 10.5: Some fungi like *Trichoderma* (yellow colour) and *Aspergillus* (white colour) growing on compost of distillery effluent and pressmud

of cellulose and it is more efficient than *P. chrysosporium* on paper mill effluents. The plant constituent that offers maximum resistance to biodegradation is lignin. Yet, in spite of its substantial microbial recalcitrance, lignin does get degraded by some fungi and a few bacteria. The most important among these are white-rot fungi belonging to Basidiomycetes. Species of *Polyporus, Pleurotus, Collybia, Poria, Fomes, Trametes, Sporotrichum, Cyathus* and *Coriolus* have also been found to degrade lignin. Temperature is one of the most important factors affecting fungal growth. Other important factors are sources of C and N and the pH. During composting, temperatures above 55°C discourage fungal growth. Fungi are excluded during the earlier high temperature stage of the composting process. A moderately high level of nitrogen is needed for fungal growth, although some fungi, mainly wood-rotting fungi, grow at low nitrogen levels. Indeed, a low nutrient nitrogen level is often a prerequisite for lignin degradation. However, low nutrient nitrogen is a rate limiting factor for the degradation of cellulose. Most fungi prefer an acidic environment but tolerate a wide range of pH, with the exception of the Basidiomycotina, which do not grow well above pH 7.5. The majority of the fungi are mesophiles, which grow between 5°C and 37°C, with an optimum temperature of 25–35°C. However, in the compost environment the elevated temperature means that the small group of thermophilic fungi is an important biodegradation agent. Thermophilic fungi that have been found growing in lignocellulose substrate or compost are *Taloromyces emersonii, T. thermophilus, Thermoascus auranticus,* and *Thermomyces lanuginosus*. The most effective lignin degraders are Basidiamycotina, but according to Cooney and Emerson (1964) all Basidiamycotina are mesophilic. However, a few Basidiamycotina grow well at elevated temperatures. *Phanerochaete chyrsoporium* (*Sporotrichum pulverulentum*) is a white-rot fungus with an optimum temperature of 36–40°C and maximum temperature of 46–49°C. *Ganoderma colosum* is another white-rot fungus that is still capable of growing at 45°C and has an optimum temperature of 40°C. In the genus Coprinus there are some species that have an optimum temperature of above 40°C. A thermophilic *Ascomycotina, Thermoascus aurantiacus,* has a high lignolytic capacity and has been isolated from compost.

10.3.2.3 Actinomycetes

Like fungi, actinomycetes also utilize complex organic material. They tend to grow in numbers in the later stages of composting and have been shown to attack polymers such as hemicellulose, lignin and cellulose. The actinomycetes that occur most frequently are Micromonospora, Streptomyces, Nocardia, and Thermoactinomyces. Actinomycetes generally show their activity at later stages of decomposition. Actinomycetes are higher forms of bacteria, which form multicellular filaments; thus they resemble fungi. They are primarily strict aerobic saprophytes and are common in many environments. Their ubiquity is a result of their ability to utilize a wide range of carbon sources and to sporulate prolifically. Actinomycetes colonize more slowly than bacteria and fungi. Colonization is minimal in areas that are poorly aerated. They appear during the thermophilic phase as well as the cooling and maturation phase of composting and can occasionally become so numerous that they are visible as a white film on the surface of the compost. The genera of the thermophilic actinomycetes isolated from compost include *Nocardia*, *Streptomyces*, *Thermoactinomyces* and *Micromonospora*. Actinomycetes are able to degrade some cellulose and solubilize lignin and they tolerate higher temperatures and pH than fungi. Thus, actinomycetes are important agents of lignocellulose degradation during peak heating, although their ability to degrade cellulose and lignin is not as high as that of fungi. The actinomycetes are thus well placed to exploit the compost environment as the piles cool in the immediate post peak heat phase. During the cooling stage of composting, actinomycetes actively degrade hemicellulose in the compost. With an optimum growth between 25–30°C and pH of 5–9, these microorganisms are the most significant group of microbes in the degradation of relatively complex, recalcitrant polymers. As actinomycetes develop more slowly than most bacteria or fungi, they are ineffective competitors when nutrient levels are high, but become more competitive as nutrient levels decrease. *Actinomycetes thermophilus*, *Streptomyces* and *Micromonospora spp.* are common in compost. Although optimum growth temperatures fall in the mesophilic range, obligate thermophiles such as Thermoactinomycetes and *Saccharomonospora spp.* have been isolated. Certain species of actinomycetes are more tolerant of high temperatures, becoming increasingly active as temperatures approach and surpass 60°C. Cellulose is not an obligate carbon source for fungi and actinomycetes, which are the microorganisms mainly responsible for cellulose degradation and the addition of readily metabolizable substances has been shown to accelerate the decomposition of cellulose. It is thought that by initially utilizing the more available C sources, the population of cellulose degraders can develop to a large size. Once the more available C source becomes limiting, the microorganisms adapt to the cellulose, with the overall effect being an increase in cellulose hydrolysis. Since cellulose degradation is largely attributed to fungi and actinomycetes, which are characterized by the formation of hyphae, it is possible that frequent turning of the compost resulted in the breaking of the hyphae and, subsequently, in reduced activity of the cellulose degraders.

10.4 ADVANTAGES OF COMPOSTING

Composting will be more cost effective and environmentally friendly than some other management options, including storage and landfill. Composting can have less impact on the environment than most alternatives. Current research aims to reduce greenhouse gas emissions. A biologically stable compost does not generate noxious odours during land application and can be stored without being a nuisance because it forms a water repellent crust. Stable compost does not provide a medium for the breeding of flies. Unlike some organic wastes (including sludge, barley, sawdust, green waste and food processing waste), mature manure compost does not contain or produce phytotoxic substances (which inhibit plant growth and seed germination). The heat generated during composting promotes moisture removal, with the result that it is less costly to store and transport the composted material than the raw manure.

10.5 DISADVANTAGES OF COMPOSTING

The effectiveness of the composting operation is usually dictated by atmospheric conditions and quality suffers during wet, cold or dry weather. A high degree of control of moisture and temperature is required to achieve a satisfactory product. The material must have a relatively high void ratio, warranting the use of low density blending agents. The markets for compost are not as well defined as those for commercial fertiliser or animal manure and the characteristics of the bulking agent can affect the quality of the compost. High application rates are required to meet crop nutrient requirements. Cartage costs are higher than for fertilisers. Composting is more demanding than the direct application of manure to land. However, compost is labour intensive and is sometimes the source of foul odors.

11

Challenges and Further Suggestions for Decolourization and Detoxification of PMDE in Indian Scenario

The major problem with the PMDE is its complex nature due to the recalcitrant constituent, inhibits the bacterial growth. The chemical structure of major colouring constituent, i.e. melanoidins is yet to be understood. Melanoidins have net negative charge and therefore, different heavy metal ions (Cu^{2+}, Cr^{3+}, Fe^{3+}, Zn^{2+} and Pb^{2+}, etc.) form large complex molecules with melanoidins, amino acids, proteins and sugars in acidic medium and get precipitated. Besides melanoidins, distillery effluent contains various phenolic compounds as gallic acid, p-coumaric acid, gentisic acid, 2-methyl-3, 4-dimethoxy-d-napthol and 3, 4-dimethoxy 2 (1,1-hydroxyethyl) α-napthol which have a high inhibitory and antimicrobial activity reducing the anaerobic digestion of distillery effluent. The sugarcane molasses, which is used as raw material for alcohol production in distilleries contain high amount of sulfate and aromatic alcoholic moieties as by products from sugar industries. Further, sulfate and heavy metals present in distillery effluent get reduced into black coloured precipitate of metal sulfides which act as competitive inhibitor for sulfate reducing bacteria (SRB) and non-SRB leading to inhibition of methanogenesis or sulphate reduction and giving toxicity to PMDE. Moreover, the interaction of melanoidin with other pollutant of PMDE (phenolics, sulfides and heavy metals) is still not known. Prior to solving the decolourization problem of effluent, it is very essential to understand the detailed chemical structure of melanoidins. One carbohydrate structure is reported as part of the melanoidin skeleton by Cammerer et al. (2002). The structural analysis of Maillard product by Yaylayan and Kaminsky (1998) in glycine/glucose model system isolated three different polymers which were further characterized through pyrolysis GC/MS and their empirical formula was obtained as $C_7H_{11}N_1O_4$. Kato and Hayase (2002) revealed that there were two pyrrolo-pyrrol rings combined with methane bridge. The degradation of melanoidin is yet to be fully understood. In addition, the detail of enzyme and its mode of action also need to be revealed. Moreover, the effect of sulfides and metals along with phenolic compounds required systematic studies to establish the mechanism of biological decolourization prior to its scope to develop an industrial scale decolourization technique for safe disposal. Recently, the sequential applications of bacteria and wetland plants have been reported to be very promising for detoxification but this has to be optimized yet with detailed microbiology of wetland plants, plant rhizospheres and detoxification mechanism. The interaction of melanoidins with other toxic compounds present in PMDE under the influence of different environmental condition is not known. Chemical structure of different types of melanoidins in PMDE is complicated and yet to be explored. Moreover, the nature of recalcitrant toxic compounds and extent of toxicity added in environment by melanoidin, phenolics, sulphates, phosphates and heavy metals is to be understood.

11.1 RESEARCH GAPS

Four decades of investigations unequivocally demonstrated that utilization of treated distillery spentwash enhances crop productivity, soil fertility besides saving considerable amount of inorganic fertilizers. However these dataset needs be proved on long term sustainability and stability basis by taking into consideration associated land and water degradation. Though research on utilization of distillery effluent was undertaken for a wide array of crops, almost all published results are restricted to short term field experiments. Researchers should focus on long term effects of spentwash on crop productivity, soil nutrient dynamics, microbial diversity and ground water quality. Some agricultural universities recommend controlled land application of spentwash once in two years. However, even in such cases, the long term assessment is lacking and it is necessary to pay attention on this direction. Moreover, some of the negative results arising after the use of distillery effluent are often not reported. Only when such information if any is published, the scientific community can resolve for effective utilization of this nutrient rich waste. On the other side, the role of distillery industries is pivotal for ensuring fair conduct of field experiments during research tie-ups with investigating organizations. Though lot of information is available in the area of distillery waste management in agriculture, data on effect of spentwash application on biogeo-chemical cycles, plant physiological adaptations to excess potassium, quality of farm produce is still lacking.

11.2 LACK OF EXTENSION SUPPORT FROM INDUSTRIES

Distilleries disposing their spentwash on land do not give adequate attention to the propagation of applicability of the spentwash for agricultural use. Most of the industries consider it as a waste which has to be disposed clandestinely either to water courses or if possible into the agricultural land. The value of the spentwash as a resource is scarcely recognised. The best strategy to remove these constraints would be to recognize the distillery spentwash as an important source of nutrients for agriculture. However, a sustainable agronomic package has to be developed which ensures increased crop yields without causing any environmental hazards. When these techniques are available to the farmers they can be convinced to accept the practice of irrigating their fields with diluted spentwash which would reduce fertilizer cost especially in rainfed agriculture and augment availability of water for irrigation. Once this nutrient rich liquid waste is recognised as a resource and finds use in agriculture, the basic problem of water pollution will be automatically solved to a great extent.

11.3 POLICIES AND INSTITUTIONAL SET-UP FOR WASTEWATER MANAGEMENT

Presently there are no separate regulations/guidelines for safe handling, transport and disposal of wastewater in the country. The existing policies for regulating wastewater management are based on certain environmental laws and certain policies and legal provisions viz. Constitutional Provisions on sanitation and water pollution; National Environment Policy, 2006; National Sanitation Policy, 2008; Hazardous waste 6

(Management and Handling) Rules, 1989; Municipalities Act; District Municipalities Act etc. Creation of sewerage infrastructure for sewage disposal is responsibility of State governments/urban local bodies, though their efforts are supplemented through central schemes, such as National River Conservation Plan, National Lake Conservation Plan, Jawaharlal Nehru National Urban Renewal Mission, and Urban Infrastructure Scheme for Small and Medium Towns (MoEF, 2012). However, operation and maintenance of sewerage infrastructure including treatment plants are responsibilities of State governments/urban local bodies and their agencies. As per Water Act 1974, State Pollution Control Boards possesses statutory power to take action against any defaulting agency. Water Act 1974 also emphasizes utilization of treated sewage in irrigation, but this issue has been ignored by the State Governments. Ministry of Environment and Forests (MoEF), Govt. of India initiated a technical and financial support scheme to promote common facilities for treatment of effluents generated from SSI units located in clusters. Under the Common Effluent Treatment Plant (CETP) financial assistance scheme, 50% subsidy on project capital cost-25% share each of Central and State Governments-was provided. As a result, 88 CETPs having total capacity of 560 MLD have been set up throughout India covering more than 10,000 polluting industries (CPCB, 2005). In addition to setting up treatment plants, Central Government, State Government and the Board have given fiscal incentives to industries/investors to encourage them to invest in pollution control. Incentives/concessions available to them are:

- Depreciation allowance at a higher rate is allowed on devices and systems installed for minimising pollution or for conservation of natural resources.
- Investment allowance at a higher rate is allowed for systems and devices listed under depreciation allowance.
- To reduce pollution and to decongest cities, industries are encouraged to shift from urban areas. Capital gains arising from transfer of buildings or lands used for the business are exempted from tax if these are used for acquiring lands or constructing building for the purpose of shifting business to a new place.
- Reduction in central excise duty for procuring the pollution control equipments.
- Subsidies to industries subject for installation pollution control devices.
- Distribution of awards to industries based on their pollution control activities.
- Amount paid by a tax payer, to any association or institution implementing programmes for conservation of natural resources, is allowed to be deducted while computing income tax. Customs duty exemption is granted by the Central Government for items imported to improve safety and pollution control in chemical industries.

11.4 FUTURE RESEARCH NEEDS IN PHYTOREMEDIATION

Modification or over expression of the enzymes that are involved in the synthesis of GSH and PCs might be a good approach to enhance heavy metal tolerance and accumulation in plants. In the process of attempting to improve rhizofiltration, it was discovered that young plant seedlings grown in aerated water (aquacultured) are often more effective than roots in removing heavy metals from water .The technology of using plant seedlings to remove toxic metals from water was termed blastofiltration (blasto is 'seedling' in Greek). Blastofiltration may represent the second generation of plant-based water treatment technology. It takes advantage of the dramatic increase in surface to volume ratio that occurs after germination and the fact that some germinating seedlings also ab/adsorb large quantities of toxic metal ions. This property makes seedlings uniquely suitable for water remediation. Seedling cultures used for blastofiltration can be produced in light or in darkness, and seeds, water and air are the only components required. Heavy metal hyper accumulators have received increased attention in recent years, due to the potential of using these plants for phytoremediation of metal contaminated sites. However, there are some limitations for this technology to become efficient and cost effective on a commercial scale, as most of the metal hyper accumulating plants identified have small biomass, and are not very adaptable to harsh environment. These limitations need to be overcome by achieving a good understanding of the mechanisms of metal hyper accumulation in plants. In the past years, most researches focusing on the physiological mechanisms of hyper accumulation have made great progress; however, the understanding of a range of molecular/cellular mechanisms will undoubtedly change our concept of metal acquisition and homeostasis in higher plants. With the completion of the Arabidopsis genome project, eventually followed by genome sequences for other plants, the full range of genes that are potentially involved in heavy metal homeostasis and accumulation will be identified. The problem of low biomass phytoremediators can be overcome by increasing plant yield and metal uptake by engineering common plants with hypera ccumulating genes. If non-native transgenic plants are used for phytoremediation, proper control of their dissemination has to be adopted to avoid the introduction of new weed species. Some key technical hurdles that must be overcome for an industry to develop and grow are:

- Identifying more species that have remediative abilities.
- Optimizing phytoremediation processes, such as appropriate plant selection and agronomic practices. Understanding more about how plants uptake, translocate and metabolize contaminants.
- Identifying genes responsible for uptake and/or degradation for transfer to appropriate high-biomass plants.
- Decreasing the length of time needed for phytoremediation to work.
- Devising appropriate methods for contaminated biomass disposal, particularly for heavy metals and radionuclides that do not degrade to harmless substances and protecting wildlife from feeding on plants used for remediation.
- In addition to technical barriers, government regulations will also determine the overall success of phytoremediation.

So one can conclude that rhizofiltration is one of the important technologies, and when such technologies are merged with existing technologies they can be proved as efficient technologies. These are natural boon of nature where natural efficiency of plants can be utilized to treat contaminated sites. Sometimes certain modifications can be done in order to make these plants resistant to toxicants. So they can be greener technology and can help in reducing pollution and hence can help us to step towards a sustainable development.

11.5 FUTURE RESEARCH NEEDS IN WETLAND TREATMENT SYSTEM

Results from research currently underway will improve our ability to design effective wetlands, but a number of questions concerning animal wastewater treatment wetlands remain. Future research must answer the following questions:

11.5.1 Are Current Design Guidelines Adequate?

Many of the animal wastewater treatment wetlands are designed according to the guidelines of the NRCS (1991). Do these procedures result in the best design for treating animal wastewater? If so, under what circumstances does it function at a maximum? Can adjustments to the guidelines be made that will improve wetland performance? With present data, it is difficult to determine whether any given design is preferable. Results are not easily comparable since wetlands of various designs treat different wastes, use different pretreatment structures, and are in different areas of the U.S. Different design methods could be used to construct a number of wetlands at a single site, to test their effectiveness in treating the same wastewater under the same conditions.

11.5.2 What are the maximum concentrations of nutrients and BOD that can be treated in a wetland?

Is it unrealistic to expect that wetlands can treat waste loads above a certain level? What loading is too high? Should the loading rate recommendations vary under different climatic or soil conditions? Do the current design equations adequately account for extremely high loads? Mesocosm experiments with a series of loading rates would help answer these questions.

11.5.3 Which pretreatment structures are most effective?

A variety of pretreatment structures have been built in conjunction with wetlands, but very few data show the effectiveness of these devices. What structures are the most effective in reducing the solid and BOD load to wetlands? Is aeration of waste to enhance nitrification feasible? Different pretreatment strategies should be tested at a single site to discern which practice best enhances wetland performance. Recommendations for pretreatment should go hand in hand with wetland construction.

11.5.4 Are site specific results applicable to other areas?

What climatic conditions must be met in order for animal wastewater treatment wetlands to be effective? Should designs be different for different regions?

11.5.5 How can the cost to the farmer be minimized?

Research concerning the use of constructed wetlands should focus on maximizing their potential for contaminant removal while minimizing their cost. The goal of research on all animal wastewater treatment structures, including wetlands, is to foster the widespread treatment of animal wastewater in order to protect the quality of surface and ground water. Therefore it is necessary to determine under what circumstances wetlands can be added to an animal farmer's waste management choices and how they can be designed to function at an optimum.

11.6 POSSIBLE SOLUTIONS OF PROBLEMS ASSOCIATED WITH THE EFFLUENTS

- To exploit the effluent as a potential source of irrigation and maintain environment, the wastewaters must be diluted either with canal or underground waters to avoid the excessive accumulation of soluble salts in the soils. It will help in maintaining the productivity of agricultural crops without any harmful effect on soil properties.

- Entry of heavy metals into food chain can be reduced by adopting soil and crop management practices, which immobilize these metals in soils and reduce their uptake by plants.
- Heavy phosphate application and also the application of kaolin/zeolite to soils can reduce the availability of heavy metals.
- Application of organic manures can mitigate the adverse effect of the toxic metals on crops. Thus in the soils contaminated with high amount of toxic metals, application of organic manures is recommended to boost the yield potentials as well as decrease the metal availability to plants.
- Raising hyper accumulator plants (mustard/trees) in toxic metals contaminated soils is recommended to avoid the entry of toxic metals in the food chain.
- The industrial effluents, sludge and the soils must be monitored continuously to avoid the excessive accumulation of toxic metals in the soils and then transfer in the food chain.
- There should be strict Government legislation that only those industrial effluents be used in the fields which are cleaned through sewage and effluent treatment plants.
- Highest priorities should be given to proper disposal of solid and liquid effluents from Industries for proper land management.

Future research needs:
- Research should be done to study the long-term effects of sewage and industrial effluents on salt and toxic metal accumulation in soils and their effect on soil biological health and crop productivity.
- Effect of sewage/industrial effluents and heavy metal pollution in soils should be studied on fixed sites.
- Bio-transpiration of the contaminates through farm forestry and the critical concentrations of toxic metals in soil and plants for better animal and human health needs to be initiated.
- To develop eco-friendly technology for the use of sewage and industrial effluents to improve crop productivity and soil quality and to protect of quality of farm produce and environment from degradation.

References

Akunna, J.C. and Clark, M. (2000). Performance of a granular-bed anaerobic baffled reactor (GRABBR) treating whisky distillery wastewater. Bioresource Technology, 74 (3), 257–261.

All India Distiller's Association, AIDA (2003). Ref No. L/2003/1690. December 18, 2003.

All India Distiller's Association, AIDA (2008). Ethanol opportunities and challenges. URL http://aidaindia.org/its08/topics_covered.html, visited on March 2008.

Allen, S.E. (1974). Chemical analysis of ecological material. Oxford (London): Blackwell Scientific Publication.

American Public Health Association, APHA, (2005). Standard method for examination of water and wastewater. 21st ed. APHA, AWWA and WEF. Washington (DC): APHA, AWWA, WEF.

AOAC, (2002). Official Methods of Analysis of AOAC International, 17th ed. AOAC International, Gaithersburg, USA.

Arnon, D.I. (1949). Copper enzymes in isolate chloroplasts, Polyphenol oxidase in Beta vulgaris. Plant Physiol., 24, 1–15.

Arora, D.S., Chander, M. and Gill, P.K. (2002). Involvement of lignin peroxidase, manganese peroxidase and laccase in degradation and selective ligninolysis of wheat straw. Int. Biodeterior. Biodegrad. 50, 115–120.

Ashoor, S.H. and Zent, J.B. (1984). Maillard browning of common Amino acids and sugars. J. Food Sci., 49, 1206–1207.

Bekedam, E.K., Schols, H.A., Van Boekel, M.A.J.S. and Smit, G. (2006). High molecular weight melanoidins from coffee brew. J. Agricultural and Food Chemistry, 54, 7658–7666.

Belsare, D.K. and Prasad, D.Y. (1988). Decolorization of effulent from the bagasse based pulp mills by white-rot fungus, Schizophyllum commune. Appl. Microbiol. Biotechnol., 28, 301–304.

Beltran, F.J., Alvarez, P.M., Rodriguez, E.M., Garcia-Araya, J.F. and Rivas, J. (2001). Treatment of high strength distillery wastewater (cherry stillage) by integrated aerobic biological oxidation and ozonation. Biotechnol. Prog. 17, 462–467.

Bhalerao, V.P., Jadhav, M.B. and Bhoi, P.G. (2006). Effect of spent wash, press mud and compost on soil properties, yield and quality of sugarcane. Indian Sugar 40(6), 57–65.

Bharagava, R.N. and Chandra, R. (2010a). Biodegradation of the major color containing compounds in distillery wastewater by an aerobic bacterial culture and characterization of their metabolites. Biodegradation, 21, 703–711.

Bharagava, R.N. and Chandra, R. (2010b). Effect of bacteria treated and untreated post-methanated distillery effluent (PMDE) on seed germination, seedling growth and amylase activity in *Phaseolus mungo* L. J. Hazardous Materials, 180, 730–734.

Bharagava, R.N., Chandra, R. and Rai, V. (2008). Phytoextraction of trance elements and physiological changes in Indian mustard plants (*Brassica nigra* L.) grown in post methanated distillery effluent (PMDE) irrigated soil. Bioresour. Technology, 99, 8316–8324.

Bilgic, H., Gokcay, C.F. and Hasirci, N. (1997). Color removal by white- rot fungi. In: Wise, D.L.Ed.), Global Environmental Biotechnology. Elsevier, UK, 211–222.

Billore, S.K., Singh, N., Ram, H.K., Sharma, J.K., Singh, V.P., Nelson, R.M. and Dass, P. (2001). Treatment of molasses based distillery effluent in a constructed wetland in central India. Water Science and Technology, 44 (11–12), 441–448.

Blonskaja, V., Menert, A. and Vilu, R. (2003). Use of two-stage anaerobic treatment for distillery waste. Advances in Environmental Research, 7(3), 671–678.

Boopathy, R. and Tilche, A. (1991). Anaerobic digestion of high strength molasses wastewater using hybrid anaerobic baffled reactor. Water Research, 25 (7), 785–790.

Borrelli, R.C. and Fogliano, V. (2005). Bread crust melanoidins as potential prebiotic ingredients. Molecular Nutrition and Food Research, 49, 673–678.

Bourbonnais, R. and Paice, M.G. (1990). Oxidation of non-phenolic substrates: An expanded role of laccase in lignin biodegradation. FEBS Letters, 267 (1), 99–102.

Brands, C.M.J., Alink, G.M., Van Boekel, M.A.J.S. and Jongen, W.M.F. (2000). Mutagenicity of heated sugar-casein systems: effect of the Maillard reaction. J. Agric. Food Chem., 48, 2271–2275.

Brooks, R. R., M. F. Chambers, L. J. Nicks and B. H. Robinsons. 1998. Phytomining. Trends in Lant and Science, 1, 359–362.

Call, H.P. and Mucke, I. (1997). History, overview and applications of mediated lignolytic systems, especially laccase-mediator-systems (Lignozym®-process). J. Biotechnology, 53(2–3), 163–202.

Cammarota, M.C., Sant Anna, J.G.L. (1992). Decolorization of kraft bleach plant E1 stage effluent in a fungal bioreactor. Environ. Technol., 13, 65–71.

Cammerer, B., Jalyschkov, V. and Kroh, L.W. (2002). Carbohydrate structures as part of the melanoidin skeleton. Int. Congr. Ser., 1245, 269–273.

Central Pollution Control Board (CPCB), 2003. Environmental Management in Selected Industrial Sectors Status and Needs, PROBES/97/2002–03, CPCB, Ministry of Environment and Forest, New Delhi.

Chandra, R. and Pandey, P.K., (2001). Decolourisation of anaerobically treated distillery effluent by activated charcoal adsorption method. Indian J. Environ. Prot., 21, 134–137.

Chandra, R., Bharagava, R.N., Rai, V. and Singh, S.K. (2009a). Characterization of sucrose-glutamic acid Maillard products (SGMPs) degrading bacteria and their metabolites. Bioresour Technol., 100, 6665–6668.

Chandra, R., Bharagava, R.N., Yadav, S and Mohan, D. (2009b). Accumulation and distribution of toxic metals in wheat (*Triticum aestivum* L.) and Indian mustard (Brassica campestris L.) irrigated with distillery and tannery effluents. J. Hazard. Mater., 162, 1514.

Chandra, R., Kumar, K. and Singh, J. (2004). Impact of anaerobically treated and untreated (raw) distillery effluent irrigation on soil microflora, growth, total chlorophyll and protein contents of *Phaseolus aureus* L. J Environ. Biol., 25(4), 381–385.

Chandra, R., Yadav, S. and Mohan, D. (2008a). Effect of distillery sludge on seed germination and growth parameters of green gram (*Phaseolus mungo* L.). J. Hazardous Material, 152, 431–439.

Chandra, R., Yadav, S., Bharagava, R.N. and Murthy, R.C. (2008b). Bacterial pretreatment enhances removal of heavy metals during treatment of post methanated distillery effluent by *Typha angustifolia* L. J. Environ. Manag., 88, 1016–1024.

Chopra, P., Singh, D., Verma, V. and Puniya, A.K. (2004). Bioremediation of melanoidin containing digested spentwash from cane-molasses distillery with white rot fungus *Coriolus versicolor*. Indian J. Microbiology, 44 (3), 197–200.

Cooney, D.G. and Emerson, R. (1964). Themophilic fungi: An account of their biology, activities and classification, W.H. Freeman, San Francisco.

Cosovic, B., Vojvodic, V., Boskovic, N., Plavsic, M. and Lee, C. (2010). Characterization of natural and synthetic humic substances (melanoidins) by chemical composition and adsorption measurements. Organic Geochemistry, 41, 200–205.

D'Souza, D.T., Tiwari, R., Sah, A.K. and Raghukumar, C. (2006). Enhanced production of laccase by a marine fungus during treatment of coloured effluents and synthetic dyes. Enzyme Microb. Technol., 38, 504–511.

Dahiya, J., Singh, D. and Nigam, P. (2001a). Decolourisation of synthetic and spentwash melanoidins using the white-rot fungus *Phanerochaete chrysosporium* JAG–40. Bioresour. Technol., 78, 9598.

Dahiya, J., Singh, D. and Nigam, P. (2001b). Decolourisation of molasses wastewater by cells of *Pseudomonas fluorescens* immobilized on porous cellulose carrier. Bioresour. Technol., 78, 111–114.

Davamani, V., Lourduraj, A.C. and Singaram, P. (2006). Effect of Sugar and distillery wastes on nutrient status, yield and quality of turmeric. Crop Research Hisar 32(3), 563–567.

David, G.F.X., Herbert, J. and Wright, G.D.S. (1973). The ultrastructure of the pineal ganglion in the ferret. J Anat., 115(1), 79–97

De Kok, P.M.T. and Rosing, E.A.E. (1994). Reactivity of peptides in the Maillard reaction. In: "Thermally generated flavors, Maillard, microwave and extrusion processes" Parliament, T.H., Morello, M.J. & MacGorrin, R.J. (Eds.) ACS symposium series 543, pp. 158–179. American Chemical Society, Washington, D.C.

Dhar, G.M., Thampli, J., Pandit, A.B., Lele, S.S. and Joshi, J.B. (1998). Overall treatment of thermally pre-treated distillery waste Part I. Indian Chemical Engineer, Section A 40 (3), 222–231.

EPA, Environmental Protection Agency (1998). Design manual constructed wetlands and aquatic plant system for municipal wastewater treatment. US Environmental Protection Agency, Office of Research and Development, Cincinnati, Ohio.

EPA, Environmental Protection Agency (2002). The environmental protection rules, 3A, Schedule-II, III. U.S. Environmental Protection Agency, Office of research and Development, Cincinnati.

Ethanol India (2008). Demand Supply for Ethanol, URL http://www.ethanolindia.net/ethanol_demand.htm, visited on 24th April 2008.

Finot, P.A. and Magnenat, E. (1981). Metabolic transit of early and advanced Maillard products. Progress in Food and Nutrition Science, 5, 193–207.

Friedrich, J. (2004). Bioconversion of distillery waste. In: Arora, D.K. (Ed.), Fungal biotechnology in agriculture, food and environmental applications. Marcel Dekker Inc., New York, pp. 431–442.

Ghosh, M., Ganguli, A., Tripathi, A.K. (2002). Treatment of anaerobically digested distillery spentwash in a two-stage bioreactor using Pseudomonas putida and Aeromonas sp. Process Biochem. 7, 857–862.

Godbole, J. (2002). Ethanol from cane molasses, Fuel Ethanol Workshop, Honululu, Hawaii, November 14, 2002. /http://www.hawaii.gov/ dbedt/ert/new-fuel/files/ethanol workshop/10-Godbole-DOE-HI.pdfS (accessed on 9.08.2006).

Gonzalez, T., Terron, M.C., Yague, S., Zapico, E., Galletti, G.C. and Gonzalez, A.E. (2000). Pyrolysis/gas chromatography/mass spectrometry monitoring of fungal-biotreated distillery wastewater using Trametes sp. I-62 (CECT 20197). Rapid Commun Mass Spec., 14, 1417–1424.

Guimaraes, C., Porto, P., Oliveira, R. and Mota, M. (2005). Continuous decolourization of a sugar refinery wastewater in a modified rotating biological contactor with Phanerochaete chrysosporium immobilized on polyurethane foam discs. Process Biochem., 40, 535–540.

Harada, H., Uemura, S., Chen, A.C. and Jayadevan, J. (1996). Anaerobic treatment of a recalcitrant wastewater by a thermophilic UASB reactor. Bioresource Technol., 55 (3), 215–221.

Hati, K.M., Biswas, A.K., Bandyopadhyay, K.K. and Mishra, A.K. (2007). Soil properties and crop yield on a vertisol in India with application of distillery effluent. Soil Till. Res., 92, 60–68.

Hiltner, L. (1904). Ueber neuere Erfahrungen und Probleme auf dem Gebiete der Bodenbakteriologie und unter besonderer BerUcksichtigung der Grundungung und Brache. Arb. Deut. Landw. Gesell, 98, 59–78.

Hiramoto, K., Nasuhara, A., Michikoshi, K., Kato, T. and Kikugawa, K. (1997). DNA strand-breaking activity of 2,3-dihydro-3,5-dihydroxy-6-methyl-4H-pyran-4-one (DDMP), a Maillard reaction product of glucose and glycine. Mutat. Res., 395, 47–56.

Hiramoto, S., Itoh, K., Shizuuchi, S., Kawachi, Y., Morishita, Y. and Nagase, M. (2004). Melanoidin, a food protein-derived advanced Maillard reaction product, suppresses helicobacter pylori in vitro and in vivo. Helicobacter, 9, 429–435.

Hoagland, D.R. and Arnon, D.I. (1938). The water culture method for growing plants without soil. Circ-347. Berkley (CA): University of California, College of Agric., Agric. Exp. Stn.

Hofmann, T. (1998a). 4-alkylidene-2-imino-5-[4-alkylidene-5-oxo-1, 3-imidazol-2-inyl] aza methylidene-1, 3- imidazolidinea novel coloured substructure in melanoidins formed by Maillard reactions of bound arginine with glyoxal and furan-2-carboxaldehyde. J. Agricultural and Food Chemistry, 46, 3896–3901.

Hofmann, T. (1998b). Studies on the relationship between molecular weight and the colour potency of fractions obtained by thermal treatment of glucose/amino acid and glucose/protein solutions by using ultracentrifugation and colour dilution techniques. J. Agricultural and Food Chemistry, 46, 3891–3895.

Hoitink, H.A.J. and Kuter, S.A. (1986). Effects of composts in growth media on soilborne pathogens. In: Y. Chen and Y. Avnimelech (Eds.), The role of organic matter in modern agriculture, Martinus Nijhoff Publishers, Dordrecht, The Netherlands, pp. 289–306.

Jadhav, M.B., Joshi, V.A., Jagtap, P.B. and Jadhav. S.B. (1992). Effect spent wash press mud cake compost on soil physicochemical. Biological properties, yield and quality of adsali sugarcane. Annual Convention DSTA II, 119–134.

Jain, N., Minocha, A.K. and Verma, C.L. (2002). Degradation of predigested distillery effluent by isolated bacterial strains. Ind. J. Exp. Bot., 40, 101–105.

Jing, H. and Kitts, D.D. (2000). Comparison of the antioxidative and cytotoxic properties of glucose-lysine and fructose-lysine Maillard reaction products. Food Research International, 33, 509–516.

Joshi, H.C. (1999). Bio-Energy potential of distillery effluent. Bioenergy News, 3 (3), 10–15.

Kalavathi, D.F., Uma, L. and Subramanian, G. (2001). Degradation and metabolization of the pigment- melanoidin in a distillery effluent by the marine cyanobacterium *Oscillatoria boryana* BDU 92181. Enzyme and Microbial Technology, 29 (4–5), 246–251.

Kaletunc, G., Lee, J., Alpas, H. and Bozoglu, F. (2004). Evaluation of structural changes induced by high hydrostatic pressure in *Leuconostoc mesenteroides*. Applied and Environmental Microbiology, 70(2), 1116–1122.

Kambe, T.N., Shimomura, M., Nomura, N., Chanpornpong, T. and Nakahara, T. (1999). Decolourization of molasses wastewater by *Bacillus sp.* under thermophilic and anaerobic conditions. J. Biosci. Bioeng., 87, 119–121.

Kato, H. and Hayase, F. (2002). An approach to estimate the chemical structure of melanoidins. Int. Congress Series. 1245, 3–7.

Kaushik, A., Nisha, R., Jagjeeta, K. and Kaushik, C.P. (2005). Impact of long and short term irrigation of a sodic soil with distillery effluent in combination with bioamendments. Biores. Technol., 96, 1860–1866.

Kim, S.J. and Shoda, M. (1999). Batch decolourization of molasses by suspended and immobilizes fungus of Geotrichum candidum Dec 1. J. Biosci. Bioeng., 88, 586–589.

Kroh, L. and Westphal, G. (1989). Die Reaktion in Lebensmittel. Chemische Gesellschaft, 35, 73–80.

Kumar, P. and Chandra, R. (2004). Detoxification of distillery effluent through *Bacillus thuringiensis* (MTCC 4714) enhanced phytoremediation potential of *Spirodela polyrrhiza* (L.) Schliden. Bull. Environ. Contam. Toxicol., 73, 903–910.

Kumar, P. and Chandra, R. (2006). Decolourisation and detoxification of synthetic molasses melanoidins by individual and mixed cultures of *Bacillus spp.* Bioresour. Technol., 7, 2096–2102.

Kumar, V., Wati, L., Nigam, P., Banat, I.M., Yadav, B.S., Singh, D. and Marchant, R. (1998). Decolorization and biodegradation of anaerobically digested sugarcane molasses spent wash effluent from biomethanation plants by white-rot fungi. Process Biochem., 33, 83–88.

Kumaresan, T., Sheriffa Begum, K.M.M., Sivashanmugam, P., Anantharaman, N. and Sundaram, S. (2003). Experimental studies ontreatment of distillery effluent by liquid membrane extraction.Chem. Eng. J., 95(1–3), 199–204.

Kwak, E.J., Lee, Y.S., Murata, M. and Homma, S. (2005). Effect of pH control on the intermediates and melanoidins of nonenzymatic browning reaction. Lebensmittel-Wissenschaft und-Technologie, 38, 1–6.

Lankinen, V.P., Inkeroinen, M.M., Pellinen, J. and Hatakka, A.I. (1991). The onset of lignin-modifying enzymes, decrease of aox and color removal by white-rot fungi grown on bleach plant effluents. Water Sci. Technol., 24(3/4), 189–198.

Lee, T.H., Aoki, H., Sugano, Y. and Shoda, M. (2000). Effect of molasses on the production and activity of dye-decolourizing peroxidase from *Geotrichum candidum* Dec 1. J. Biosci. Bioeng., 89, 545–549.

Lewis, S., Handy, R.D., Cordi, B., Billinghurst, Z. and Depledge, M.H. (1999). Stress proteins (HSPs): Methods of detection and then use as an Enviromental biomarker. Ecotoxicol., 8, 351–368.

Lin, Y. and Tanaka, S. (2006). Ethanol fermentation from biomass resources: current state and prospects. Applied Microbiology Biotechnol., 69, 627–642.

Lowry, O.H., Rosenbrough, N.J., Farr, A.L. and Randall, R.J. (1951). Protein measurement with folin-phenol reagent. J. Biol. Chem., 193, 265–275.

Mahimaraja, S. and Bolan, N.S. (2004). Problems and prospects of agricultural use of distillery spentwash in India. SuperSoil 2004. 3rd Australian New Zealand Soils Conference. 5–9 December'2004. University of Sydney, Australia

Mansur, M., Suarez, T., Fernandez-Larrea, J., Brizuela, M.A. and Gonzalez, A.E. (1997). Identification of a Laccase gene family in the new lignin degrading basidiomycete CECT 20197. Appl. Environ. Microbiol., 63, 2637–2646.

Marschner H (1995). Mineral nutrition of higher plants. 2nd edn. Academic Press, London.

Mayer, A.M. and Harel, E. (1979). Polyphenol oxidases in plants. Phytochemistry, 18(2), 193–215.

Mayer, A.M. and Staples, R.C. (2002). Laccase: New functions for an old enzyme. Phytochemistry, 60(6), 551–565.

Migo, V.P., Del Rosario, E.J. and Matsumura, M. (1997). Flocculation of melanoidins induced by inorganic ions. J. Fermentation and Bioengeering, 83, 287–291.

Migo, V.P., Matsumara, M., Rosario, E.J.D. and Kataoka, H. (1993). Decolorization of molasses wastewater using an inorganic flocculant. J. Fermentation and Bioengineering, 75 (6), 438–442.

Mishra, S., Srivastava, S., Tripathi, R.D., Kumar, R., Seth, C.S and Gupta, D.K. (2006). Lead detoxification by coontail (*Ceratophyllum demersum* L.) involves induction of phytochelatins and antioxidant system in response to its accumulation. Chemosphere, 65, 1027.

Miyata, N., Iwahori, K. and Fujita, M. (1998). Manganese independent and dependent decolourisation of melanoidin by extracellular hydrogen peroxide and peroxidases from *Coriolus hirsutus* pellets. J. Ferment. Bioeng., 85, 550–553.

Miyata, N., Mori, T., Iwahori, K. and Fujita, M. (2000). Microbial decolorization of melanoidins containing wastewaters: combined use of activated sludge and the fungus *Coriolus hirsutus*. J. Biosci. Bioeng. 89, 145–50.

Mohana, S., Desai, C. and Madamwar, D. (2007). Biodegradation and decolourization of anaerobically treated distillery spent wash by a novel bacterial consortium. Bioresour. Technol., 98, 333–339.

Morales, F.J., Fraguas, C.F. and Perez, S.J. (2005). Iron-binding ability of melanoidins from food and model systems. Food Chem., 90, 821–827.

Motai, H. (1974). Relationship between the molecular weight and colour intensity of colour components of melanoidin from glycine-xylose systems. Agric. Biol. Chem., 38, 2299–2304.

Mundt, S. and Wedzicha, B.L. (2003). A kinetic model for the glucose-fructose-glycine browning reaction. J. Agricultural and Food Chemistry, 51, 3651–3655.

Nussbaum, S., Schmutry, D. and Brunold, C. (1988). Regulation of assimilatory sulfate reduction by cadmium in Zea mays L. Plant Physiol., 88, 1407–1410.

Nweke, C.O., Alisi, C.S., Okolo, J.C. and Nwanyanwu, C.E. (2007). Toxicity of zinc to heterotrophic bacteria from a tropical river sediment. Applied Ecology and Environmental Research, 5, 123–132.

O'Brien, J. and Morrissey, P.A. (1989). Nutritional and toxicological aspects of the Maillard browning reaction in foods. Crit. Rev. Food Sci. Nutr., 28, 211–248.

Ohmomo, S., Daengsabha, W., Yoshikawa, H., Yui, M., Nozaki, K., Nakajima, T. and Nakamura, I. (1988b). Screening of anaerobic bacteria with the ability to decolourize molasses melanoidin. Agric. Biol. Chem., 57, 2429–2435.

Ohmomo, S., Itoh, N., Wantanabe, Y., Kaneko, Y., Tozawa, Y. and Udea, K. (1985). Continuous decolorization of molasses wastewater with mycelia of *Coriolus versicolor* Ps4a. Agric. Biol. Chem., 49, 2551–2555.

Ohmomo, S., Kainuma, M., Kamimura, K., Sirianuntapiboon, S., Oshima, I. and Atthasumpunna, P. (1988a). Adsorption of melanoidin to the mycelia of *Aspergillus oryzae* Y-2–32. Agric. Biol. Chem., 52, 381–386.

Ohmomo, S., Kaneko, Y., Sirianuntapiboon, S., Somachi, P., Atthasumpunna, P. and Nakamura, I. (1987). Decolourization of molasses wastewater by a thermophilic strain *Aspergillus fumigatus* G-2-6. J. Agric. Biol. Chem., 52 (12), 3339–3346.

Painter, T.J. (1998). Carbohydrate polymers in food preservation: an integrated view of the Maillard reaction with special reference to discoveries of preserved foods in Sphagnum dominated peat bogs. Carbohyd. Polym., 36, 335–347.

Pant, D. and Adholeya, A. (2007). Enhanced production of ligninolytic enzymes and decolorization of molasses distillery wastewater by fungi under solid state fermentation. Biodegradation, 18, 647– 659.

Pant, D. and Adholeya, A. (2009). Nitrogen removal from biomethanated spent wash using hydroponic treatment followed by fungal decolorization. Environ. Eng. Sci., 26, 559–65.

Pant, D., Reddy, U.G. and Adholeya, A. (2006). Cultivation of oyster mushrooms on wheat straw and bagasse substrate amended with distillery effluent. World J. Microbiol. Biotechnol., 22, 267–275.

Pathak, H., Joshi, H.C., Chaudhary, A., Kalra, N. and Dwivedi, M.K. (1999). Soil amendment with distillery effluent for wheat and rice cultivation. Water Air Soil Poll., 113, 133–140.

Patil, P.U., Kapadnis, B.P. and Dhammankar, V.S. (2003). Decolorization of synthetic melanoidin and biogas effluent by immobilized fungal isolated of Aspergillus niger UM2. All India Distiller's Association (AIDA) Newsletter, pp. 53–56.

Peleg Z. and Blumwald E. (2011). Hormone balance and abiotic stress tolerance in crop plants, Current Opinion in Plant Biology, 2011, 14(3), 290–295.

Petruccioli, M., Duarte, J.C. and Fedrerici, F. (2000). High rate aerobic treatment of winery wastewater using bioreactors with free and immobilized activated sludge. J. Biosci. Bioeng., 90, 381–386.

Plavsic, M., Cosovic, B. and Lee, C. (2006). Copper complexing properties of melanoidins and marine humic material. Sci. Total Environ., 366, 310–319.

Prasad, D.Y. and Joyce, T.W. (1991). Color removal from kraft blech-plant effluents by *Trichoderma spectroscope*. Tappi. J., 74, 165–169.

Raghukumar, C., Chandramohan, D., Michel, F.C. and Reddy, C.A. (1996). Degradation of lignin and decolorization of paper mill bleach plant effluent by marine fungi. Biotechnol. Lett., 18 (1), 105–106.

Raghukumar, C., Mohandass, C., Kamat, S. and Shailaja, M.S. (2004). Simultaneous detoxification and decolorization of molasses spent wash by the immobilized white-rot fungus *Flavadon flavus* isolated from the marine habitat. Enzyme Microb. Technol., 35, 197–202.

Ramakritinan, C.M., Kumaraguru, A.K. and Balasubramanian, M.P. (2005). Impact of distillery effluent on carbohydrate metabolism of freshwater fish *Cyprinus carpio*. Ecotoxicol., 14, 693–707.

Ramana, S., Biswas, A.K. and Singh, A.B. (2002b). Effect of distillery effluents on some physiological aspects in maize. Bioresour. Technol., 84, 295–297.

Ramana, S., Biswas, A.K., Kundu, K., Saha, J.K. and Yadav, R.B.R (2002c). Effect of distillery effluent on seed germination in some vegetable crops. Bioresour. Technol., 82, 273–275.

Ramana, S., Biswas, A.K., Kundu, S., Saha, J.K. and Yadava, R.B.R. (2002a). Effect of distillery effluent on seed germination in some vegetable crops. Bioresour. Technol., 82, 273–275.

Rascio, N., Mariani, P., Tommasini, E., Bodner, M. and Larcher, W. (1991). Photosynthetic strategies in leaves and stem of *Egeria densa*. Planta, 185, 297–303.

Romheld, V. and Marschner, H. (1983). Mechanism of iron uptake by peanut plants. Plant Physiol., 71, 949–954.

Rout, G.R., Samantaray, S. and Das, P. (2001). Aluminum toxicity in plants: a review, Agronomie, 21, 3.

Rubia, T.D.L., Linares, A., Perez, J., Dorado, J.M., Romera, J. and Martinez, J. (2002). Characterization of Manganese-Dependent Peroxidase isoenzymes from the ligninilytic fungus *Phanerochaete flavido-alba*. Research in Microbiol., 153–547.

Rufian-Henares, J.A. and Morales, F.J. (2007). Functional properties of melanoidins: In vitro antioxidant, antimicrobial and antihypertensive activities. Food Research International, 40, 995–1002.

Sangave, P.C. and Pandit, A.B. (2006a). Enhancement in biodegradability of distillery wastewater using enzymatic pretreatment. J. Environ. Manag., 78, 77–85.

Sangave, P.C. and Pandit, A.B. (2006b). Ultrasound and enzyme assisted biodegradation of distillery wastewater. J. Environ. Manag., 80, 36–46.

Sangeeta, Y. and Chandra R. (2011). Heavy metals accumulation and ecophysiological effect on *Typha angustifolia* L. and *Cyperus esculentus* L. growing in distillery and tannery effluent polluted natural wetland site, Unnao, India. Environ. Earth Sci., 62, 1235–1243.

Sangeeta, Y. and Chandra, R. (2013). Effect of pH on melanoidin extraction from post methanated distillery effluent (PMDE) and its decolorization by potential bacterial consortium. International Journal of Recent Scientific Research., 4(10), 1492–1496

Sangeeta, Y. and Chandra, R. (2012). Biodegradation of organic compounds of molasses melanoidin (MM) from biomethanated distillery spent wash (BMDS) during the decolourisation by a potential bacterial consortium. Biodegradation, 23(4), 609–620.

Sangeeta, Y., Chandra, R. and Vibhuti, R. (2011). Characterization of potential MnP producing bacteria and its metabolic products during decolourisation of synthetic melanoidins due to biostimulatory effect of D-xylose at stationary phase. Process Biochemistry, 46, 1774–1784.

Sayadi, S. and Ellouz, R. (1995). Roles of lignin peroxidase and manganese peroxidase from Phanerochaete chrysosporium in the decolorization of olive mill wastewaters. Appl Environ Microbiol., 61(3), 1098–1103.

Schliephake, K., Lonergan, G.T., Jones, C.L. and Mainwaring, D.E. (1993). Decolorisation of a packed-bed bioreactor. Biotechnol. Lett., 15 (11), 1185–1188.

Schnickels, R.A., Warmbier, H.C. and Labuza, T.P. (1976). Effect of protein substitution on nonenzymatic browning in an intermediate moisture food system. J. Agric. Food Chem., 24, 901–903.

Shen, G., Lu, Y. and Hong, J. (2006). Combined effect of heavy metals and polycyclic aromatic hydrocarbons on urease activity in soil. Ecotoxicol. Environ. Safety, 63, 474–480.

Shepherd HL, Grsimer ME, Tchobanoglous G (2001). Treatment of high-strength winery wastewater using a subsurface-flow constructed wetland. Water Environ. Res. 73: 394–403.

Singh, S., Melo, J.S., Eapen, S. and D'Souza, S.F. (2006). Phenol removal by *Brassica juncea* hairy roots: role of inherent peroxidase and H2O2. J Biotechnol., 123, 43–49.

Singh, S., Melo, J.S., Eapen, S. and D'Souza, S.F. (2008). Potential of vetiver (Vetiveria zizanoides L.) for phytoremediation of phenol. Ecotoxicol. Environ. Saf., 71, 671–676.

Sirianuntapiboon, S., Phothilangka, P. and Ohmomo, S. (2004). Decolourization of molasses wastewater by a strain no. BP103 of Acetogenic bacteria. Bioresour. Technol., 92, 31–39.

Sirianuntapiboon, S., Sihanonth, P., Somchai, P., Atthasampunna, P. and Hayashida, S. (1995). An adsorption mechanism for melanoidin decolourization by *Rhizoctonia* sp. Biosci. Biotechnol. Biochem., 59, 1185–1189.

Sirianuntapiboon, S., Somchai, P., Ohmomo, S. and Atthasampunna, P. (1988). Screening of filamentous fungi having the ability to decolourize molasses pigments. Agric. Biol. Chem., 52, 387–392.

Sirianuntapiboon, S., Zohsalam, P. and Ohmomo, S. (2003). Decolourization of molasses wastewater by *Citeromyces sp.* WR–43–6. Process Biochem., 39, 917–924.

Smarrelli, J. and Castignetri, D. (1986). Iron acquisition by plants: the reduction of fernisiderophores by higher plant NADH:nitrate reductase. Biochim. Biophys. Acta., 882, 337–342

Somoza, V., Wenzel, E., Lindenmeier, M., Grothe, D., Erbersdobler, H.F. and Hofmann, T. (2005). Influence of feeding malt, bread crust, and a pronylated protein on the activity of chemopreventive enzymes and antioxidative defense parameters in vivo. J. Agricultural and Food Chemistry, 53, 8176–8182.

Sridhar BBM, Diehl SV, Han FX, Monts DL, Su Y. 2005. Anatomical changes due to uptake and accumulation of Zn and Cd in Indian mustard (*Brassica juncea*). Environ Exp Bot. 54: 131–141.

Stallwood, B., Shears, J., Williams, P.A. and Hughes, K.A. (2005). Low temperature bioremediation of oil-contaminated soil using biostimulation and bioagumentation with a *Pseudomonas sp.* from maritime Antartica. J. Appl. Microbiol., 99, 794–802.

Takahama, U. and Hirota, S. (2008). Reduction of nitrous acid to nitric oxide by coffee melanoidins and enhancement of the reduction by thiocyanate: Possibility of its occurrence in the stomach. J. Agricultural and Food Chemistry, 56, 4736–4744.

Tano F, Valenti L, Failla O, Beltrame E (2005). Effects of distillery vinasses on vineyard yield and quality in the D.O.C. "Oltrepo' Pavese Pinot Nero"-Lombardy, Italy. Water Sci. Technol. 51: 199–204.

Terry, N., Carlson, C. Raab, T.K. and Zayed, A.M. (1992). Rates of selenium volatilization among crop species. J. Environ. Quality, 21, 341–344.

Thakkar, A.P., Dhamankar, V.S. and Kapadnis, B.P. (2006). Biocatalytic decolourisation of molasses by *Phanerochaete chrysosporium*. Bioresour. Technol., 97, 1377–1381.

Tressl, R., Wondrak, G.T., Kruger, R.P. and Rewicki, D. (1998). New melanoidin-like Maillard polymers from 2-deoxypentoses. J. Agricultural and Food Chemistry, 46, 104–110.

Uppal, J. (2004). Water utilization and effluent treatment in the Indian alcohol industry - an overview. In: Liquid Assets, Proceedings of Indo-EU workshop on Promoting Efficient Water Use in Agro-based Industries. TERI Press, New Delhi, India, pp. 13–19.

US EPA. (2001). A Citizen's Guide to Phytoremediation. US Environmental Protection Agency, Office of Solid Waste and Emergency Response. EPA-542-F-01-002.

Vlissidis, A. and Zouboulis, A.I. (1993). Thermophilic anaerobic digestion of alcohol distillery wastewaters. Bioresour. Technol. 43 (2), 131–140.

Vlyssides, A.G., Israilides, C.J., Loizidou, M., Karvouni, G. and Mourafeti, V. (1997). Electrochemical treatment of vinasse from beet molasses. Water Science and Technol., 36 (2–3), 271–278.

Volesky, B., May, H. and Holan, Z.R. (1993). Cadmium biosorption by *Saccharomyces cerevisiae*. Biotechnol. Bioeng., 41, 826–829.

Watanabe, Y., Sugi, R., Tanaka, Y. and Hayashida, S. (1982). Enzymatic decolourization of melanoidin by *Coriolus sp.* Agric. Biol. Chem., 46 (20), 1623–1630.

Wen, X., Enokizo, A., Hattori, H., Kobayashi, S., Murata, M. and Homma, S. (2005). Effect of roasting on properties of the zinc-chelating substance in coffee brews. J. Agricultural and Food Chemistry, 53, 2684–2689.

WHO. (2004). Guidelines for Drinking water quality, vol. 1. World Health Organization Press, Geneva, Switzerland.

WHO. (2006). Guidelines for the Safe Use of Wastewater, Excreta and Greater. Vol. 3. World Health Organisation Press, Geneva, Switzerland.

Wiegant, W.M., Claassen, J.A. and Lettinga, G. (1985). Thermophilic anaerobic digestion of high strength wastewaters. Biotechnology and Bioengineering, 27, 1374–1381.

Xie, X.H, Fu, J., Wang, H.P. and Liu, J.S. (2010). Heavy metals resistance by two bacteria strains isolated from a copper mine tailing in China. J. Biotechnol., 9(26), 4056–4066.

Yamaoka, Y., Inoue, H., Takimura, O. and Oata, S. (2002). Effect of iron on the degradation of triphenyltin by pyoverdins isolated from *Pseudomonas chlororaphis*. Applied Organometallic Chemistry, 16, 277–279.

Yaylayan, V.A. and Kaminsky, E. (1998). Isolation and structural analysis of Maillard polymers: caramel and melanoidins formation in glycine/glucose model system. Food Chem. 63, 25–31.Yilmaz, Y. and Toledo, R. (2005). Antioxidant activity of water-soluble Maillard reaction products. Food Chem., 93, 273–278.

Yilmaz, Y. and Toledo, R. (2005). Antioxidant activity of water-soluble Maillard reaction products. Food Chem., 93, 273–278.

Zhang, T., Zhao, Q.X., Huang, H., Li, Q., Zhang, Y. and Qi, M. (1998). Kinetic study on the removal of toxic phenol and chlorophenol from waste water by horseradish peroxidase, Chemosphere, 37, 1571–1577.

Abbreviations

ABTS	:	2.2-azino-bis-(3-ethylbenzothiazoline-6-sulphonic acid
ABC	:	ATP-binding cassette
AGP	:	Arabinogalactan proteins
AIDA	:	All India Distillers Association
ATP	:	Adenosine Triphosphate
aw	:	Water activity
BCF	:	Bioconcentration Factor
BIS	:	Bureau of Indian Standards
BMCs	:	Bounded Melanoidin Compounds
BOD	:	Biological Oxygen Demand
BODST	:	Biological Oxygen Demand Short Term Test
BSED	:	Back-Scattered Electron Detector
BTX	:	Benzene, Toluene and Xylene
Cd	:	Cadmium
CEC	:	Cation Exchange Capacity
CETP	:	Common Effluent Treatment Plant
CFU	:	Colony Forming Unit
Chl	:	Chlorophyll
COD	:	Chemical Oxygen Demand
C	:	Carbon—Carbon
Cr	:	Chromium
Cu	:	Copper
CW	:	Constructed Wetland
d	:	Day
DMPD	:	N, N-dimethyl phenylenediamine dihydrochloride
DWAF	:	Department of Water Affairs and Forestry
EC	:	Electrical Conductivity
EGSB	:	Expanded Granular Sludge Bed
ELM	:	Emulsion Liquid Membrane
EMP	:	Embden-Meyerhof-Parnas
EPR	:	Electron Paramagnetic Resonance
EPS	:	Extracellular Polymeric Substances
ESI-MS	:	Electrospray Ionization-Mass Spectrum
ETS	:	Electron Transport System
FAD	:	Flavin Adenine Dinucleotide

FBR	:	Fluidized bed reactor
FC	:	Ferrichrome
Fe	:	Iron
FOB	:	Ferrioxamine B
GAA	:	Glucose-Aspartic-Acid Maillard Product
GAC	:	Granular Activated Carbon
GC	:	Gas Chromatography
GGA	:	Glucose-Glutamic-acid Maillard Product
GPYM	:	Glucose-Peptone-Yeast Extract-Mineral salt broth
GSH	:	Growth Stimulating Hormone
h	:	Hours
HA	:	Humic Acid
HABR	:	Hybrid Anaerobic Baffled Reactor
HFS	:	Horizontal Flow System
HMF	:	Hydroxymethylfurfural
HMW	:	High Molecular Weight
HRP	:	Horse Radish Peroxidase
HRT	:	Hydraulic Retention Time
HSF	:	Horizontal Subsurface Flow
HSP	:	Heat Shock Proteins
K	:	Potassium
kDa	:	Killo Dalton
LC-MS	:	Liquid Chromatography Mass Spectrometry
LiP	:	Lignin Peroxidase
LM	:	Light Microscopy
LME	:	Lignin Modifying Enzymes
LMW	:	Low Molecular Weight
MAP	:	Magnesium Ammonium Phosphate
MDA	:	Malondialdehyde
MFG	:	Monofructoseglycine
Mg	:	Magnesium
Mn	:	Manganese
MnIP	:	Manganese Independent Peroxidase
MnP	:	Manganese Peroxidase
MR	:	Maillard Reaction
MRPs	:	Maillard Reaction Products
MTs	:	Metallothioneins
MWW	:	Molasses Waste Water
N	:	Nitrogen
NAD	:	Nicotinamide Adenine Dinucleotide
NF	:	Nanofiltration
Ni	:	Nickel
nm:	:	Nano meter
NMR	:	Nuclear Magnetic Resonance
NPK + FYM	:	nitrogen, phosphate, potassium and farmyard manure
OM	:	Organic Matter
P	:	Phosphorous

PAHs	:	Polycyclic Aromatic Hydrocarbons
Pb	:	Lead
PBH	:	Poly-beta hydrobutyric acid
PCS	:	Phytochelatin Synthase
PCs	:	Phytochelatins
PHA	:	Polyhydroxy-alkanoates
PHB	:	Polyhydroxybutyrate
PI	:	Isoelectric point
PIGE	:	Particle-Induced Gamma-Ray Emission
PIXE	:	Particle-Induced X-ray Emission
PMDE	:	Post Methanated Distillery Effluent
POPs	:	Persistent Organic Pollutants
PPM	:	Parts Per Million
PVC	:	Polyvinyl Chloride
Py/GC/MS	:	Pyrolysis/gas chromatography/mass spectrometry
QTOF-MS	:	Quadrupole Time-of-Flight Mass Spectrometer
RBC	:	Rotating Biological Contactor
RBR	:	Rotating Biological Reactor
RCFB	:	Recirculating Fluidized Bed
RDX	:	1, 3, 5-trinitro-1, 3, 5-hexahydrotriazine
RGR	:	Relative Growth Rate
RO	:	Reverse Osmosis
ROS	:	Reactive Oxygen Species
SAA	:	Sucrose-Aspartic-acid Maillard Product
SAR	:	Sodium Adsorption Ratio
SBR	:	Aerobic Sequencing Batch Reactor
SD	:	Standard Deviation
SDS-PAGE	:	Sodium Dodecyl Sulfate Polyacrylamide Gel Electrophoresis
SEM	:	Scanning Electron Microscopy
SGA	:	Sucrose-Glutamic acid Maillard Product
SRB	:	Sulfate Reducing Bacteria
SRT	:	Solids Retention Time
TCE	:	Trichloroethylene
TEM	:	Transmission Electron Microscopy
TF	:	Translocation Factor
TKN	:	Total Kjeldahl Nitrogen
TNT	:	Trinitrotoluene
TOC	:	Total Organic Carbon
TS	:	Total Solid
TSS	:	Total Suspended Solids
UASB	:	Upflow anaerobic sludge blanket
USDA	:	United States Department of Agriculture
VFS	:	Vertical Flow System
VP	:	Versatile Peroxidases
WDCS	:	Waste Discharge Charge System
WHO	:	World Health Organization
Zn	:	Zinc

Glossary

Acedogenesis: A biological reaction where simple monomers are converted into volatile fatty acids.

Acitogenesis: A biological reaction where volatile fatty acids are converted into acetic acid, carbon dioxide, and hydrogen.

Activated carbon: Activated carbon, also called activated charcoal, activated coal, or carbo activatus, is a form of carbon processed to be riddled with small, low-volume pores that increase the surface area available for adsorption or chemical reactions.

Adaptation: Changes in an organism or population through which they become more suited for living in the current environment.

Bacterial Cell Growth: Bacterial growth is the division of one piece of bacteria into two daughter cells in a process called binary fission. Providing no mutational event occurs the resulting daughter cells are genetically identical to the original cell. Hence, "local doubling" of the bacterial population occurs. In the laboratory, under favorable conditions, a growing bacterial population doubles at regular intervals. Growth is by geometric progression: 1, 2, 4, 8, etc. or $2^0, 2^1, 2^2, 2^3.........2^n$ (where n = the number of generations). This is called exponential growth.

Bifidobacterium: Bifidobacterium is a genus of Gram-positive, non-motile, often branched anaerobic bacteria. They are ubiquitous, endosymbiotic inhabitants of the gastrointestinal tract, vagina and mouth (*B. dentium*) of mammals, including humans. Bifidobacteria are one of the major genera of bacteria that make up the colonflora in mammals. Some bifidobacteria are used as probiotics.

Bioaccumulation: The accumulation of substances, such as pesticides, or other organic chemicals in an organism. Bioaccumulation occurs when an organism absorbs a toxic substance at a rate greater than that at which the substance is lost. Thus, the longer the biological half-life of the substance the greater the risk of chronic poisoning, even if environmental levels of the toxin are not very high.

Bioaugmentation: Bioaugmentation is the practice of adding actively growing, specialized microbial strains into a microbial community in an effort to enhance the ability of the microbial community to respond to process fluctuations or to degrade certain compounds, resulting in improved treatment. By changing the microbial community to include specific microbes, the characteristics of the microbial community can be improved.

Bioavailability: The availability of chemicals to degradative microorganisms.

Biodegradation: The breakdown of organic substances by microorganisms.

Biological Oxygen Demand (BOD): The amount of Oxygen (mg/1) is taken up by micro-organism that decomposes organic waste matter in water. It is therefore used as the measure of the amount of certain types of organic pollutants in water.

Bioremediation: Bioremediation is the use of micro-organism metabolism to remove pollutants. Technologies can be generally classified as in situ or ex situ. In situ bioremediation involves treating the contaminated material at the site, while ex situ involves the removal of the contaminated material to be treated elsewhere. Some examples of bioremediation related technologies are phytoremediation, bioventing, landfarming, bioreactor, composting, bioaugmentation, rhizofiltration and biostimulation.

Biosorption: Biosorption is a physiochemical process that occurs naturally in certain biomass which allows it to passively concentrate and bind contaminants onto its cellular structure. Though using biomass in environmental cleanup has been in practice for a while, scientists and engineers are hoping this phenomenon will provide an economical alternative for removing toxic heavy metals from industrial wastewater and aid in environmental remediation.

Biostimulation: Biostimulation involves the modification of the environment to stimulate existing bacteria capable of bioremediation. This can be done by addition of various forms of rate limiting nutrients and electron acceptors, such as phosphorus, nitrogen, oxygen, or carbon.

Biotransformation: The metabolic alteration to the chemical structure of a compound by a living organism or enzyme.

Brix: Measurement of sugar concentration in a solution. One degree Brix is 1 gram of sucrose in 100 grams of solution and represents the strength of the solution as percentage by weight (% w/w).

Bulk density: Bulk density is an indicator of soil compaction. It is calculated as the dry weight of soil divided by its volume. This volume includes the volume of soil particles and the volume of pores among soil particles. Bulk density is typically expressed in g/cm^3.

Cation-Exchange Capacity (CEC): Cation-exchange capacity (CEC) is the maximum quantity of total cations, of any class, that a soil is capable of holding, at a given pH value, available for exchange with the soil solution. CEC is used as a measure of fertility, nutrient retention capacity, and the capacity to protect groundwater from cation contamination. It is expressed as milliequivalent of hydrogen per 100 g of dry soil(meq+/100g), or the SI unit centi-mol per kg (cmol+/kg).

Chemical Oxygen Demand (COD): A parameter of water quality which measures the amount of oxygen in parts per million required to oxidize organic and oxidizable inorganic compounds in the water sample.

Consortium: A microbial consortium is two or more microbial groups living symbiotically

Distillation: Distillation is a method of separating mixtures based on differences in volatility of components in a boiling liquid mixture. Distillation is a unit operation, or a physical separation process, and not a chemical reaction.

Distillery: A plant where winery by-products are distilled

Electrical Conductivity (EC): a measure of electrical conductance due to dissolved salts. EC can be converted to an approximate salt content. Its SI unit is siemens per metre (S/m)

Electrostatic Interactions: These are interactions between cations and anions, which are ions and functional groups with formal charge. Electrostatic interactions can be either attractive or repulsive, depending on the charges of the interacting species. Electrostatic interactions can be very strong, and fall off slowly with distance (1/r).

Ellman's reagent: Ellman's reagent (5, 5'-dithiobis-(2-nitrobenzoic acid) or DTNB) is a chemical used to quantify the number or concentration of thiol groups in a sample

Enzyme: Enzymes are biological catalysts that facilitate the conversion of substrates into products by providing favorable conditions that lower the activation energy of the reaction.

Eutrofication: The process by which a body of water acquires a high concentration of nutrients, especially phosphates and nitrates. These typically promote excessive growth of algae. As the algae die and decompose, high levels of organic matter and the decomposing organisms deplete the water of available oxygen, causing the death of other organisms, such as fish. Eutrophication is a natural, slow-aging process for a water body, but human activity greatly speeds up the process.

Fermentation: Fermentation is a metabolic process converting sugar to acids, gases and/or alcohol using yeast or bacteria. In its strictest sense, fermentation is the absence of the electron transport chain and takes a reduced carbon source, such as glucose, and makes products like lactic acid or acetate. No oxidative phosphorylation is used, only substrate level phosphorylation, which yields a much lower amount of ATP.

Hazardous waste: Hazardous waste is waste that poses substantial or potential threats to public health or the environment

Heavy Metals: The term heavy metal refers to any metallic chemical element that has a relatively high density and is toxic or poisonous at low concentrations. Group of metallic elements with, generally, a specific gravity greater than 5.

Hydraulic conductivity: Hydraulic conductivity is a soil property that describes the ease with which the soil pores permit water (not vapor) movement. It depends on the type of soil, porosity, and the configuration of the soil pores. In saturated soils, the hydraulic conductivity is represented as Ksat and in unsaturated soils, the hydraulic conductivity is represented as K.

Hydrolysis: Hydrolysis comes from Greek hydro, meaning "water" and lysis, meaning "separation") usually means the cleavage of chemical bonds by the addition of water. Generally, hydrolysis or saccharification is a step in the degradation of a substance.

Incineration: Incineration is a waste treatment process that involves the combustion of organic substances contained in waste materials. Incineration and other high temperature waste treatment systems are described as "thermal treatment". Incineration of waste materials converts the waste into ash, flue gas, and heat. The ash is mostly formed by the inorganic constituents of the waste, and may take the form of solid lumps or particulates carried by the flue gas. The flue gases must be cleaned of gaseous and particulate pollutants before they are dispersed into the atmosphere.

Malondialdehyde: Malondialdehyde is the organic compound with the formula $CH_2(CHO)^2$. The structure of this species is more complex than this formula suggests. This reactive species occurs naturally and is a marker for oxidative stress

Melanoidins: Melanoidins are natural condensation products of sugar and amino acids produced by non-enzymatic Maillard amino-carbonyl reaction taking place between the amino and carbonyl groups in organic substances.

Methanogenesis: A biological reaction where acetates are converted into methane and carbon dioxide, while hydrogen is consumed.

Nanofiltration (NF): It is a cross-flow filtration technology which ranges somewhere between ultrafiltration (UF) and reverse osmosis (RO). The nominal pore size of the membrane is typically about 1 nanometre.

Phyotvolatilisation: The use of plants to volatilise pollutants from polluted soils and water.

Phytodegradation: The process where plants are able to metabolically degrade organic pollutants.

Phytoextraction: The use of plants to extract contaminants from the environment.

Phytomining: Use of plants to extract metal compounds of high economic value.

Phytoremediation: Phytoremediation, meaning "plant" and Latin remedium, meaning "restoring balance" describes the treatment of environmental problems (bioremediation) through the use of plants that mitigate the environmental problem without the need to excavate the contaminant material and dispose of it elsewhere.

Phytoremediation: Use of plants to remediate polluted soil and/or groundwater.

Phytostabilisation: Use of plants to reduce bioavailability and migration of contaminants.

pI: The isoelectric point (pI), sometimes abbreviated to IEP, is the pH at which a particular molecule or surface carries no net electrical charge.

Prebiotic activity: Prebiotics are non-digestible food ingredients that stimulate the growth and/or activity of bacteria in the digestive system in ways claimed to be beneficial to health. They were first identified and named by Marcel Roberfroid in 1995. As a functional food component, prebiotics, like probiotics, are conceptually intermediate between foods and drugs. Depending on the jurisdiction, they typically receive an intermediate level of regulatory scrutiny, in particular of the health claims made concerning them.

Press mud: It is a compressed sugar industries waste. It is a soft, spongy, amorphous and dark brown solid containing sugar, fiber and coagulated colloid including cane wax, albuminoids, inorganic salts, soil particles and mineral elements in varying amount

Probiotic: Probiotics are generally the live micro-organisms in foods such as yoghurts; they survive passage through the gut and temporarily bring the benefits of the normal gut flora. Probiotics have been used to treat or prevent diarrhoea and to improve symptoms in lactose intolerance.

Putriciable waste: Solid waste that contains organic matter capable of being decomposed by microorganisms and of such a character and proportion as to cause obnoxious odors and to be capable of attracting or providing food for birds or animals.

Pyorchar: Activated carbon both in granular and powdered form, manufactured from paper mill sludge.

Reactive oxygen species (ROS): ROS are chemically reactive molecules containing oxygen. Examples includeoxygen ions and peroxides. ROS form as a natural byproduct of the normal metabolism of oxygen and have important roles in cell signaling and homeostasis.

Reverse Osmosis (RO): It is a water purification technology that uses a semipermeable membrane. This membrane-technology is not properly a filtration method. In RO, an applied pressure is used to overcome osmotic pressure, a colligative property that is driven by chemical potential, a thermodynamic parameter. RO can remove many types of molecules and ions from solutions and is used in both industrial processes and in producing potable water. The result is that the solute is retained on the pressurized side of the membrane and the pure solvent is allowed to pass to the other side. To be "selective," this membrane should not allow large molecules or ions through the pores (holes), but should allow smaller components of the solution (such as the solvent) to pass freely. In the normal osmosis process, the solvent naturally moves from an area of low solute concentration (High Water Potential), through a membrane, to an area of high solute concentration (Low Water Potential).

Rhizofiltration: The uptake of contaminants by the roots of plants which are immersed in water.

Rhizosphere: The soil area immediatley surrounding the plant root surface. Typically up to a few millimetres from the root surface.

Scanning Electron Microscopy: A scanning electron microscope (SEM) is a type of electron microscope that produces images of a sample by scanning it with a focused beam ofelectrons. The electrons interact with electrons in the sample, producing various signals that can be detected and that contain information about the sample's surface topography and composition. The electron beam is generally scanned in a raster scan pattern, and the beam's position is combined with the detected signal to produce an image. SEM can achieve resolution better than 1 nanometer. Specimens can be observed in high vacuum, low vacuum and in environmental SEM specimens can be observed in wet conditions.

Siderophores: Siderophores are small, high-affinity iron chelating compounds secreted by microorganisms such as bacteria, fungi and grasses. Siderophores are amongst the strongest soluble Fe^{3+} binding agents known.

Sludge: Material that has settled to the bottom of a wastewater collection, treatment or storage device.

Sodium adsorption ratio (SAR): SAR is a measure of the suitability of water for use in agricultural irrigation, as determined by the concentrations of solids dissolved in the water. It is also a measure of the sodicity of soil, as determined from analysis of water extracted from the soil. The formula for calculating sodium adsorption ratio is:

$$S.A.R = \frac{Na^+}{\sqrt{\frac{1}{2}\left(Ca^2 + Mg^{2+}\right)}}$$

Spentwash: The wastewater from distilleries, major portion of which is spentwash, is nearly 15 times the total alcohol production.

Synergistic: Producing or capable of producing synergy. Synergy is the interaction of multiple elements in a system to produce an effect different from or greater than the sum of their individual effects.

Vinasse: Vinasse is a byproduct of the sugar industry. Sugarcane or Sugar beet is processed to produce crystalline sugar, pulp and molasses. The latter are further processed by fermentation toethanol, ascorbic acid or other products. After the removal of the desired product (alcohol, ascorbic acid, etc.) the remaining material is called vinasse.

Wastewater Storage Lagoon: Lagoon any dam, pond or lagoon constructed and used for the purpose of holding wastewater; does not include a sediment retention basin.

Wetland plants: Plants grow in wetland is called wetland plants. A wetland is a land area that is saturated with water, either permanently or seasonally, such that it takes on the characteristics of a distinct ecosystem.

Xenobiotic: Xenobiotics are any chemical compounds that are found in a living organism, but which are foreign to that organism, in the sense that it does not normally produce the compound or consume it as part of its diet.

Zeta potential: Zeta potential is the potential difference between thedispersion medium and the stationary layer of fluid attached to the dispersed particle.

Index

A

% Accumulation 158
Acetobacter acetii 199
Acinetobacter sp. 198
Activated sludge 180
Active Efflux Pumping 122
Adsorption 172
Aerated static pile process 212
Aeromonas sp. 198
Alcaligens faecalis 198
Alcohol Manufacture 10
Algal
    blooms 92
    treatment 200
Aliphatic acids 113
All Indian Distillers Association
    (AIDA) 3, 9
Alzheimer disease 107
Amadori rearrangement 24
Aminolevulinic acid 132
Amylase 98
Anabolism 188
Anaerobic process 175
Analytical techniques 45
Anatomical changes 164
Antihypertensive activity 41
Antimicrobial activity 39
Antioxidant defense 123
Antioxidants 99, 102, 123
Ascorbic acid 161
Aspergillus fumigates 195
Aspergillus niger UM2 196
Aspergillus oryzae 195
Aspergillus 221

B

Bacillus
    brevis 199
    megaterium 198
    smithii 198
    species 57, 73, 77, 198
    thuringiensis 199

Bacterial
    flocs 184
    growth 191–193
    treatment 197
Basidiomycete 196
Bioaugmentation 7, 49
Biochemical changes 132
Biocomposting 194, 208–223
Biological Oxygen Demand (BOD)
    90, 181
Biological treatment 175
Biosparging 5
Biostimulation 6, 49
Biotic Factors 218
Bioventing 5

C

Caramel 20
Carbon sources 57
Carbon/Nitrogen ratio 213–214, 218
Catabolism 188
Catalase 160
Cellulase 94
Chelate 147
Chlorophyll 99, 137
Citeromyces sp. WR-43-6 196
Coagulation and flocculation 172
COD/BOD 14, 176
Combustion 174
Conventional treatment techniques 169
Coriolus
    hirsutus 196
    versicolor sp no. 20 196
    versicolor 196
    *versicolor*Ps4a 195
Cyperus esculentus 109, 127, 135–140
Cytotoxicity 37

D

Degradable and non-degradable
    carbon 181
Dendroremediation 147

Distillation 12
Distillery sludge 96
D-xylose 61

E

Electrical Conductivity (EC) 91, 217
Enclosed systems 213
Enrichment factor 95
Enterobacter sakazakii 57, 73, 77
Enzymes 47, 50, 113, 203
Ethylinic and azomethine linkage 58
Eutrophication 89
Evaporation/combustion 174
Ex-situ 153

F

Fermentation 10
Ferti-irrigation 91, 93
Fixed bed reactor 178
Flavobacterium sp. 198
Flavodon flavus 196
Fluidized bed reactor (FBR) 179
Fungal treatment 194
Furfural 20, 26

G

Genotoxicity 37
Geotrichum candidum 196
Glycosylamine 25, 29

H

Health hazards 87
Heat Shock Proteins (HSP) 123
Heavy Metals 20, 72, 73–86, 89, 107
High molecular weight 31
High Performance Liquid Chromatography (HPLC) 65, 81
Hodge Diagram 29
Horizontal flow system 205
Hormones 113, 124
Hybrid reactors 180

I

Impact of distillery effluent 89–104
Inceptisols 95
Infiltration 95
In-situ bioremediation 4
Intracellular accumulation 114

L

Laccase 54
Lactobacillus hilgardii 198
Laughing gas 211
Light microscopy 164
Lignin Peroxidases (LiP) 52
Lipid peroxidase 102
Liquid Chromatography-Mass Spectrometry (LC-MS) 64
Low molecular weight 31

M

Maillard reaction 23, 60
Major element for bacterial growth 58
Malondialdehyde 102
Manganese peroxidise (MnP) 49, 53, 54, 62, 85
Maturation phase 211
Melanoidine 4, 14, 72
Membrane treatment 174
Mesophilic phase 211
Metabolism of bacteria 185
Metabolites characterization 64–70
Metal
    accumulation 127
    analysis 158
    deficient soil 107–108
Metallothioneins (MTs) 123
Methanobacillus 16
Methanobacterium 16
Methanococcus 16
Methanosarcina 16
Methemoglobinemia 92
Microbial Biotechnology 124
Microorganism associated with composting 219
Molasses 11
Mycelia sterilia 195
Mycorrhizal Association 114

N

NBD-desferrioxamine B (NBD-DFO) 118
Nicotinamide Adenine Dinucleotide (NAD) 15
Nitrifying bacteria 190

O

Oxidation-reduction reactions 124
Oxidation process 173

P

Particle-induced X-ray emission or proton-induced X-ray emission (PIXE) 141
Permissible limit for Discharge 19
Peroxidise activity 160
Persistent organic pollutants (POPs) 22
Phanerochaete chrysosporium JAG-40 196
Phanerochaete chrysosporium 196
Phenolics 20, 113
Photo-degradation 175
Phragmites cummunis 109, 127–128, 133
Phytochelatins (PCs) 123
Phytodegradation 5
Phytoextraction 5, 145–147
Phytofiltration/rhizofilteration 149
Phytohydraulics 5
Phytoremediation 5, 107, 167, 201
Phytosequestration 5
Phytostabilization 148
Phytovolatilization 5, 145, 146, 148
Plant root exudates 112
Pleurotus florida 196
Pollution and pollutant 87
Prebiotic activity 40
Pressmud 208, 209
Protein 101
Proteolytic enzymes 15
Proteus mirabilis 51, 199
Pseudomonas fluorescens 198
Pseudomonas putida 198
Pseudomonas sp. 198
Pyrolysis 22

Q

Quadrupole time-of-flight mass spectrometer (QTOF-MS) 64

R

Reactive Oxygen Species (ROS) 100, 123
Recycling Process 202
Reductone 26
Relative Growth Rate (RGR) 160
Respiration 187–189
Rhizoctonia sp. 195.
Rhizodegradation 5
Roultella planticola 57, 73, 77

S

Scanning Electron Microscopy (SEM) 82–84
Schiff base 24, 29, 30
SDS-PAGE 62, 99
Seed germination 96
Siderophore 115, 153
Soft and Hard BOD 182
Spent Wash 1, 13, 14, 23–46
Stecker degradation 25, 28
Stenotrophomonas maltophila 199
Sterols 113
Sucrose-aspartic acid Maillard product (SAA-MP) 56
Sugar-Amine condensation 24
Sugar dehydration 25
Sugar fermentation 25
Sulfate-reducing bacteria (SRB) 17, 224
Suspended bed reactor 176

T

Thermophilic phase 211
Thin Layer Chromatography (TLC) 65
Tolerance 120
Trametes sp. I-62 196
Transmission Electron Microscopy (TEM) 133, 164
Trichoderma 209–221
Typha anguistifolia 109, 133, 135–140, 155, 161, 206

U

Upflow Anaerobic Sludge Blanket (UASB) 176, 206
Urease 94

V

Versatile Peroxidases (VP) 55
Vertical flow system 205
Vitamins 113

W

Wetland plants 109–112, 127, 153, 205
Windrow process 213
World Health Organisation (WHO) 87

X

Xanthomonas fragariae 198